Ultrarelativistic Heavy-Ion Collisions

Ultrarelativistic Heavy-Ion Collisions

Ramona Vogt

Lawrence Berkeley Laboratory
Berkeley, CA, USA

ELSEVIER

Amsterdam – Boston – Heidelberg – London – New York – Oxford – Paris
San Diego – San Francisco – Singapore – Sydney – Tokyo

Elsevier
Radarweg 29, PO Box 211, 1000 AE Amsterdam, The Netherlands
Linacre House, Jordan Hill, Oxford OX2 8DP, UK

First edition 2007

Library of Congress Cataloging-in-Publication Data
A catalog record for this book is available from the Library of Congress

British Library Cataloguing in Publication Data
A catalogue record for this book is available from the British Library

ISBN: 978-0-444-52196-5

For information on all Elsevier publications
visit our website at books.elsevier.com

Transferred to Digital Print 2009

Printed and bound in the United Kingdom

Preface

This book is an extension of a course for advanced undergraduate and graduate students that I have taught at the University of California, Davis over the last ten years. It represents an attempt to gather useful information for new practitioners in the field of relativistic heavy-ion collisions in one place. As a resource for students, the text is on a rather basic level of theoretical depth. There is not a strong focus on data because the most recent data from the Relativistic Heavy Ion Collider is in a state of flux and no final conclusions have been drawn.

There are some excellent references that go beyond the scope of this book which can be used for more in-depth studies of theory. Cross sections and perturbative QCD is covered well in Halzen and Martin 'Quarks and Leptons: An Introductory Course in Modern Particle Physics', the CTEQ 'Handbook of Perturbative QCD' and, on a more phenomenological level, Perkins 'Introduction to High Energy Physics'. For more on ultraperipheral collisions, see the reviews by Baur *et al.* and Bertulani *et al.*. The review by Cleymans, Gavai and Suhonen 'Quarks And Gluons At High Temperatures And Densities' was very useful in preparation of the chapters on thermodynamics and hydrodynamics. The book 'Thermal Physics' by Kittel and Kroemer gives a good basic introduction to thermodynamics. The books 'Finite Temperature Field Theory' by Kapusta and 'Introduction to Relativistic Heavy Ion Collisions' by Csernai go further in depth in on these topics. Both Ramond 'Field Theory: A Modern Primer' and Cheng and Li 'Gauge Theory of Elementary Particle Physics' give good discussions of Grassman variables. Cheng and Li also have an excellent discussion of symmetries. A good early book on lattice gauge theory is 'Quarks,

Gluons and Lattices' by Creutz. The earlier textbook, 'Introduction to High-Energy Heavy-Ion Collisions' by Wong has an understandable section on lattice gauge theory. For finite temperature on the lattice, there are a number of very good reviews by Karsch and by Gupta.

For a snapshot of the conclusions reached from the early RHIC runs with Au+Au, d+Au and *pp* collisions at 200 GeV in the center of mass, see Volume 757 of Nuclear Physics A containing reviews by all of the first four RHIC experiments. A compendium of new data is available in the proceedings of the Quark Matter conference series, the major meeting in this field.

This book is organized into two parts. The first part covers the basic physics of heavy ion collisions with chapters on kinematics, cross sections, geometry, thermodynamics, hydrodynamics and lattice gauge theory. It includes examples in the text and exercises at the end of each chapter. The second part includes several chapters that are more like extended examples using concepts developed in the first part. The chapters on high mass thermal dileptons and quarkonium cover two high energy probes in some depth. The third discusses fragmentation and hadronization of final state particles in proton-proton collisions.

Finally, I would like to thank my colleagues at UC Davis and the students in the class that have helped make this course better. In particular I would like to mention Daniel Cebra, Mike Anderson and Brooke Haag. Roppon Picha was great help, putting all the equations in my hand-written lecture notes into LaTeX. I would like to thank Frithjof Karsch for discussions about the lattice chapter. I also thank Jean Cleymans, Vesa Ruuskanen and Bengt Friman for discussions on other sections and Joe Kapusta for useful remarks. I thanks Carl Schwarz at Elsevier for originally suggesting that I do this book and enthusiasm about the project. Last but not least I thank Jørgen and Kristina for their patience during the completion of the book.

Contents

Part I

Basics

Chapter 1

Kinematics and invariants

1.1 Introduction

Over the years, nuclear collision energies have increased from beam kinetic energies of a few MeV/nucleon on fixed targets in small university laboratories to, at present, collider energies at large laboratories with international collaborations. As the energy is increased, the relevant degrees of freedom change. At the lowest energies, the nucleus may remain intact or be broken up into light nuclear fragments. As various thresholds for particle production are reached, some of the energy of the system may go into producing new particles, such as pions or kaons. At high enough energies, the relevant degrees of freedom are expected to be quarks and gluons rather than hadrons, forming the quark-gluon plasma.

The modern era of heavy-ion collisions arrived with beam energies of 10-200 GeV/nucleon at fixed-target facilities: the Alternating Gradient Synchrotron (AGS) at Brookhaven National Laboratory (BNL) and the Super Proton Synchrotron (SPS) at the European Center for Nuclear Research (CERN). Both the AGS and the SPS accelerated protons and several types of ions onto fixed targets of heavy nuclei. While the AGS accelerated silicon, Si, and gold, Au, ions, the SPS provided oxygen, O, sulfur, S, and lead, Pb, beams as well as, more recently, indium, In. Proton-proton (pp) and proton-nucleus (pA) interactions are used as baseline comparison measurements to distinguish true dense matter effects in nucleus-nucleus (AB) collisions from those already present in pp and pA collisions. For gold ions (nuclear mass number, A, of 197), the maximum energy at the BNL AGS was 10 GeV/nucleon while the maximum lead ion ($A = 208$) energy at the CERN SPS was 158 GeV/nucleon. Lower energy ion beams were also used to perform energy scans, down to 2 GeV/nucleon at the AGS and 20 GeV/nucleon at the SPS.

Note that the maximum energy per nucleon of the heaviest ion beam is not as high as the maximum possible proton beam energy, $E_{\rm max}$. The maximum energy for ions is $E_{\rm max}Z/A$ where Z is the proton number (nuclear charge). Protons and lighter ions can be accelerated to higher energies per nucleon, E/A, due to their larger charge-to-mass ratio, Z/A. The Z/A ratio determines the acceleration capability because while the uncharged neutrons are unaffected by the electromagnetic fields, they remain bound in the nucleus. The maximum possible center-of-mass energy for these fixed-target machines is rather low, 4.4 GeV/nucleon for Au+Au collisions at the AGS and 16.8 GeV/nucleon for Pb+Pb collisions at the SPS.

Now, two nuclear colliders take heavy-ion physics to the next level. In a collider, both collision partners, the 'projectile' and 'target' are accelerated, leading to much higher energies than those possible at fixed-target facilities. The Relativistic Heavy-Ion Collider (RHIC) at BNL and the Large Hadron Collider (LHC) at CERN produce Au+Au and Pb+Pb collisions at energies up to 200 GeV/nucleon and 5500 GeV/nucleon respectively in the center of mass. Diagrams of the RHIC and LHC accelerator complexes are shown in Figs. 1.1 and 1.2. These high energies, far above production threshold for most particles, make it possible to study the production of rare particles not accessible at lower energy facilities.

At the RHIC complex, the atoms are stripped of some of their electrons, leaving a positive charge which is accelerated by an electric field in the Tandem van de Graaff. The ions are then sent through a beam line in vacuum at 5% the velocity of light via a magnetic field. Protons are accelerated by the Linac and then sent to the Booster. Both ions and protons are further accelerated by the Booster and then

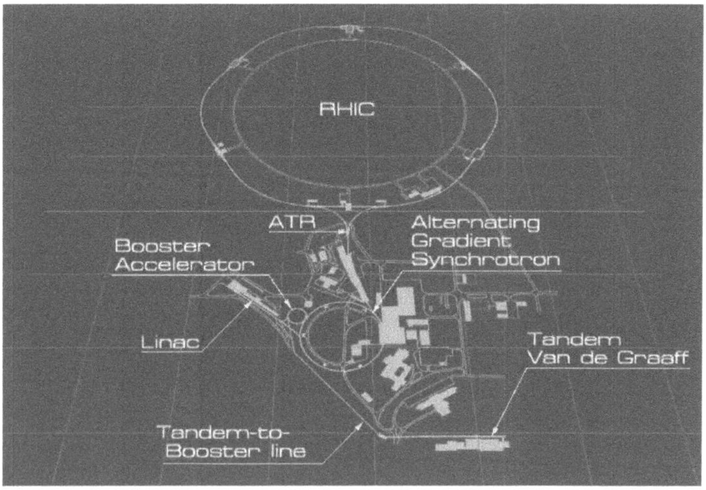

Figure 1.1: Schematic diagram of the RHIC accelerator complex [1]. Image courtesy of Brookhaven National Laboratory.

fed into the AGS where the ions are further accelerated from 37% the velocity of light to 99.7% the velocity of light. After reaching 99.7% the velocity of light, the ion beam is sent via a transfer line to the RHIC ring. A magnetic field that switches field direction sends the ions either left to travel clockwise around the ring or right to travel counterclockwise in the second ring. The ring is 2.4 miles (3.8 km) in circumference. The two oppositely-directed beams can collide at one of six interaction points. The first experiments at RHIC (STAR, PHENIX, PHOBOS and BRAHMS) occupied four of these interaction points. Together, these four experiments included 970 physicists from around the world working to build the detectors, take and analyze the data.

At the LHC, both protons and lead ions start their acceleration chain using Linacs which then feed into the Proton Synchrotron (PS) and the SPS before being transferred to the LHC rings. Note that on Fig. 1.2 there are several tangential lines lines off the PS and SPS rings (East Area on the PS and North and West Areas on the SPS) which end in fixed-target experimental areas. In addition there is a

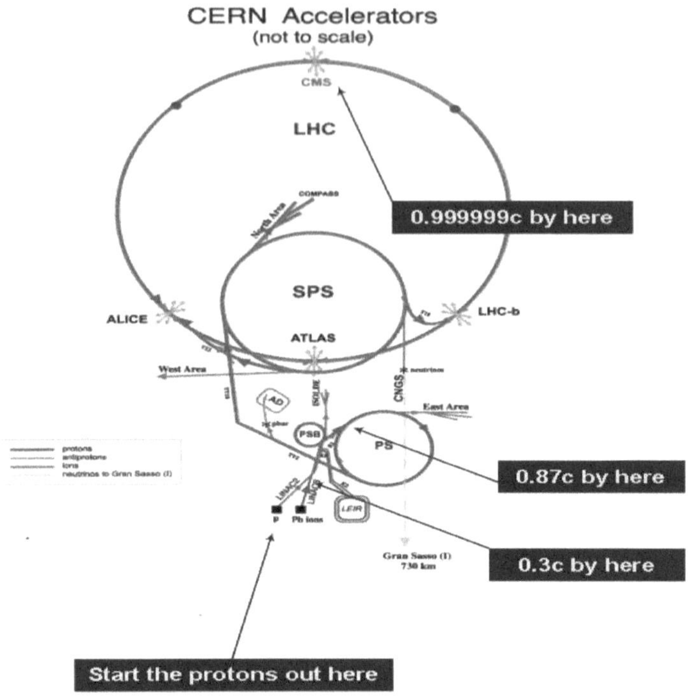

Figure 1.2: Schematic diagram of the CERN accelerators [2]. CERN copyright.

beam line from the SPS to a neutrino production area. The neutrinos are detected at the Gran Sasso laboratory in Italy, 456 miles (730 km) away. The LHC rings are 16.9 miles (27 km) in circumference. The CERN complex straddles the borders of Switzerland and France. There are four interaction points for experiments (ALICE, ATLAS, CMS and LHCb). To get an idea of the geography, ATLAS is the only detector located in Switzerland. The others are all in France. The LHC tunnel is 50-175 m underground. The varying distance below ground is due to the fact that the tunnel runs under the Jura mountain range. The ALICE experiment is dedicated to heavy-ion studies. CMS and ATLAS were mainly conceived as proton-proton experiments but CMS is also an approved heavy-ion experiment and ATLAS also has a heavy-ion group. The LHCb detector is designed to study CP violation in bottom quark-antiquark systems. The ALICE experiment alone has more than 1000 collaborators, greater than all the RHIC experiments combined.

We will not go into detail here about the acceleration process or particle detectors but will instead concentrate on the theory behind understanding these collisions.

Here we introduce some of the important kinematic variables and notation used to describe particle interactions in heavy-ion collisions. Since heavy-ion experiments now are at energies approaching those of fundamental interactions in particle physics, it is appropriate to use relativistic kinematics and four-vector notation. The next section covers this notation while the following section deals with boost-invariant quantities.

1.2 Four-vectors and kinematic variables

We first introduce some kinematic variables that are useful to describe a particle's position and momentum. We will discuss how these variables transform under special relativity and introduce quantities invariant under Lorentz transformations.

It is useful to describe a particle's position and momentum in terms of four-vectors. In the case of position, the four-vector is represented as $\mathbf{x}^\mu$ where $\mu = 0, \cdots, 3$ and

$$\mathbf{x}^\mu = (x^0, x^1, x^2, x^3) = (ct, x, y, z) \ . \tag{1.1}$$

The zeroth coordinate is the time coordinate while $\mu = 1$, 2 and 3 are the spatial coordinates. We will use boldface type to distinguish four-vectors more explicitly here but note that this is not typically done. The right-hand side of Eq. (1.1) identifies the zeroth component of the four-vector with the time variable while the components with $\mu = 1$, 2 and 3 are the positions on the x, y and z spatial axes. The spatial four vector is often collapsed to two components with the introduction of a transverse coordinate, $x_T = \sqrt{x^2 + y^2}$, neglecting the azimuthal angle, so that

$$\mathbf{x}^\mu = (t, x_T, z) \tag{1.2}$$

where we have now taken $c = 1$, as is often done.

The momentum four-vector can be defined similarly so that

$$\mathbf{p}^\mu = (p^0, p^1, p^2, p^3) = (E/c, p_x, p_y, p_z) \tag{1.3}$$
$$= (E, p_T, p_z) \tag{1.4}$$

Note that now the zeroth component corresponds to the particle energy and $p_T = \sqrt{p_x^2 + p_y^2}$, again neglecting the azimuthal angle. We have again taken $c = 1$ between Eqs. (1.3) and (1.4).

To multiply four-vectors, we need to be able to raise and lower the indices, essentially to change a row vector to a column vector. We use the metric tensor $g_{\mu\nu}$ to raise and lower the indices where

$$g_{\mu\nu} = \begin{pmatrix} 1 & 0 & 0 & 0 \\ 0 & -1 & 0 & 0 \\ 0 & 0 & -1 & 0 \\ 0 & 0 & 0 & -1 \end{pmatrix} . \tag{1.5}$$

When the metric tensor operates on a four-vector, it changes the sign of the three-vector components, $\mu = 1, 2, 3$. The four-vector with the lowered index is written as a column vector rather than a row vector so that *e.g.*,

$$\mathbf{x}_\mu = g_{\mu\nu}\mathbf{x}^\nu = \begin{pmatrix} t \\ -x_T \\ -z \end{pmatrix} \tag{1.6}$$

$$\mathbf{p}_\mu = g_{\mu\nu}\mathbf{p}^\nu = \begin{pmatrix} E \\ -p_T \\ -p_z \end{pmatrix} . \tag{1.7}$$

Then, when we multiply four-vectors, we multiply a row vector with four components by a column vector with four components, resulting in a scalar. We have

$$\mathbf{a} \cdot \mathbf{b} = \mathbf{a}^\mu \mathbf{b}_\mu = g^{\mu\nu}\mathbf{a}_\mu \mathbf{b}_\nu = a^0 b^0 - \vec{a} \cdot \vec{b} \tag{1.8}$$

where **a** and **b** are any pair of four-vectors. Two cases are particularly useful in our further studies. They are the multiplication of the position and momentum four-vectors of a particle, $\mathbf{p} \cdot \mathbf{x}$, and the multiplication of two momentum four-vectors. The first is denoted by

$$\mathbf{p} \cdot \mathbf{x} = Et - \vec{p} \cdot \vec{x} \tag{1.9}$$

while the second is

$$\mathbf{p}_1 \cdot \mathbf{p}_2 = E_1 E_2 - \vec{p}_1 \cdot \vec{p}_2 \tag{1.10}$$

which, when $1 = 2$, collapses to

$$\mathbf{p} \cdot \mathbf{p} = \mathbf{p}^2 = E^2 - |\vec{p}|^2 = m^2 \ . \tag{1.11}$$

Then, since the square of the particle mass, m, is always greater than or equal to zero, $\mathbf{p}^2 = m^2 \geq 0$ and $\mathbf{p}^2$ is a 'time-like' quantity since the zeroth components of the four-vectors dominate. If, on the other hand, the square of a four-vector is negative, $\mathbf{a}^2 < 0$, it is referred to as a 'space-like' quantity since the spatial (3-vector) contribution is largest. For example, in deep-inelastic scattering, the square of the momentum transferred from the lepton to the proton, $\mathbf{q}^2$, is negative but is usually referred to as $\mathbf{Q}^2 = -\mathbf{q}^2 \geq 0$, as we discuss later.

Figure 1.3 shows the collision of two particles at ultrarelativistic energies. The vertical axis is the time direction where the lower half plane is before the collision while the upper half plane is after the collision. The projectile comes from the left ($z < 0$) and goes off to the right. The target comes from the right ($z > 0$) and goes off to the left. The diagonal lines where $t^2 - z^2 = 0$ along the path of the 'projectile' and the 'target' define the 'light cone'. The collision occurs at $t = z = 0$. The upper part of the light cone, where $t^2 - z^2 > 0$, is the time-like region. Particle production in a real collision occurs in the upper half plane within the light cone. Outside the light cone, where $t^2 - z^2 < 0$, is the space-like region.

Most experiments are either done with either a beam on a fixed target (laboratory frame) or in a collider. We use special relativity to write the laboratory quantities in the center-of-mass frame by

$$\begin{pmatrix} E^* \\ p_z^* \end{pmatrix} = \begin{pmatrix} \gamma & -\beta\gamma \\ -\beta\gamma & \gamma \end{pmatrix} \begin{pmatrix} E \\ p_z \end{pmatrix} \tag{1.12}$$

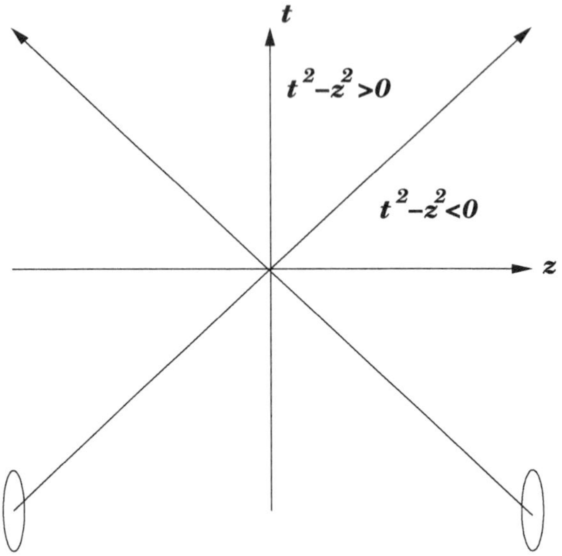

Figure 1.3: The collision axis representing one space (z) and one time dimension. The time-like ($t^2 - z^2 > 0$) and space-like ($t^2 - z^2 < 0$) regions after the collision at $t = z = 0$ are indicated.

where the variables in the new frame are 'starred'. Note that only the energy and longitudinal momentum are changed under the Lorentz transformation so that $\vec{p}_T^* = \vec{p}_T$. The velocity of the particle is denoted by β so that $\vec{\beta} = \vec{p}/E$. When transforming the projectile and target momenta from the laboratory to the center-of-mass frame, the particle has only a longitudinal velocity and we can write $\beta = p_z/E$. The gamma factor, γ, is related to the velocity by $\gamma = (1 - \beta^2)^{-1/2}$ so that, after substituting for β, we have $\gamma = E/\sqrt{E^2 - p_z^2} = E/m$ since $p_T = 0$. Then, defining β and γ as the velocity and boost factor from the laboratory frame to the center-of-mass frame, we have

$$E^* = \gamma(E - \beta p_z) , \quad (1.13)$$
$$p_z^* = \gamma(p_z - \beta E) . \quad (1.14)$$

We now define two variables that are very convenient for describing particle kinematics, the transverse mass and the rapidity. The transverse mass, m_T, is related to the difference between the squares of the energy and longitudinal momentum,

$$m_T^2 = E^2 - p_z^2 = p_T^2 + m^2 \; . \tag{1.15}$$

It is an invariant under Lorentz transformations.

EXAMPLE: By substituting Eqs. (1.13) and (1.14) into Eq. (1.15), show that m_T^2 is an invariant.
By substitution,

$$\begin{aligned} (E^*)^2 - (p_z^*)^2 &= (\gamma(E - \beta p_z))^2 - (\gamma(\beta E - p_z))^2 \\ &= \gamma^2(E^2(1-\beta^2) - p_z^2(1-\beta^2)) \; . \end{aligned} \tag{1.16}$$

Then, using the definition of γ and β,

$$(E^*)^2 - (p_z^*)^2 = E^2 - p_z^2 \; . \tag{1.17}$$

The rapidity of a particle can also be defined in terms of its energy and longitudinal momentum, as in Eq. (1.15) above. In this case,

$$y = \frac{1}{2} \ln \frac{E + p_z}{E - p_z} \; . \tag{1.18}$$

The rapidity of the particle can be written in terms of its velocity, β, instead of the momentum and energy if we define its direction of motion to be along the z axis since $\beta = p_z/E$. Then

$$y = \frac{1}{2} \ln \frac{1+\beta}{1-\beta} \; . \tag{1.19}$$

We can make use of Eq. (1.18) to define the energy and longitudinal momentum in terms of the transverse mass and rapidity, using the definitions of the hyperbolic sine and cosine, $\sinh y = [\exp(y) - \exp(-y)]/2$ and $\cosh y = [\exp(y) + \exp(-y)]/2$. Then, using the definition of rapidity in Eq. (1.18), we have

$$E = m_T \cosh y \; , \tag{1.20}$$

$$p_z = m_T \sinh y \; . \tag{1.21}$$

EXAMPLE: Using Eq. (1.18), prove these equalities.

We can exponentiate the left- and right-hand sides of Eq. (1.18) to find

$$\exp(y) = \sqrt{\frac{E+p_z}{E-p_z}}\,, \qquad \exp(-y) = \sqrt{\frac{E-p_z}{E+p_z}}\,.$$

Then we form the sum and difference of the two exponentials,

$$\begin{aligned} \exp(y)+\exp(-y) &= \frac{E+p_z}{\sqrt{E^2-p_z^2}} + \frac{E-p_z}{\sqrt{E^2-p_z^2}} = \frac{2E}{m_T}\,, \\ \exp(y)-\exp(-y) &= \frac{E+p_z}{\sqrt{E^2-p_z^2}} - \frac{E-p_z}{\sqrt{E^2-p_z^2}} = \frac{2p_z}{m_T}\,. \end{aligned}$$

Rearranging terms and using the definitions of $\cosh y$ and $\sinh y$, we have Eqs. (1.20) and (1.21).

An advantage of rapidity over velocity is that it transforms more straightforwardly under Lorentz boosts. If y is defined as in Eq. (1.18), then, in the boosted frame,

$$y^* = \frac{1}{2}\ln\left(\frac{E^*+p_z^*}{E^*-p_z^*}\right) = y - \frac{1}{2}\ln\left(\frac{1+\beta}{1-\beta}\right)\,. \tag{1.22}$$

EXAMPLE: Prove Eq. (1.22).
Substituting the definitions of E^* and p_z^* from Eqs. (1.13) and (1.14) and rearranging terms, we have

$$\begin{aligned} y^* &= \frac{1}{2}\ln\left(\frac{E^*+p_z^*}{E^*-p_z^*}\right) = \frac{1}{2}\ln\left(\frac{\gamma(E-\beta p_z)+\gamma(p_z-\beta E)}{\gamma(E-\beta p_z)-\gamma(p_z-\beta E)}\right) \\ &= \frac{1}{2}\ln\left(\left(\frac{1-\beta}{1+\beta}\right)\left(\frac{E+p_z}{E-p_z}\right)\right) = y - \frac{1}{2}\ln\left(\frac{1+\beta}{1-\beta}\right)\,. \end{aligned}$$

We now examine two extreme cases and look at the particle rapidity. The first case is when the velocity, β, is very small, $\beta \ll 1$. Then we can expand the numerator and denominator of Eq. (1.19) to find

$$y \approx \beta\,. \tag{1.23}$$

EXAMPLE: Prove Eq. (1.23).

Using $x = 1 \pm \beta$ in Eq. (1.19), we make a Taylor expansion of $\ln x$ around $x_0 = 1$. Then

$$\ln x \approx \frac{x - x_0}{x_0} - \frac{(x - x_0)^2}{2x_0^2} + \frac{(x - x_0)^3}{3x_0^3} + \cdots . \tag{1.24}$$

We write Eq. (1.19) as

$$y = \frac{1}{2}[\ln(1 + \beta) - \ln(1 - \beta)] .$$

and substitute the values of x and x_0 ($x - x_0 = \pm\beta$) into Eq. (1.24) to find

$$\begin{aligned} y &= \frac{1}{2}[\beta - \beta^2/2 + \beta^3/3 - (\beta - \beta^2/2 - \beta^3/3)] = \frac{1}{2}[2\beta + 2\beta^3/3] \\ &\approx \beta + \mathcal{O}(\beta^3) . \end{aligned}$$

On the other hand, when the particle momentum is so high that the mass can be ignored, $p \gg m$, we can relate the rapidity to the pseudorapidity, η, and then to the angle of emission, θ. The pseudorapidity is often a more useful experimental measure, especially if the particles detected are not identified and their masses are thus unknown. We define the angle of particle emission relative to the z axis as $\cos\theta = p_z/p$. Then $E + p_z = \sqrt{p^2 + m^2} + p_z$ and

$$\sqrt{p^2 + m^2} = p\sqrt{1 + (m/p)^2} \approx p(1 + m^2/2p^2 + \cdots) \tag{1.25}$$

when $p \gg m$. Then

$$\begin{aligned} E + p_z &\approx p([1 + p_z/p] + m^2/2p^2) = p(1 + \cos\theta) + m^2/2p^2)) \\ &= 2p(\cos^2(\theta/2) + m^2/4p^2) \end{aligned} \tag{1.26}$$

using the definition $2\cos^2(\theta/2) = 1 + \cos\theta$. Similarly, $E - p_z = 2p(\sin^2(\theta/2) + m^2/4p^2)$ since $2\cos^2(\theta/2) = 1 - \cos\theta$. Because $p \gg m$, we can drop the second term in each equality so that

$$y \approx -\ln[\tan(\theta/2)] \equiv \eta . \tag{1.27}$$

It is often more convenient to use rapidity for phenomenological calculations since the mass of the desired particle is known. On the other hand, pseudorapidity is typically used by experimentalists for reasons of convenience. As stated above, if the measured particles are unidentified, then rapidity cannot be defined as in Eq. (1.18). In addition, a typical detector or detector component covers some well-defined θ region with respect to the beam axis, making pseudorapidity a natural variable to use.

1.3 Invariants

The momentum and position of a particle are not frame invariant. However, many of the physical quantities, such as cross sections and rates, measured in experiments should be independent of the frame in which the measurement is made. Thus, for convenience, the cross sections are normally written in terms of invariants. These invariants, based on four-momenta, are used to describe interactions between particles and thus appear in calculations of the scattering cross sections. A typical two-body scattering diagram is shown in Fig. 1.4. The invariants are

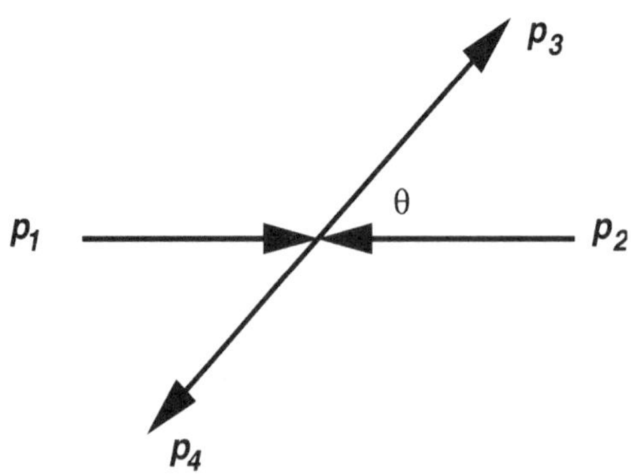

Figure 1.4: A typical $2 \to 2$ scattering diagram with particles 1 and 2 incoming while 3 and 4 are outgoing.

the squares of sums or differences of four-momenta. There are three invariants that characterize Fig. 1.4,

$$S = (\mathbf{p}_1 + \mathbf{p}_2)^2 = (\mathbf{p}_3 + \mathbf{p}_4)^2 , \tag{1.28}$$
$$T = (\mathbf{p}_1 - \mathbf{p}_3)^2 = (\mathbf{p}_2 - \mathbf{p}_4)^2 , \tag{1.29}$$
$$U = (\mathbf{p}_1 - \mathbf{p}_4)^2 = (\mathbf{p}_2 - \mathbf{p}_3)^2 . \tag{1.30}$$

They are also known as Mandelstam invariants. The first, S, depends only on the initial or final state while the other two mix the initial and final states. Thus the square root of the first, $\sqrt{S}$, is usually used to define the total 'energy' available in a collision. (We say 'energy' here because, while this terminology is standard, the units of $\sqrt{S}$ are that of momentum, *e.g.* GeV/c. Energy units are typical because $c = 1$ units are used.)

EXAMPLE: Show that S is a Lorentz invariant.
We replace the energy and the longitudinal momentum by their boosted counterparts. The transverse momentum remains unchanged by the boost. Then

$$\begin{aligned} S' &= (E_1' + E_2')^2 - (\vec{p}_1' + \vec{p}_2')^2 \\ &= (\gamma(E_1 - \beta p_{1\,z}) + \gamma(E_2 - \beta p_{2\,z}))^2 \\ &\quad - (\vec{p}_{1\,T} + \vec{p}_{2\,T})^2 - (\gamma(p_{1\,z} - \beta E_1) + \gamma(p_{2\,z} - \beta E_2))^2 . \end{aligned} \tag{1.31}$$

Rearranging terms and using $\gamma^2 - \beta^2\gamma^2 = 1$, we find $S' = S$.

It is straightforward to calculate $\sqrt{S}$, known as the center-of-mass energy, in two special cases, the center-of-mass frame and a frame in which the 'target' particle, 2, is at rest, called the fixed-target frame. In general,

$$\begin{aligned} S &= (\mathbf{p}_1 + \mathbf{p}_2)^2 = (E_1 + E_2)^2 - (\vec{p}_1 + \vec{p}_2)^2 \\ &= m_1^2 + m_2^2 + 2E_1E_2 - 2\vec{p}_1 \cdot \vec{p}_2 . \end{aligned} \tag{1.32}$$

In the last equality we have used the definition, $E^2 - p^2 = m^2$.

In the center-of-mass frame, $\vec{p}_1 + \vec{p}_2 = 0$ so that $\sqrt{S} = E_1 + E_2$. It also follows from this that, if $m_1 = m_2$, $E_1 = E_2 = E$ so that $\sqrt{S} = 2E$.

EXAMPLE: Starting from $S = m_1^2 + m_2^2 + 2E_1E_2 - 2\vec{p}_1 \cdot \vec{p}_2$, calculate the center-of-mass momentum $p_{\rm cm}$ and the center-of-mass energy, $E_{\rm cm}$. Let $|\vec{p}_1| = |\vec{p}_2| \equiv p_{\rm cm}$ so that $E_1 = \sqrt{p_{\rm cm}^2 + m_1^2}$, $E_2 = \sqrt{p_{\rm cm}^2 + m_2^2}$ and $\vec{p}_1 \cdot \vec{p}_2 = p_{\rm cm}^2$. Expanding and solving for $p_{\rm cm}$, we find

$$p_{\rm cm} = \frac{1}{2\sqrt{S}}[(S - (m_1 - m_2)^2)(S - (m_1 + m_2)^2)]^{1/2} . \tag{1.33}$$

Note that $p_{\rm cm}$ is symmetric with respect to 1 and 2. If $m_1 = m_2 = m$, then

$$p_{\rm cm} = \frac{1}{2}\sqrt{S - 4m^2} . \tag{1.34}$$

The center-of-mass energy of *e.g.* particle 1 with mass m_1 is $E_{1\,\rm cm} = \sqrt{p_{\rm cm}^2 + m_1^2}$. Replacing $p_{\rm cm}$ by the expression derived in Eq. (1.33), we find

$$E_{1\,\rm cm} = \frac{1}{2\sqrt{S}}(S + m_1^2 - m_2^2) . \tag{1.35}$$

If, instead, we wanted $E_{2\rm cm}$, then m_1 and m_2 are interchanged in Eq. (1.35) so that

$$E_{2\,\rm cm} = \frac{1}{2\sqrt{S}}(S + m_2^2 - m_1^2) . \tag{1.36}$$

Note also that if $m_1 = m_2$, $E_{1\,\rm cm} = E_{2\,\rm cm} = \sqrt{S}/2$. Finally, similar relations to Eqs. (1.33)-(1.36) can be determined for the final state in $2 \to 2$ scatteringif $m_1 \to m_3$ and $m_2 \to m_4$.

In the fixed-target frame, $\vec{p}_2 = 0$ and $E_2 = m_2$, so that, using $E^2 - |\vec{p}|^2 = m^2$,

$$S = m_1^2 + m_2^2 + 2E_1 m_2 . \tag{1.37}$$

Typically, the beam kinetic energy is given for an accelerator. If the energy is low, on the order of a few tens of MeV, as was the case for the earliest nuclear accelerators, the kinetic energy of the beam is significantly smaller than the total energy since

$$E_1 = E_{1\,\rm kin} + m_1 \tag{1.38}$$

where m_1, the particle rest mass, is its potential energy. At collider energies, $E_{1\,\rm kin} \gg m_1$ and $E_1 \approx E_{1\,\rm kin}$.

EXAMPLE: Find the relationship between the beam momentum, p_1, in the fixed-target frame and the center-of-mass momentum.
Because two beams in collision can make more use of the available energy than by deposition on a fixed target, the equivalent fixed-target beam energy for the center-of-mass energy of a collider is very large. We see this by using the definition $S = m_1^2 + m_2^2 + 2E_1 m_2$ in Eq. (1.33) which is reduced to

$$p_{\rm cm} = \frac{1}{2\sqrt{S}}[4m_2^2(E_1^2 - m_1^2)]^{1/2} = \frac{p_1 m_2}{\sqrt{S}} \ . \tag{1.39}$$

This gives us

$$p_1 = \frac{p_{\rm cm}\sqrt{S}}{m_2} \approx \frac{S}{2m_2} \tag{1.40}$$

where the approximation holds if $p_{\rm cm} \gg m_1, m_2$ since then $p_{\rm cm} \approx \sqrt{S}/2$. For a collider with center-of-mass energy of 200 GeV, the beam momentum needed for a fixed-target experiment with the same center-of-mass energy would be 21300 GeV. The energy of the target particle in the center-of-mass frame may be obtained from the value of $p_{\rm cm}$ in Eq. (1.39). Then

$$E_{2\,\rm cm} = \sqrt{p_{\rm cm}^2 + m_2^2} = \frac{m_2}{\sqrt{S}}(E_1 + m_2) \ . \tag{1.41}$$

We can transform between the laboratory and center-of-mass frames using the results in Eqs. (1.39) and (1.41). Note that, in its own rest frame, the particle velocity is $\beta = p/E$. However, if we want to transform from the laboratory frame to the center-of-mass frame, we need the velocity of the center of mass in the laboratory frame. For example, in the laboratory frame, $\beta_{2\,\rm lab} = 0$ since the target is at rest. In the center of mass, $\beta_{2\,\rm cm} = p_{\rm cm}/E_{2\,\rm cm}$. With Eq. (1.39) for the numerator and Eq. (1.41) for the denominator, we find

$$\beta_{2\,\rm cm} = \frac{p_1}{E_{1\,\rm lab} + m_2} \ . \tag{1.42}$$

Likewise, the Lorentz boost of the center of mass in the laboratory frame is

$$\gamma_{2\,\rm cm} = \frac{E_{1\,\rm lab} + m_2}{\sqrt{S}} \ . \tag{1.43}$$

The rapidity of a particle at rest in the laboratory frame, boosted to the center-of-mass system, is related to the particle velocity in the center-of-mass frame, as seen by setting $\vec{p} = 0$ in Eqs. (1.13) and (1.14).

EXAMPLE: Show the relationship between Eqs. (1.18) and (1.19).
When $\vec{p} = 0$, $E_{2\,\mathrm{lab}} = m_2$ and $p_{z_{2\,\mathrm{lab}}} = 0$ in the target frame. Then, from Eqs. (1.13) and (1.14), we have $E^* = E_{2\,\mathrm{cm}} = \gamma_{2\,\mathrm{cm}} m_2$ and $p_z^* = p_{\mathrm{cm}} = \gamma_{2\,\mathrm{cm}} \beta_{2\,\mathrm{cm}} m_2$. Then

$$\begin{aligned} y_{\beta_{2\,\mathrm{cm}}} &= \frac{1}{2} \ln \left(\frac{\gamma_{2\,\mathrm{cm}} m_2 - \gamma_{2\,\mathrm{cm}} \beta_{2\,\mathrm{cm}} m_2}{\gamma_{2\,\mathrm{cm}} m_2 + \gamma_{2\,\mathrm{cm}} \beta_{2\,\mathrm{cm}} m_2} \right) \\ &= -\frac{1}{2} \ln \left(\frac{1 + \beta_{2\,\mathrm{cm}}}{1 - \beta_{2\,\mathrm{cm}}} \right) . \end{aligned} \tag{1.44}$$

since the factors $\gamma_{2\,\mathrm{cm}} m_2$ cancel.

From Eq. (1.44), we can see that the center-of-mass rapidity of the projectile will be boosted backward from the laboratory frame, essentially by $y_{\beta_{2\,\mathrm{cm}}}$ so that, in Eq. (1.22), $y^* = y_{\mathrm{cm}} = y_{\mathrm{lab}} + y_{\beta_{2\,\mathrm{cm}}}$. The center-of-mass rapidity of the target is $y_{\beta_{2\,\mathrm{cm}}} = -y_{\mathrm{cm}}$. Thus the sum of the projectile and target rapidities in the laboratory frame is $y_1 + y_2 = y_{\mathrm{lab}}$ since $y_1 = y_{\mathrm{lab}}$ and $y_2 = 0$ while in the center-of-mass frame this sum is zero since $y_{1\,\mathrm{cm}} = y_{\mathrm{cm}}$ and $y_{2\,\mathrm{cm}} = -y_{\mathrm{cm}}$, as shown schematically in Fig. 1.5. Here $y_{\mathrm{proj}} \equiv y_1$ and $y_{\mathrm{targ}} = y_2$. This is also the case in symmetric energy colliders where the projectile and target particles are moving toward each other with equal and opposite rapidity since $y_1 = -y_2$. The rapidity difference between the beams at a collider, however, grows with energy since the difference is $y_1 - y_2 = 2y_1$ and y_1 increases with energy.

EXAMPLE: For a proton beam with 450 GeV kinetic energy, calculate $\beta_{2\,\mathrm{lab}}$, $\beta_{2\,\mathrm{cm}}$, the rapidity of the beam in the laboratory frame and the rapidity of the beam in the center-of-mass frame. (Use the Particle Data Book [3] value of the proton mass, $m = 0.93827203$ GeV.)
The laboratory energy, $E_{1\,\mathrm{lab}}$, is the sum of the kinetic energy and proton rest mass, $E_{1\,\mathrm{lab}} = 450 + 0.93827203\,\mathrm{GeV} = 450.93827203$ GeV. The laboratory momentum is $p_1 = (E_{1\,\mathrm{lab}}^2 - m_1^2)^{1/2} = 450.937296$ GeV. In the laboratory frame $\beta_{2\,\mathrm{lab}} = 0$ while, according Eq. (1.42), $\beta_{2\,\mathrm{cm}} = 0.99792145$, smaller

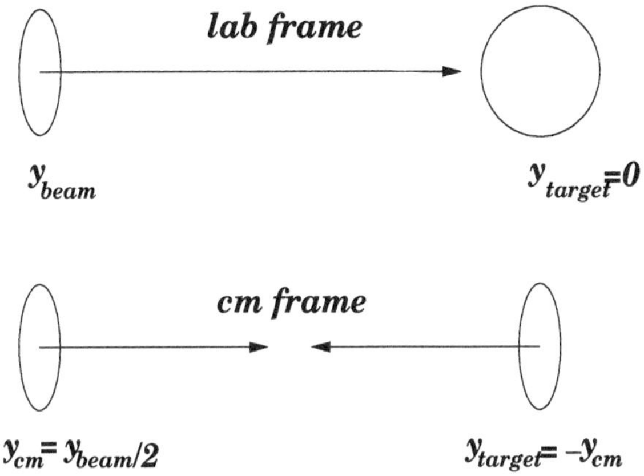

Figure 1.5: The rapidities are shown for the laboratory (top) and center-of-mass (bottom) frames. In the laboratory frame, the projectile is shown as Lorentz contracted while the target is not. In the center-of-mass frame, both the projectile and target are boosted.

than β_1 for the beam in the laboratory frame, $\beta_1 = 0.999997835$. Using Eq. (1.19), the rapidity of the beam in the laboratory frame is

$$y_{\rm lab} = \frac{1}{2} \ln \left(\frac{1+\beta_1}{1-\beta_1} \right) = 6.86819192$$

while the rapidity of the beam in the center-of-mass frame is

$$y_{\beta_{\rm cm}} = \frac{1}{2} \ln \left(\frac{1+\beta_{2\,{\rm cm}}}{1-\beta_{2\,{\rm cm}}} \right) = 3.43409596 \ .$$

Then $y_{\rm cm} = y_{\rm lab} - y_{\beta_{\rm cm}} = 3.43409596 = y_{\beta_{\rm cm}} = y_{\rm lab}/2$. Note that we have kept full accuracy here since, while $\beta \approx 1$, rounding off would mean $y \to \infty$, a situation that would be impossible in a real accelerator.

A third option, neither fixed target nor the collision of two beams at the same energy, exists at the LHC. It is most likely only possible to collide asymmetric systems such as proton-nucleus, pA, or nucleus$_1$-nucleus$_2$, AB, with two beams of different energies. Going back to the original definition of S and neglecting the masses at these high energies,

$$S = (E_1 + E_2)^2 - (\vec{p}_1 + \vec{p}_2)^2 \approx 4E_1E_2 \qquad (1.45)$$

For the realistic case of proton-lead collisions at the LHC, where the maximum proton beam energy is 7000 GeV while the maximum lead beam energy is 2750 GeV, $\sqrt{S} = 8775$ GeV. The difference in available energy is due to the charge-to-mass ratio, Z/A. The more neutrons in the nucleus, the more difficult it is to accelerate to higher energies. Note that in this case the two beams will also not have the same rapidity due to the different energies. The absolute value of the rapidity of the proton beam is $y_p = 9.61$ while the absolute value of the nuclear beam is $y_{\rm Pb} = 8.67$. Thus the center of the collision is shifted away from $y_{\rm cm} = 0$ by $\Delta y_{\rm cm} = (y_p - y_{\rm Pb})/2 = 0.47$.

The discussion in Eqs. (1.32)-(1.41) above concentrates on the initial state where the situation is typically fairly well under control and one knows the beam momenta rather exactly. When we have $2 \to 2$ scattering, the center-of-mass momentum and energy of the final state may be found by replacing 1 and 2 with 3 and 4 in Eqs. (1.33) and (1.35).

This is typically not done, however, since not all the interaction products may be observed in a single experiment. For example, when the massive gauge bosons W and Z^0 were discovered at CERN in the early 1980's, the detectors were designed to measure all hadrons and charged leptons. Neutral hadrons, which may not be visible to a detector that can only track charged particles, are observed through charged decay products. The detectors were also made as hermetic as possible to cover all regions of phase space. Even with all these precautions, the detector is still blind to neutrinos which pass through all material without a trace. The detector could compensate for the escape of the neutrinos because the events were very clean and it was possible to observe the neutrinos through 'missing energy' in the detector. One way to characterize the missing energy was to observe a high energy electron or muon track with no opposing track to balance it and conserve energy and momentum in decays such as $W^- \to e^- \nu_e$. These 'exclusive' measurements, where all decay products are accounted for, are ideal but generally not feasible in the environment of a high energy heavy-ion collision. Thus the measurements are more 'inclusive' where perhaps only certain types of particles are recorded in an event and the momenta of the unobserved particle or particles are integrated away.

We now turn to the invariant T. Because it involves both an initial and final-state four-momentum, it is often referred to as the momentum transfer between the initial and final states. For example, in deep-inelastic scattering of an electron off a proton, the interaction is characterized by the change in the energy of the electron after it scatters and the angle θ through which it scatters.

Following the calculation in Eq. (1.32), it is easy to show that T is also Lorentz invariant. We can calculate T in the center-of-mass frame, using the angle θ shown in Fig. 1.4. We have

$$\begin{aligned} T &= (E_{1\,\mathrm{cm}} - E_{3\,\mathrm{cm}})^2 - (\vec{p}_{1\,\mathrm{cm}} - \vec{p}_{3\,\mathrm{cm}})^2 \\ &= (E_{1\,\mathrm{cm}} - E_{3\,\mathrm{cm}})^2 - (p_{1\,\mathrm{cm}}^2 + p_{3\,\mathrm{cm}}^2 - 2p_{1\,\mathrm{cm}}p_{3\,\mathrm{cm}}\cos\theta) \\ &= (E_{1\,\mathrm{cm}} - E_{3\,\mathrm{cm}})^2 - (p_{1\,\mathrm{cm}} - p_{3\,\mathrm{cm}})^2 - 4p_{1\,\mathrm{cm}}p_{3\,\mathrm{cm}}\sin^2(\theta/2) \end{aligned} \tag{1.46}$$

Since $0 \leq \theta \leq \pi$, the extremes of T can be defined as

$$T_{\mp} = (E_{1\,\mathrm{cm}} - E_{3\,\mathrm{cm}})^2 - (p_{1\,\mathrm{cm}} \mp p_{3\,\mathrm{cm}})^2 \tag{1.47}$$

where the '$-$' sign corresponds to $\theta = 0$ and the '$+$' to $\theta = \pi$. By substituting the values of $E_{1,3\,\mathrm{cm}}$ and $p_{1,3\,\mathrm{cm}}$ into Eq. (1.47), an expression for $T_{\mp}$ can be found in terms of S and the particle masses for $2 \rightarrow 2$ scattering. We note that the expression for T in Eq. (1.46) is fully general and it is only when making the last substitution in Eq. (1.47) that we assume $2 \rightarrow 2$ scattering. If the energy is sufficiently high that the masses can be ignored, $\sqrt{S} \gg m$ and we find $T_- = 0$ and $T_+ = -S$. Thus T_- corresponds to the upper limit on T while T_+ is the minimum value. Since T is negative, scattering is a space-like process.

The sum of the Mandelstam invariants is also an invariant,

$$\begin{aligned} S + T + U &= (\mathbf{p}_1 + \mathbf{p}_2)^2 + (\mathbf{p}_1 - \mathbf{p}_3)^2 + (\mathbf{p}_1 - \mathbf{p}_4)^2 \\ &= 3m_1^2 + m_2^2 + m_3^2 + m_4^2 + 2\mathbf{p}_1(\mathbf{p}_2 - \mathbf{p}_3 - \mathbf{p}_4) \\ &= \Sigma_i m_i^2 \,, \end{aligned} \tag{1.48}$$

since conservation of energy and momentum dictates $-\mathbf{p}_1 = \mathbf{p}_2 - \mathbf{p}_3 - \mathbf{p}_4$ and $\mathbf{p}_1^2 = m_1^2$.

As discussed above, not all the final state particles produced in the interaction of particles 1 and 2 may be observed. For example, if the detector can only measure charged pions, then this is an 'inclusive' process. All other final-state particles are ignored. Then, if $\mathbf{p}_3$ is the four-momentum of the final-state pion produced in a hadron-hadron interaction, T and U can be defined only in terms of the detected pion momenta and the center-of-mass energy so that $T = (\mathbf{p}_1 - \mathbf{p}_3)^2$ and $U = (\mathbf{p}_2 - \mathbf{p}_3)^2$.

For collisions of two equal-mass particles, as in a pp interaction, we can define the four-momenta as

$$\mathbf{p}_1 = (E_{\rm cm}, 0, p_{\rm cm}) \approx (\sqrt{S}/2, 0, \sqrt{S}/2) \tag{1.49}$$
$$\mathbf{p}_2 = (E_{\rm cm}, 0, -p_{\rm cm}) \approx (\sqrt{S}/2, 0, -\sqrt{S}/2) \; . \tag{1.50}$$

Here we have assumed that 1 and 2 are traveling in equal but opposite directions along the z axis with no momentum component in the transverse direction. By convention, the 'projectile' particle, 1, is defined to have positive longitudinal momentum while the 'target' particle, 2, has equal but opposite longitudinal momentum. When the center-of-mass energies of particles 1 and 2 are much larger than their masses, the masses may be neglected and the approximate equalities on the right-hand sides of Eqs. (1.49) and (1.50) hold.

If we assume that the pion is produced in a $2 \rightarrow 2$ scattering as in Fig. 1.4, then the final-state particles have four-momenta

$$\mathbf{p}_3 = (E_3, \vec{p}_{T_3}, p_{z_3}) = (m_{T_3} \cosh y_3, \vec{p}_{T_3}, m_{T_3} \sinh y_3) \tag{1.51}$$
$$\mathbf{p}_4 = (E_4, \vec{p}_{T_4}, p_{z_4}) = (m_{T_4} \cosh y_4, \vec{p}_{T_4}, m_{T_4} \sinh y_4) \; . \tag{1.52}$$

For two-body kinematics, assuming that $\mathbf{p}_1$ and $\mathbf{p}_2$ only have momentum in the longitudinal direction, momentum conservation in the transverse direction gives

$$\vec{p}_{T_3} = -\vec{p}_{T_4} \tag{1.53}$$

where $|p_{T_3}| = |p_{T_4}| \equiv p_T$. If only particle 3 can be detected,

$$\begin{aligned} T &= m_1^2 + m_3^2 - 2m_{T_3}(E_{\rm cm} \cosh y_3 - p_{\rm cm} \sinh y_3) \\ &\approx -p_T \sqrt{S} \exp(-y_3) \end{aligned} \tag{1.54}$$
$$\begin{aligned} U &= m_1^2 + m_3^2 - 2m_{T_3}(E_{\rm cm} \cosh y_3 + p_{\rm cm} \sinh y_3) \\ &\approx -p_T \sqrt{S} \exp(y_3) \; . \end{aligned} \tag{1.55}$$

The last expressions on the right-hand sides of Eqs. (1.54) and (1.55) result if the hadron masses are neglected. We will return to this discussion on the parton (quark and gluon) level in the next chapter.

Exercises

1. For fixed target experiments, given a kinetic energy of the beam, calculate the beam energy and momentum in the laboratory frame, the rapidity covered in the lab, $y_{\rm proj} - y_{\rm targ}$, the center-of-mass rapidity, the velocity of the beam in the laboratory frame, $\sqrt{S}$, $E_{\rm cm}$, $p_{\rm cm}$, and the velocity of the beam in the center-of-mass frame. Write the expression used to calculate each quantity. Assume pp interactions with a proton mass of 0.938272 GeV. The velocities should be in $c = 1$ units so that $v \equiv \beta$. The values of $E_{\rm kin}$ are 10, 100 and 500 MeV, 1, 2, 10, 14, 60, 160 and 200 GeV.

2. For collider experiments, given a value of $\sqrt{S}$, calculate $E_{\rm cm}$, $p_{\rm cm}$, total rapidity coverage, $y_{\rm proj} - y_{\rm targ}$, the velocity of the center-of-mass frame and the equivalent value of $E_{\rm lab}$ for a fixed-target experiment. Write the expression used to calculate each quantity. Assume pp interactions with a proton mass of 0.938272 GeV. The velocities should be in $c = 1$ units so that $v \equiv \beta$. The values of $\sqrt{S}$ are 60, 200, 500, 1800, 5500 and 14000 GeV.

3. At the LHC, the ion species list for acceleration is as follows: O ($Z = 8$, $A = 16$), Ar ($Z = 18$, $A = 40$), Kr ($Z = 36$, $A = 84$), Sn ($Z = 50$, $A = 119$) and Pb ($Z = 82$, $A = 208$). For each species, calculate the maximum beam energy per nucleon, $7Z/A$ TeV. Then, for a proton-nucleus collision, calculate the nucleon-nucleon center-of-mass energy, $\sqrt{S_{NN}}$, the rapidity of the nuclear beam, y_A, the rapidity difference between the two beams, $y_p - y_A$, and the shift in the center-of-mass rapidity, $\Delta y_{\rm cm}$. Do the same calculations assuming that, instead of a proton beam, a deuteron beam ($Z = 1$, $A = 2$) is used. Assume that the proton mass is 0.938272 GeV.

Chapter 2

Cross sections

2.1 Introduction

When two particles, be they leptons (electrons, muons or neutrinos), photons, or hadrons (protons, pions, hyperons, nuclei $\cdots$) collide, they can either suffer a glancing collision that changes their momentum but otherwise leaves them unaltered or they can produced other particles by a more direct hit when energy is transferred from the initial state to the new particles in the final state. In this last case, the particles that originally collided may not appear at all in the final state. Either scenario is allowed as long as no conservation laws such as energy, momentum, charge and baryon number are violated. The first situation of a glancing collision where no new particles are produced, is known as an elastic collision, while the second, where new particles are produced, is referred to as an inelastic collision. Together, these two components make up the total scattering cross section.

To describe what happens in a heavy-ion collision, we measure as many outgoing particles as possible and describe their behavior in terms of kinematic variables such as rapidity and transverse momentum, as discussed in Chapter 1. The best way to characterize these results is in terms of a 'cross section'. In this chapter, we describe, very generally, how to calculate particle interactions and relate them to the cross section, which we then define. We will show that the cross section has to be a Lorentz invariant and relate it to the wave-optical picture of particle interactions. Finally, we introduce quark and hadron notation and discuss the parton (quark, antiquark and gluon) constituents of hadrons.

2.2 Derivation of the cross section from nonrelativistic perturbation theory

We first start with nonrelativistic quantum mechanics to get our feet wet and then move to the relativistic picture more appropriate for high energy heavy-ion collisions. The classical nonrelativistic kinetic energy of a free particle with mass m,

$$E = \frac{p^2}{2m} , \tag{2.1}$$

may be written in terms of differential operators by the substitutions

$$E \equiv i\hbar\frac{\partial}{\partial t} \qquad \vec{p} \equiv -i\hbar\frac{\partial}{\partial \vec{x}} = -i\hbar\vec{\nabla} . \tag{2.2}$$

We let $\hbar \equiv 1$, as before, and allow the differential operators to act on a complex wavefunction, $\psi(\vec{x}, t)$, so that

$$i\frac{\partial \psi}{\partial t} + \frac{1}{2m}\nabla^2\psi = 0 , \tag{2.3}$$

the Schrödinger equation for zero potential.

EXAMPLE: By substituting Eq. (2.2) into Eq. (2.1), obtain Eq. (2.3). Note that $p^2 = (-i\hbar\vec{\nabla})^2 = -\hbar^2\nabla^2$. Replacing this in Eq. (2.1), moving $p^2/2m$ to the left-hand side to change the sign, setting $\hbar = 1$ and multiplying by the wavefunction ψ on the right-hand side, we obtain Eq. (2.3).

The square of the wavefunction, $|\psi|^2$, we take to be the probability density, ρ, so that $\rho = |\psi|^2 = \psi^*\psi$ is the probability of finding a particle in the volume element $dxdydz \equiv d^3x \equiv dV$ where ψ^* is the complex conjugate of the wavefunction. In a collision, we need to keep track of the density flux, $\vec{j}$, in *e.g.* a beam of particles. To conserve probability, we need the decrease of particles in volume V with time,

$$-\frac{\partial}{\partial t}\int_V \rho dV \ , \tag{2.4}$$

to be equal to the total flux of particles out of the same volume,

$$\int_v \vec{\nabla}\cdot\vec{j}dV \ , \tag{2.5}$$

so that

$$-\frac{\partial}{\partial t}\int_V \rho dV = \int_v \vec{\nabla}\cdot\vec{j}dV \ . \tag{2.6}$$

Dropping the integral over V and the dV from both sides, we obtain the continuity equation,

$$\frac{\partial\rho}{\partial t} + \vec{\nabla}\cdot\vec{j} = 0 \ . \tag{2.7}$$

To find the flux through the volume element, we take the operator equation, Eq. (2.3), and multiply by $-i\psi^*$ on the left-hand side and subtract it from the complex conjugate equation, multiplied by $-i\psi$ on the left-hand side. We can then identify the current as

$$\vec{j} = \frac{-i}{2m}\left[\psi^*\vec{\nabla}\psi - \psi\vec{\nabla}\psi^*\right] \ . \tag{2.8}$$

EXAMPLE: Show this.
First multiply the left-hand side of Eq. (2.3) by $-i\psi^*$ to obtain

$$(-i\psi^*)\left(i\frac{\partial\psi}{\partial t}\right) + \frac{-i\psi^*}{2m}\nabla^2\psi = 0 \ . \tag{2.9}$$

Next, take the complex conjugate of Eq. (2.3) and multiply the left-hand side of this conjugate equation by $-i\psi$ to find

$$(-i\psi)\left(-i\frac{\partial\psi^*}{\partial t}\right) + \frac{-i\psi}{2m}\nabla^2\psi^* = 0 \ . \tag{2.10}$$

Then subtract Eq. (2.10) from Eq. (2.9) to obtain

$$\left(\psi^*\frac{\partial\psi}{\partial t}+\psi\frac{\partial\psi^*}{\partial t}\right)+\frac{-i}{2m}\left(\psi^*\nabla^2\psi-\psi\nabla^2\psi^*\right)=0\,. \tag{2.11}$$

Equation (2.11) can be recast as the continuity equation, Eq. (2.7), if the parenthetical expression on the left-hand side can be identified as

$$\psi^*\frac{\partial\psi}{\partial t}+\psi\frac{\partial\psi^*}{\partial t}=\frac{\partial}{\partial t}(\psi^*\psi)\equiv\frac{\partial\rho}{\partial t} \tag{2.12}$$

where we have used the definition $\rho=|\psi|^2$. Then the second parenthetical expression in Eq. (2.11) can be identified with $\vec{\nabla}\cdot\vec{j}$,

$$\frac{-i}{2m}\left(\psi^*\nabla^2\psi-\psi\nabla^2\psi^*\right)\equiv\vec{\nabla}\cdot\vec{j}\,. \tag{2.13}$$

Since $\nabla^2=\vec{\nabla}\cdot\vec{\nabla}$, we can drop one $\vec{\nabla}$ from Eq. (2.13), leaving us with Eq. (2.8) as the definition of the current. This can be seen by operating on both sides of Eq. (2.8) with $\vec{\nabla}$ since the cross terms of the derivatives, $\nabla\psi\cdot\nabla\psi^*$, drop out.

A free particle solution of the nonrelativistic operator equation is

$$\psi=N\exp(i\vec{p}\cdot\vec{x}-iEt) \tag{2.14}$$

where the particle has energy E and momentum $\vec{p}$.

EXAMPLE: Prove Eq. (2.14) and determine ρ and $\vec{j}$ for this value of ψ.

Taking the derivatives of ψ with respect to time and space, we obtain $\partial\psi/\partial t=-iE\psi$ and $\nabla^2\psi=-p^2\psi$. Substituting these into Eq. (2.3), we find the nonrelativistic expression for the kinetic energy, Eq. (2.1). Then using the definitions of ρ and $\vec{j}$ in terms of ψ, we find

$$\rho = |N|^2 \tag{2.15}$$

$$\vec{j} = \frac{\vec{p}}{m}|N|^2\,. \tag{2.16}$$

This is the nonrelativistic solution. However, Eq. (2.1) is not Lorentz covariant.

EXAMPLE: Show this.
Write $p^2 = p_T^2 + p_z^2$ to separate the transverse and longitudinal momenta. Then, from Chapter 1, substitute Eq. (1.13) for E and Eq. (1.14). It is clear that the boosted equation cannot reduce to Eq. (2.1).

Since it is not Lorentz covariant, Eq. (2.1) is not appropriate for relativistic particles. Instead, we use the invariant definition,

$$E^2 = p^2 + m^2 \ , \tag{2.17}$$

to try to obtain a covariant wave equation. In Chapter 1 we have already shown that $m_T^2 = E^2 - p_z^2 = p_T^2 + m^2$ is covariant so we know that this is a better starting point. We can substitute Eq. (2.2) into Eq. (2.17) and let it operate on a covariant wavefunction ϕ to obtain (again with $\hbar \equiv 1$),

$$-\frac{\partial^2 \phi}{\partial t^2} = -\nabla^2 \phi + m^2 \phi \rightarrow (\Box^2 + m^2)\phi = 0 \ . \tag{2.18}$$

It is clear from Eq. (2.2) that covariant four-momenta $\mathbf{p}_\mu$ transforms like $i\partial_\mu$. Rearranging terms on the left-hand side of Eq. (2.18) and using $\partial^\mu \partial_\mu \equiv \Box^2$, we are left with the right-hand side of Eq. (2.18). This wave equation is known as the Klein-Gordon equation, the relativistic version of the Schrödinger equation.

We can also obtain relativistic forms for the probability density and current,

$$\mathbf{j}^\mu = (\rho, \vec{j}) = i(\phi^* \partial^\mu \phi - \phi \, \partial^\mu \phi^*) \ , \tag{2.19}$$

so that the covariant continuity equation becomes

$$\partial_\mu \mathbf{j}^\mu = 0 \ . \tag{2.20}$$

EXAMPLE: Prove Eq. (2.19).
As in the nonrelativistic case, we multiply Eq. (2.18) by $-i\phi^*$ on the left-hand side and subtract it from the complex conjugate equation, multiplied by $-i\phi$ on the left-hand side. Before subtraction, we have

$$-i\phi^* \left(\frac{\partial^2 \phi}{\partial t^2} - \nabla^2 \phi + m^2 \phi \right) = 0 \ , \tag{2.21}$$

$$-i\phi \left(\frac{\partial^2 \phi^*}{\partial t^2} - \nabla^2 \phi^* + m^2 \phi^* \right) = 0 \ . \tag{2.22}$$

The term proportional to the mass cancels in the subtraction, leaving

$$-i\phi^*\frac{\partial^2\phi}{\partial t^2}+i\phi\frac{\partial^2\phi^*}{\partial t^2}+i\phi^*\nabla^2\phi-i\phi\nabla^2\phi^*=0\ . \tag{2.23}$$

We pull one derivative out from the temporal and spatial derivatives and a factor of -1 to obtain

$$\frac{\partial}{\partial t}\left[i\left(\phi^*\frac{\partial\phi}{\partial t}-\phi\frac{\partial\phi^*}{\partial t}\right)\right]+\vec{\nabla}\cdot\left[-i\left(\phi^*\vec{\nabla}\phi-\phi\vec{\nabla}\phi^*\right)\right]=0\ . \tag{2.24}$$

Equation (2.24) is suggestive of the nonrelativistic continuity equation in Eq. (2.7). The first term in square brackets can be identified as ρ while the second term corresponds to the current, $\vec{j}$.

A free particle, plane-wave solution of the Klein-Gordon equation is

$$\phi=N\exp(i\vec{p}\cdot\vec{x}-iEt)=N\exp(-i\mathbf{p}\cdot\mathbf{x}) \tag{2.25}$$

where the last equality is written in covariant form as a four-vector product.

EXAMPLE: Find the covariant form of the current for ϕ.
Substituting ϕ and its complex conjugate ϕ^* into Eq. (2.19), we find

$$\rho = 2E|N|^2\ , \tag{2.26}$$
$$\vec{j} = 2\vec{p}|N|^2\ , \tag{2.27}$$
$$\mathbf{j}^\mu = 2\mathbf{p}^\mu|N|^2\ . \tag{2.28}$$

Given the covariant definition of ϕ, we can plug this solution into Eq. (2.18) to regain Eq. (2.17). Thus the energy eigenvalues of the Klein-Gordon equation are

$$E=\pm\sqrt{p^2+m^2}\ . \tag{2.29}$$

We then have a negative energy solution as well as one with positive energy. This means that the probability density, ρ, in Eq. (2.26) is also negative.

These negative energy solutions were rather remarkable at first and difficult to interpret. They could not be discarded or explained away and disappear. Dirac derived his famous equation,

$$i\partial_\mu \gamma^\mu \psi - m\psi = 0 \ , \tag{2.30}$$

where γ^μ are 4×4 matrices, to obtain an alternative relativistic wave equation linear in ∂_μ rather than quadratic, as in Eq. (2.18). A bonus was being able to describe relativistic spin 1/2 particles, the fermions. However, he did not succeed in getting rid of the negative energy solutions. He then postulated that there was an infinite sea of negative energy states that was all filled up (the Dirac sea) so that positive energy particles would not collapse into it. The negative energy states could, however, be excited into a positive energy state. The way to get around this problem is to let the positive energy solutions be particles and the negative energy solutions be antiparticles.

Stuckelberg and Feynman developed an elegant solution to this problem of negative energy states which is how they are now treated. If the electric charge e is inserted into the current, $\mathbf{j}^\mu$, then it can be described as a charge current density rather than a probability density,

$$\mathbf{j}^\mu = -ie(\phi^* \partial^\mu \phi - \phi \partial^\mu \phi^*) \ . \tag{2.31}$$

Then the four-vector current for an electron is

$$\mathbf{j}^\mu(e^-) = -2e|N|^2(E, \vec{p}) \ . \tag{2.32}$$

Let the antiparticle of the electron, the positron, have the same energy and momentum. It has a positive electric charge, so that

$$\mathbf{j}^\mu = 2e|N|^2(E, \vec{p}) \tag{2.33}$$

$$= -2e|N|^2(-E, -\vec{p}) \ . \tag{2.34}$$

Note that the second equality, Eq. (2.34), is the same as an electron with negative four-momentum. Thus we can describe the propagation of an antiparticle forward in time by a negative energy particle moving backward in time since $\exp(-i(-E)(-t)) = \exp(-iEt)$.

Thus, for example, we can draw a diagram for the production of an e^-e^+ pair by a virtual photon two ways, as shown in Fig. 2.1. On the

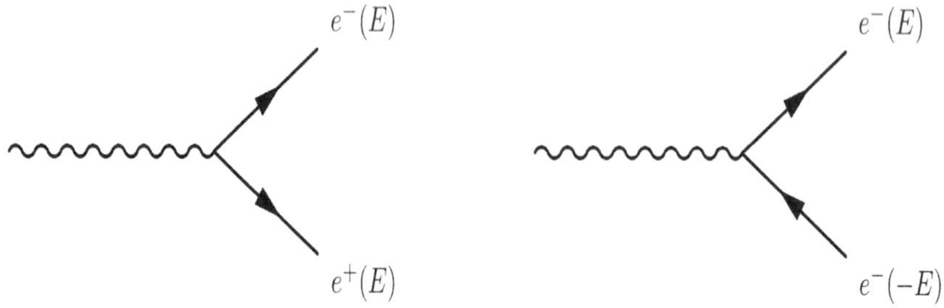

Figure 2.1: Diagram for a virtual photon splitting into an electron-positron pair. The left-hand side shows both the electron and positron as propagating forward in time while the right-hand side represents the positron as an electron propagating backward in time toward the vertex.

left-hand side, the e^-e^+ pair is "popped out of the vacuum" and both the electron and the positron are propagating outward from the vertex forward in time. The way this is actually calculated is on the right-hand side. A negative energy electron moves backward in time to the vertex where it scatters with the virtual photon and then propagates forward in time. (The time direction is indicated by the arrows: $\longrightarrow$ is forward in time while $\longleftarrow$ is backward in time.)

So far, all we have discussed are the free particle solutions to the wave equation. However, to get a cross section, we need to discuss interactions. We will go back to nonrelativistic perturbation theory to describe the basic framework for calculating the cross section, starting from free-particle solutions to the Schrödinger equation.

We solve the Schrödinger equation for the behavior of a particle in the presence of an interaction potential, $V(\vec{x}, t)$,

$$(H_0 + V(\vec{x}, t))\psi(\vec{x}, t) = i\frac{\partial\psi(\vec{x}, t)}{\partial t} \,. \tag{2.35}$$

The solutions,

$$\psi = \sum_n a_n(t)\phi_n(\vec{x}) \exp(-iE_n t) \tag{2.36}$$

are expressed in terms of a time-independent part, $\phi_n(\vec{x})$ and a time-dependent coefficient, $a_n(t)$. Let H_0 be a time-independent Hamiltonian so that

$$H_0\phi_n(\vec{x},t) = E_n\phi_n(\vec{x},t) \ , \tag{2.37}$$

where

$$\int_V d^3x\phi_m^*\phi_n = \delta_{mn} \ . \tag{2.38}$$

The normalization δ_{mn} requires only one particle per volume V and that, unless $m = n$, the box is empty. This is the orthonormality condition.

To find an expression for the coefficient $a_n(t)$, we substitute Eq. (2.36) into Eq. (2.35),

$$\begin{aligned} i\sum_n \left(\frac{da_n(t)}{dt} - iE_na_n(t)\right)\phi_n(\vec{x})\exp(-iE_nt) = \\ (H_0 + V(\vec{x},t))\sum_n a_n(t)\phi_n(\vec{x})\exp(-iE_nt) \end{aligned} \tag{2.39}$$

leading to, with Eq. (2.37),

$$\begin{aligned} i\sum_n \frac{da_n(t)}{dt}\phi_n(\vec{x})\exp(-iE_nt) = \\ V(\vec{x},t)\sum_n a_n(t)\phi_n(\vec{x})\exp(-iE_nt) \ . \end{aligned} \tag{2.40}$$

Now we can get $da_n(t)/dt$ on one side by itself by multiplying both sides of Eq. (2.40) by ϕ_f^*, integrating over the volume and using Eq. (2.38),

$$\begin{aligned} \int d^3x\phi_f^*(\vec{x})i\sum_n \frac{da_n(t)}{dt}\phi_n(\vec{x})\exp(-iE_nt) = \\ \int d^3x\phi_f^*(\vec{x})V(\vec{x},t)a_n(t)\phi_n(\vec{x})\exp(-iE_nt) \ . \end{aligned} \tag{2.41}$$

(Note that by calling the spatial function ϕ_f, we are biasing toward a final state.) Since $da_n(t)/dt$ and $\exp(-iE_nt)$ are independent of the volume, they can be moved out of the integral on the left-hand side of Eq. (2.41). The orthonormality condition then gives us $n \to f$, so that

$$i\frac{da_f(t)}{dt}\exp(-iE_f t) = \int d^3x \phi_f^*(\vec{x})V(\vec{x},t)a_n(t)\phi_n(\vec{x})\exp(-iE_n t)\ . \tag{2.42}$$

Multiplying both sides of Eq. (2.42) by $-i\exp(iE_f t)$, we obtain a coupled differential equation for $a_f(t)$:

$$\frac{da_f(t)}{dt} = -i\sum_n a_n(t)\int d^3x \phi_f^*(\vec{x})V(\vec{x},t)\phi_n(\vec{x})\exp[i(E_f - E_n)t]\ . \tag{2.43}$$

Before V interacts on the particle, at some initial eigenstate i of the unperturbed Hamiltonian, $a_i(-T/2) = 1$. For other eigenstates, $n \neq i$, $a_n(-T/2) = 0$. Thus we take $n = i$ on the right-hand side of Eq. (2.43) to obtain

$$\frac{da_f(t)}{dt} = -i\int d^3x \phi_f^*(\vec{x})V(\vec{x},t)\phi_i(\vec{x})\exp[i(E_f - E_i)t]\ . \tag{2.44}$$

If the potential $V(\vec{x},t)$ is not too large and acts over a short time interval, we can assume it is independent of time, $V(\vec{x},t) \to V(\vec{x})$. We then integrate both sides over time up to $t = T/2$ when the interaction is over. The resulting left-hand side, $a_f(T/2)$, is called T_{fi} and is written suggestively as

$$\begin{aligned} T_{fi} &\equiv -i\int_{-T/2}^{T/2} dt \int d^3x\,[\phi_f(\vec{x})\exp(-iE_f t)]^*\,V(\vec{x})\,[\phi_i(\vec{x})\exp(-iE_i t)] \\ &= -i\int d^4x \phi_f^*(\mathbf{x})V(\mathbf{x})\phi_i(\mathbf{x})\ . \end{aligned} \tag{2.45}$$

The second equality is written covariantly with $\phi(\mathbf{x}) = \phi(\tilde{\mathbf{x}})\exp(-\mathbf{iEt})$.

Note that $|T_{fi}|^2$ is not quite the probability for a particle to be scattered from the initial state i into the final state f. Since $V(x)$ is time-independent,

$$V_{fi} = \int d^3x \phi_f^*(\vec{x})V(\vec{x})\phi_i(\vec{x})\ , \tag{2.46}$$

so that then

$$\begin{aligned} T_{fi} &= -iV_{fi}\int_{-\infty}^{\infty} dt\, \exp[i(E_f - E_i)t] \\ &= -2\pi i\, V_{fi}\, \delta(E_f - E_i)\ . \end{aligned} \quad (2.47)$$

The delta function, $\delta(E_f - E_i)$, implies energy conservation. Since infinite time separates the initial and final states, $|T_{fi}|^2$ is not a probability. Instead we define a transition probability per unit time

$$W_{fi} = \lim_{T\to\infty} \frac{|T_{fi}|^2}{T} \quad (2.48)$$

$$\begin{aligned} &= \lim_{T\to\infty} \int_{-\infty}^{\infty} dt\, V_{fi}\, \exp[-i(E_f - E_i)t] \\ &\quad \times \int_{-T/2}^{T/2} dt'\, V_{fi}^{*}\, \exp[i(E_f - E_i)t']\ . \end{aligned} \quad (2.49)$$

Note that we have kept the limits of the first time integral as $\pm\infty$ while the second integral over t' has limits $\pm T/2$. We can pull out the V_{fi}'s from the integrals and use the definition of the delta function in the first integration to set $E_f = E_i$ in the second integral. Then

$$W_{fi} = \lim_{T\to\infty} \frac{|V_{fi}|^2}{T} 2\pi\delta(E_f - E_i) \int_{-T/2}^{T/2} dt' \exp[i(E_f - E_i)t'] \quad (2.50)$$

$$= 2\pi |V_{fi}|^2 \delta(E_f - E_i)\ . \quad (2.51)$$

After we use $\delta(E_f - E_i)$ in Eq. (2.50), the integral over t' is just equal to T which then cancels the T in the denominator, giving us a time-independent result in Eq. (2.51). This is meaningful only after integration over the initial and final states.

In a collision environment, we usually know the initial state, *e.g.*, two particle beams. In the case of a secondary beam, produced by a proton acting on an intermediate target, the actual particle that interacts may not be completely clear since the secondary beam is an admixture of several particles. For example, a hyperon beam may primarily be Σ^- but also contains pions and kaons.) The final state, when not exclusively specified, is denoted X, as in $pp \to X$. We can call $\rho(E_f)$ the density of final states so that $\rho(E_f)\,dE_f$ is the number of states in the interval $E_f < E < E_f + dE_f$. Integrating Eq. (2.51) over $dE_f\rho(E_f)$, we get the transition rate

$$\begin{aligned} W_{fi} &= 2\pi \int dE_f\, \rho(E_f)|V_{fi}|^2\delta(E_f - E_i) \\ &= 2\pi\rho(E_i)|V_{fi}|^2 \,. \end{aligned} \tag{2.52}$$

This is the first term in the perturbative expansion of the amplitude. This result can be iterated over any number of intermediate states. More terms can be added. If we have one particle in and one particle out, what happens in the intermediate region can be said to occur within a region centered around the interaction point V_{fi}, as shown in Fig. 2.2(a) for the case of a single scattering

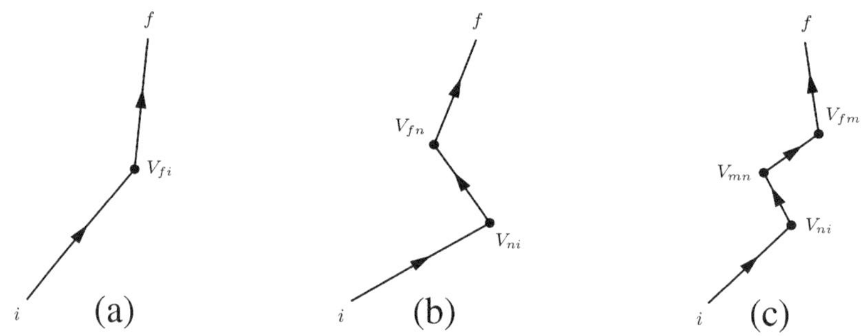

Figure 2.2: Single particle scattering with (a) one, (b) two, and (c) three intermediate states.

The intermediate state V_{fi} is the simplest possible case but any number of intermediate steps can be added involving sums over unknown intermediate states so that

$$\begin{aligned} V_{fi} &\rightarrow V_{fi} + \sum_{n \neq i} V_{fn} \frac{1}{E_i - E_n + i\epsilon} V_{ni} \\ &+ \sum_{n\neq i, m\neq n} V_{fn} \frac{1}{E_m - E_n + i\epsilon} V_{nm} \frac{1}{E_i - E_m + i\epsilon} V_{mi} + \cdots \,. \end{aligned} \tag{2.53}$$

where now more than one interaction point is present, see Figs. 2.2(b) and (c). The energy differences in the denominators, $(E_m - E_n + i\epsilon)^{-1}$, represent the propagation of the particle from state n to state m.

So far, this is not covariant because there really isn't an interaction, the potential is static, independent of x and t. We can make it covariant though by *e.g.* considering the motion of a charged particle, like the electron, with charge $-e$, through an electromagnetic four-vector potential $\mathbf{A}^\mu$ so that in Eq. (2.17), $\mathbf{p}^\mu \to \mathbf{p}^\mu + e\mathbf{A}^\mu$ or $i\partial^\mu \to i\partial^\mu + e\mathbf{A}^\mu$. Then the Klein-Gordon equation becomes

$$(\Box^2 + m^2)\phi = -V\phi \, . \tag{2.54}$$

EXAMPLE: Show this.

Start with Eq. (2.17) but now make the substitution $\mathbf{p}^\mu = i\partial^\mu \to i\partial^\mu + e\mathbf{A}^\mu$. Then

$$E^2 = \left(i\frac{\partial}{\partial t} + eA_0\right)^2 \tag{2.55}$$

$$p^2 = (-i\vec{\nabla} - e\vec{A})^2 \, . \tag{2.56}$$

After squaring, these become

$$E^2 = -\frac{\partial^2}{\partial t^2} + ie\frac{\partial A_0}{\partial t} + ieA_0\frac{\partial}{\partial t} + e^2A_0^2 \tag{2.57}$$

$$p^2 = -\nabla^2 + ie\vec{\nabla}\cdot\vec{A} + ie\vec{A}\cdot\vec{\nabla} + e^2\vec{A}^2 \tag{2.58}$$

Substituting this into Eq. (2.17) and rearranging terms, we have

$$\begin{aligned}\frac{\partial^2}{\partial t^2} - \nabla^2 + m^2 &= +ie\left(\frac{\partial A_0}{\partial t} + A_0\frac{\partial}{\partial t} - \vec{\nabla}\cdot\vec{A} - \vec{A}\cdot\vec{\nabla}\right) \\ &\quad + e^2(A_0^2 - \vec{A}^2)\end{aligned} \tag{2.59}$$

Writing Eq. (2.59) in covariant form and operating on ϕ gives

$$(\Box^2 + m^2)\phi = (ie(\partial_\mu\mathbf{A}^\mu + \mathbf{A}^\mu\partial_\mu) + e^2\mathbf{A}^2)\phi = -V\phi \, . \tag{2.60}$$

Thus we obtain Eq. (2.54) after making the identification

$$V = -ie(\partial_\mu\mathbf{A}^\mu + \mathbf{A}^\mu\partial_\mu) - e^2\mathbf{A}^2 \, . \tag{2.61}$$

Equation (2.61) is the electromagnetic potential. The charge e can be related to the electromagnetic coupling constant by $\alpha = e^2/4\pi \approx 1/137$, small enough to expand V in powers of α. Keeping only the lowest order term in e in the potential, the amplitude, Eq. (2.45), is

$$T_{fi} = i \int d^4x \phi_f^* \left[ie(\mathbf{A}^\mu \partial_\mu + \partial_\mu \mathbf{A}^\mu)\right] \phi_i \ . \tag{2.62}$$

Since the partial derivative in the second term of Eq. (2.62) acts on both $\mathbf{A}^\mu$ and ϕ_i, we can integrate by parts to get it to act on ϕ_f^* instead. Then

$$\begin{aligned} T_{fi} &= -i \int d^4x \mathbf{A}^\mu \left[(-ie)(\phi_f^* \partial_\mu \phi_i - \phi_i \, \partial_\mu \phi_f^*)\right] \\ &\equiv -i \int d^4x \mathbf{j}_\mu^{fi} \mathbf{A}^\mu \ , \end{aligned} \tag{2.63}$$

where we have multiplied the definition of the current in Eq. (2.19) by $-e$ to obtain the electromagnetic current, called the current for 'spinless' electron scattering in Halzen and Martin [4].

EXAMPLE: If $\phi_i(x) = N_i \exp(-i\mathbf{p}_i \cdot \mathbf{x})$ and $\phi_f(x) = N_f \exp(-i\mathbf{p}_f \cdot \mathbf{x})$, calculate $\mathbf{j}_\mu^{fi}$.
We assume N is real. We also have $\partial_\mu \phi_i = (-i\mathbf{p}_i)_\mu \phi_i$, $\partial_\mu \phi_f^* = (i\mathbf{p}_f)_\mu \phi_f^*$ and $\phi_f^* \phi_i = N_i N_f \exp[i(\mathbf{p}_f - \mathbf{p}_i) \cdot \mathbf{x}]$. Then, by substitution,

$$\mathbf{j}_\mu^{fi} = -eN_i N_f (\mathbf{p}_f + \mathbf{p}_i)_\mu \exp[i(\mathbf{p}_f - \mathbf{p}_i) \cdot \mathbf{x}] \ . \tag{2.64}$$

Thus if we have two spinless particles, say two spinless electrons, scattering off each other elastically, let 1 and 2 be incoming (the i state) while 3 and 4 are outgoing (the f state), as shown in Fig. 2.3. Each electron has a current associated with it. Elastic scattering occurs via exchange of a photon between the two electrons. If initial-state 1 scatters to final-state 3 and initial-state 2 scatters to final-state 4, the two currents are

$$\mathbf{j}_\mu^{31} = -eN_1 N_3 (\mathbf{p}_1 + \mathbf{p}_3)_\mu \exp[i(\mathbf{p}_3 - \mathbf{p}_1) \cdot \mathbf{x}] \ , \tag{2.65}$$

$$\mathbf{j}_\mu^{42} = -eN_2 N_4 (\mathbf{p}_2 + \mathbf{p}_4)_\mu \exp[i(\mathbf{p}_4 - \mathbf{p}_2) \cdot \mathbf{x}] \ . \tag{2.66}$$

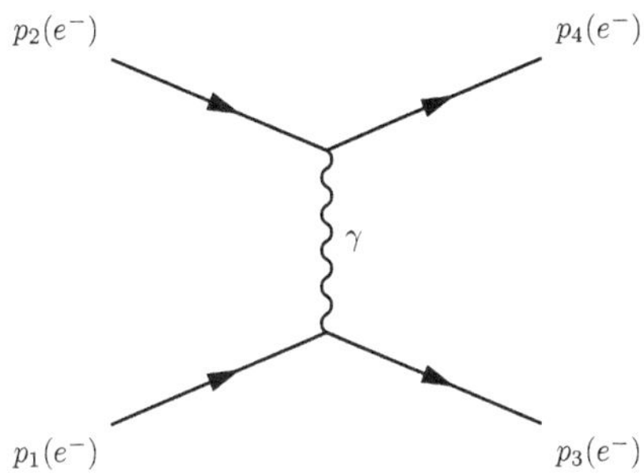

Figure 2.3: Diagram for spinless electron scattering. The incoming electrons have four-momenta $\mathbf{p}_1$ and $\mathbf{p}_2$ while the outgoing electrons have four-momenta $\mathbf{p}_3$ and $\mathbf{p}_4$.

Maxwell's equations can be written covariantly as

$$\Box^2 \mathbf{A}^\mu = \mathbf{j}^\mu \ . \tag{2.67}$$

Since $\Box^2 \exp(i\mathbf{q}\cdot\mathbf{x}) = -\mathbf{q}^2 \exp(i\mathbf{q}\cdot\mathbf{x})$, we can make the identification $\mathbf{A}^\mu = (-1/\mathbf{q}^2)\mathbf{j}^\mu_{42}$ with $\mathbf{q} = \mathbf{p}_4 - \mathbf{p}_2$. Using this in Eq. (2.63), we find

$$T_{fi} = -i \int d^4x \, \mathbf{j}^{31}_\mu \mathbf{A}^\mu = i \int d^4x \, \mathbf{j}^{31}_\mu \frac{1}{\mathbf{q}^2} \mathbf{j}^\mu_{42} \ . \tag{2.68}$$

After substitution of the currents, Eqs. (2.65) and (2.66), and integration over d^4x, we find

$$T_{fi} = -iN_1N_2N_3N_4(2\pi)^4\delta^4(\mathbf{p}_1 + \mathbf{p}_2 - \mathbf{p}_3 - \mathbf{p}_4)\mathcal{M} \ , \tag{2.69}$$

where the invariant amplitude $\mathcal{M}$, the matrix element, is defined as

$$-i\mathcal{M} = [ie(\mathbf{p}_1 + \mathbf{p}_3)^\mu] \left(-i\frac{g_{\mu\nu}}{\mathbf{q}^2} \right) [ie(\mathbf{p}_2 + \mathbf{p}_4)^\nu] \tag{2.70}$$

where $(-ig_{\mu\nu}/\mathbf{q}^2)$ is the photon propagator. The four-dimensional delta function arises from the integration

$$\int d^4x \exp[i(\mathbf{p}_1 + \mathbf{p}_2 - \mathbf{p}_3 - \mathbf{p}_4)\cdot\mathbf{x}] = (2\pi)^4\delta^4(\mathbf{p}_1 + \mathbf{p}_2 - \mathbf{p}_3 - \mathbf{p}_4) \ . \tag{2.71}$$

Now we have an expression for T_{fi} that is completely general for $2 \to 2$ scattering. The matrix element can be written down for any process.

From here we can define the cross section. First we have to fix N. We can do this from the probability density recalling $\rho = 2E|N|^2$. If we fix $2E$ particles in volume V, $\int_V \rho dV = 2E$ and $N = 1/\sqrt{V}$. The $1+2 \to 3+4$ transition rate per unit volume and unit time is

$$W_{fi} = \frac{|T_{fi}|^2}{TV} \,. \tag{2.72}$$

Substituting Eq. (2.69) and squaring, we have

$$W_{fi} = \frac{(N_1 N_2 N_3 N_4)^2}{TV}(2\pi)^8[\delta^4(\mathbf{p}_1 + \mathbf{p}_2 - \mathbf{p}_3 - \mathbf{p}_4)]^2|\mathcal{M}|^2 \,. \tag{2.73}$$

We can cancel the TV in the denominator by one power of $(2\pi)^4\delta^4(\mathbf{p}_1 + \mathbf{p}_2 - \mathbf{p}_3 - \mathbf{p}_4)$. Then putting in the definition of N, we are left with

$$W_{fi} = \frac{(2\pi)^4}{V^4}\delta^4(\mathbf{p}_1 + \mathbf{p}_2 - \mathbf{p}_3 - \mathbf{p}_4)|\mathcal{M}|^2 \,. \tag{2.74}$$

This is proportional to the cross section after we sum over all final states and average over the initial states:

$$\text{cross section} = \frac{W_{fi}}{\text{initial flux}}\text{number of final states} \,. \tag{2.75}$$

The cross section is usually denoted σ and has dimensions of area. For most particle interactions, the area is expressed in units of barns, b, as in "can't hit the side of a barn". The total cross section for *e.g.* proton-proton interactions is in units of millibarns, mb.

The number of final-states is calculated in a periodic box of volume $L \times L \times L = V$. In the z direction, the momentum ranges from p_z to $p_z + dp_z$ so that the number of states within the momentum interval dp_z is $Ldp_z/2\pi$. If the same is true in the x and y directions, the number of states in the box is $Vd^3p/(2\pi)^3$. Normalizing to $2E$ particles in V, as defined by the density of states, the number of states per particle is

$$\frac{V\, d^3p}{(2\pi)^3 2E} \,. \tag{2.76}$$

Then in the reaction $1 + 2 \to 3 + 4$, the number of final states is

$$\frac{V\,d^3p_3}{(2\pi)^3 2E_3}\frac{V\,d^3p_4}{(2\pi)^3 2E_4} \ . \tag{2.77}$$

The cross section must be Lorentz invariant.

EXAMPLE: Show that d^3p/E is Lorentz invariant.
We start with the definition of the longitudinal boost. Then the differential of the boost is

$$dp_z^* = \gamma\,dp_z - \gamma\beta\,dE \ . \tag{2.78}$$

Taking the derivative of Eq. (2.17), we have

$$E\,dE = p_z\,dp_z \ . \tag{2.79}$$

Substituting for dE in Eq. (2.78) and factoring out dp_z/E, we have

$$dp_z^* = \frac{dp_z}{E}(\gamma E - \beta p_z) = \frac{dp_z}{E}E^* \ . \tag{2.80}$$

The last expression comes from the definition of the boosted energy. Using the invariance of the transverse momentum, we find that the differential is boost invariant,

$$\frac{d^3p^*}{E^*} = \frac{d^2p_\perp\,dp_z^*}{E^*} = \frac{d^2p_\perp\,dp_z}{E} = \frac{d^3p}{E} \ , \tag{2.81}$$

as it should be.

The initial flux is also invariant but it is most convenient to calculate it in the laboratory frame for a fixed target. The number of target particles per volume is $2E_2/V$ since the number of particles is normalized to $2E$. The number of beam particles is not computed per volume but per area per time. Thus the number of particles per volume is multiplied by $|\vec{v}_1|$ to obtain $2E_1|\vec{v}_1|/V$. Then in the laboratory frame we have

$$\text{initial flux} = |\vec{v}_1|\frac{2E_1}{V}\frac{2E_2}{V} \ . \tag{2.82}$$

If we consider only the numerator of the flux, defined as F, for the case when both the 'beam' and 'target' particles are in motion with relative velocity $|\vec{v}_1 - \vec{v}_2|$,

$$F = |\vec{v}_1 - \vec{v}_2| 2E_1 2E_2 \tag{2.83}$$

Inserting the definition $\vec{v}_1 = \vec{p}_1/E_1$ and noting that $|\vec{v}_1 - \vec{v}_2| = |\vec{v}_1| + |\vec{v}_2|$ for a collinear collision where particles 1 and 2 are moving toward each other, we can write F as

$$F = 4(|\vec{p}_1|E_2 + |\vec{p}_2|E_1) . \tag{2.84}$$

We can finally write F in the invariant form,

$$F = 4\sqrt{(\mathbf{p}_1 \cdot \mathbf{p}_2)^2 - m_1^2 m_2^2} , \tag{2.85}$$

equivalent to the definition in Eq. (2.84).

EXAMPLE: Show that Eqs. (2.84) and (2.85) are equivalent.
We expand the square of the four-vector product, $(\mathbf{p}_1 \cdot \mathbf{p}_2)^2$ so that, for $F = 4\sqrt{R}$,

$$R \equiv (\mathbf{p}_1 \cdot \mathbf{p}_2)^2 - m_1^2 m_2^2 = (E_1 E_2 - \vec{p}_1 \cdot \vec{p}_2)^2 - m_1^2 m_2^2 .$$

We substitute $E^2 = p^2 + m^2$ for the squared energy terms. We also use the collinear momentum to write $\vec{p}_1 \cdot \vec{p}_2 = -|\vec{p}_1||\vec{p}_2|$. Then

$$R = |\vec{p}_1|^2|\vec{p}_2|^2 + m_1^2|\vec{p}_2|^2 + m_2^2|\vec{p}_1|^2 + 2|\vec{p}_1||\vec{p}_2|E_1E_2 + |\vec{p}_1|^2|\vec{p}_2|^2$$

after canceling the $m_1^2 m_2^2$ terms. Factoring, we have

$$\begin{aligned} R &= |\vec{p}_1|^2(|\vec{p}_2|^2 + m_2^2) + |\vec{p}_2|^2(|\vec{p}_1|^2 + m_1^2) + 2|\vec{p}_1||\vec{p}_2|E_1E_2 \\ &= |\vec{p}_1|^2 E_2^2 + 2|\vec{p}_1||\vec{p}_2|E_1E_2 + |\vec{p}_2|^2 E_1^2 \\ &= (|\vec{p}_1|E_2 + |\vec{p}_2|E_1)^2 . \end{aligned}$$

After inserting R into Eq. (2.85), we find Eq. (2.84).

In the center-of-mass frame, $|\vec{p}_1| = |\vec{p}_2| \equiv p_1$ so that

$$F = 4p_1(E_1 + E_2) . \tag{2.86}$$

Since $S = (\mathbf{p}_1 + \mathbf{p}_2)^2 = (E_1 + E_2)^2 - (\vec{p}_1 + \vec{p}_2)^2$ and $\vec{p}_1 = -\vec{p}_2$, we have

$$F = 4p_1\sqrt{S} \ . \tag{2.87}$$

We can now obtain a more quantitative expression for the two-body scattering cross section. Inserting the number of final states, Eq. (2.77), the initial flux, Eq. (2.82), and the transition rate, Eq. (2.74), into Eq. (2.75), the differential cross section is

$$\begin{aligned} d\sigma &= \frac{V^2|\mathcal{M}|^2}{|\vec{v}_1|2E_1 2E_2}\frac{(2\pi)^4}{V^4}\delta^4(\mathbf{p}_1 + \mathbf{p}_2 - \mathbf{p}_3 - \mathbf{p}_4) \\ &\quad \times \frac{V d^3p_3}{(2\pi)^3 2E_3}\frac{V d^3p_4}{(2\pi)^3 2E_4} \ . \end{aligned} \tag{2.88}$$

Note that the volume cancels so that the cross section is independent of volume, as it should be. The cross section may be written schematically as

$$d\sigma = \frac{|\mathcal{M}|^2}{F}\, d\text{Lips} \tag{2.89}$$

where dLips is the Lorentz invariant phase space,

$$d\text{Lips} = (2\pi)^4\delta^4(\mathbf{p}_1 + \mathbf{p}_2 - \mathbf{p}_3 - \mathbf{p}_4)\frac{d^3p_3}{(2\pi)^3 2E_3}\frac{d^3p_4}{(2\pi)^3 2E_4} \ . \tag{2.90}$$

We now calculate the phase space. By canceling factors of 2π in Eq. (2.90) and splitting the four-dimensional delta function into its energy and momentum components, we have

$$\begin{aligned} d\text{Lips} &= \frac{1}{4\pi^2}\delta(E_1 + E_2 - E_3 - E_4)\frac{d^3p_3}{2E_3} \\ &\quad \times \frac{d^3p_4}{2E_4}\delta(\vec{p}_1 + \vec{p}_2 - \vec{p}_3 - \vec{p}_4) \ . \end{aligned} \tag{2.91}$$

In the center-of-mass frame of $1 + 2 \to 3 + 4$, $\vec{p}_1 + \vec{p}_2 = 0$ and $\sqrt{S} = E_1 + E_2$. Thus the momenta of the final-state particles must also be equal and opposite. We can integrate over p_4 using $\delta(\vec{p}_3 + \vec{p}_4)$. The invariant phase space is reduced to

$$d\text{Lips} = \frac{1}{4\pi^2}\delta(\sqrt{S} - E_3 - E_4)\frac{p_3^2\, dp_3\, d\Omega}{4E_3 E_4} \ . \tag{2.92}$$

On the final-state side, the remaining energy delta function gives $\sqrt{S} = E_3 + E_4$. We change the energy delta function to a delta function in p_3 using the property

$$\delta(f(p_3)) = \delta(p_3 - p_{3_0})/|\partial f(p_3)/\partial p_3|_{p_3=p_{3_0}} \; . \tag{2.93}$$

We find

$$\delta(\sqrt{S} - E_3 - E_4) = \frac{1}{p_{3_0}} \frac{E_3 E_4}{\sqrt{S}} \delta(p_3 - p_{3_0}) \; . \tag{2.94}$$

EXAMPLE: Show this.
Let $f(p_3) = \sqrt{S} - E_3 - E_4$. To determine p_{3_0}, we let $f(p_3) = 0$ so that

$$\sqrt{S} = E_3 + E_4 = \sqrt{p_3^2 + m_3^2} + \sqrt{p_3^2 + m_4^2} \; .$$

After some algebra, $p_{3_0} = p_{\text{cm}}$ in Eq. (1.33). Then $\partial f(p_3)/\partial p_3$ is

$$\begin{aligned} \frac{\partial f(p_3)}{\partial p_3}|_{p_3=p_{3_0}} &= -p_{3_0}\left(\frac{1}{E_3} + \frac{1}{E_4}\right) \\ &= p_{3_0}\left(\frac{E_3 + E_4}{E_3 E_4}\right) = -p_{3_0}\frac{\sqrt{S}}{E_3 E_4} \; . \end{aligned} \tag{2.95}$$

After substituting Eq. (2.94) into Eq. (2.92), the factors of $E_3 E_4$ cancel along with one power of p_3, leaving us with

$$d\text{Lips} = \frac{1}{4\pi^2} \frac{\delta(p_3 - p_{3_0})}{4\sqrt{S}} p_3 dp_3 \, d\Omega = \frac{1}{4\pi^2} \frac{p_{3_0}}{4\sqrt{S}} \, d\Omega \; . \tag{2.96}$$

Using the result for dLips with the flux calculated in the center-of-mass frame, Eq. (2.87), the differential cross section in solid angle $d\Omega$ is

$$\frac{d\sigma}{d\Omega} = \frac{|\mathcal{M}|^2}{4p_1\sqrt{S}} \frac{p_{3_0}}{16\pi^2\sqrt{S}} = \frac{p_{3_0}}{p_1} \frac{|\mathcal{M}|^2}{64\pi^2 S} \; . \tag{2.97}$$

We can change variables from solid angle to momentum transfer T to obtain

$$\frac{d\sigma}{dT} = \frac{|\mathcal{M}|^2}{64\pi p_1^2 S} \; . \tag{2.98}$$

EXAMPLE: Show the conversion from $d\Omega$ to dT.
Using Eq. (1.46) in Chapter 1,

$$dT = 2p_1 p_{3_0} d(\cos\theta) \ . \tag{2.99}$$

Since $d\Omega = 2\pi d(\cos\theta)$, we have

$$d\Omega/dT = \pi/p_1 p_{3_0} \ . \tag{2.100}$$

Finally, we note that more than one diagram may lead to the same final state. For example, in electron scattering, since the scattered particles are identical in both the initial and final states, the matrix element is symmetric under the interchange of labels $3 \leftrightarrow 4$ and $1 \leftrightarrow 2$. Thus we actually have two different diagrams that describe $p_1(e^-) + p_2(e^-) \to p_3(e^-) + p_4(e^-)$, shown in Fig. 2.4. The left-hand

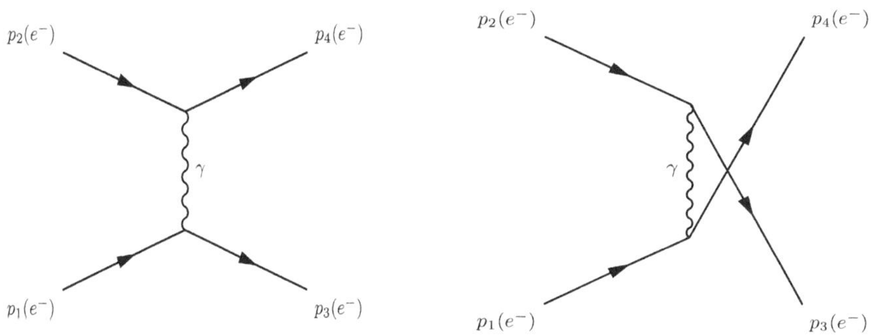

Figure 2.4: Diagrams for electron scattering. The T-channel (left-hand side) and U-channel (right-hand side) processes are shown.

diagram, shown also in Fig. 2.3, is known as a T-channel diagram and the right-hand as a U-channel diagram from the corresponding Mandelstam invariants. The matrix elements from the two diagrams have to be added and squared to get the complete result.

We do not have to have identical particle scattering to have symmetry under interchange. In the case of heavy quark-antiquark ($Q\overline{Q}$) pairs, the dominant contribution at leading order is from $gg \to Q\overline{Q}$.

Here, $p_1(g) + p_2(g) \to p_3(Q) + p_4(\overline{Q})$, and the identical particles, the gluons, are in the initial state but we can draw two diagrams similar to those above for electron scattering. In addition, the fact that gluons can interact with each other leads to a third diagram with a three-gluon vertex. All three diagrams are shown in Fig. 2.5. The left-hand diagram shows the three-gluon vertex (the s channel) while the middle and right-hand diagrams correspond to the t and u channel processes respectively. The u channel diagram is the result of interchanging p_1

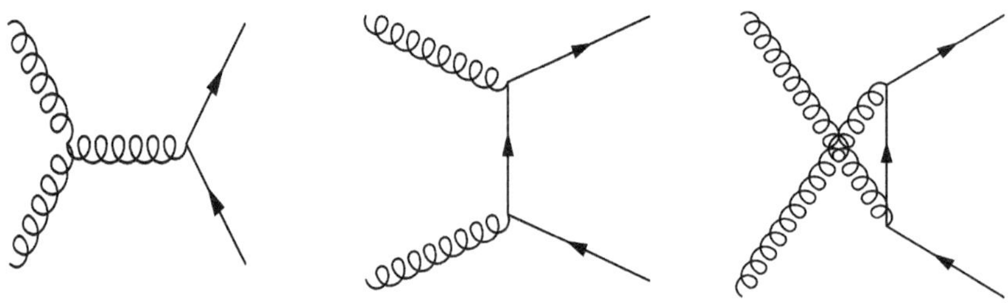

Figure 2.5: Leading order contributions to $gg \to Q\overline{Q}$ in the s (left), t (middle) and u (right) channels.

and p_2 (rather than the outgoing particles as in Fig. 2.4) since quarks and antiquarks are not indistinguishable. The three matrix elements are added before squaring. For completeness, we note that, at leading order, there is another contribution to heavy quark-antiquark pair production, $q\overline{q} \to Q\overline{Q}$, referred to as $q\overline{q}$ annihilation, shown in Fig. 2.6. This process is also in the s channel. Note that when referring to the invariants in the gg and $q\overline{q}$ processes we have used lower case letters, s, t and u to refer to the Mandelstam invariants. We do this because the quarks and gluons carry only a fraction of the parent hadron momentum, as discussed later in this chapter, and are thus on the parton level rather than on the hadron level. These partonic Mandelstam invariants are written in lower case letters.

We can use these properties to translate between different types of scattering also. This is known as crossing. The electron scattering matrix elements can be used to arrive at those for e^+e^- scattering

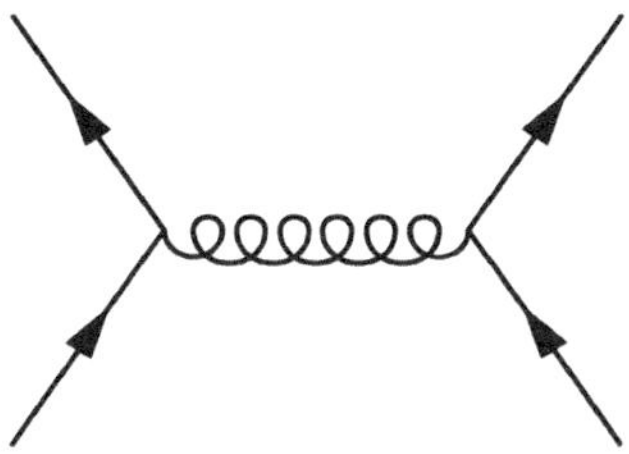

Figure 2.6: Leading order contributions to $q\bar{q} \to Q\overline{Q}$ in the s channel.

instead, as shown in Fig. 2.7, so that

$$\mathcal{M}_{e^+e^-\to e^+e^-}(p_1,p_2,p_3,p_4) = \mathcal{M}_{e^-e^-\to e^-e^-}(p_1,-p_2,p_3,-p_4)\ . \quad (2.101)$$

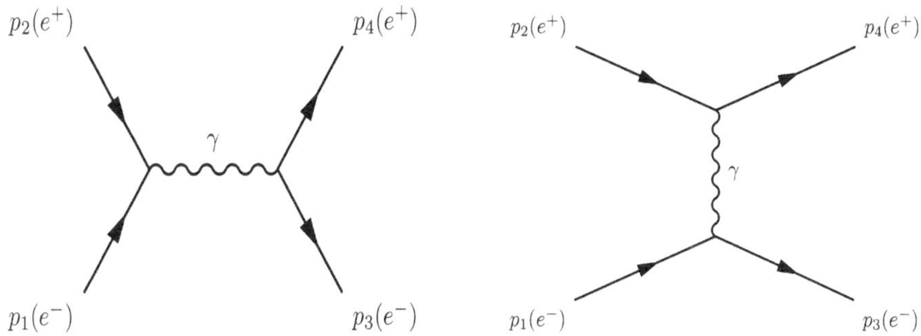

Figure 2.7: The electron scattering diagrams of Fig. 2.4 crossed to obtain those for electron-positron scattering. The left-hand annihilation diagram is an S-channel process while the right-hand scattering diagram is a T-channel process.

2.3 The wave-optical model and total cross sections

We can use the picture of plane-wave scattering discussed previously to obtain the total and elastic cross sections for interactions of two hadrons. This technique is referred to as the wave-optical model and provides a rather general description of scattering.

Consider a plane wave, normalized to unity, $\psi_i = \exp(ikz)$, where $k = 1/\lambdabar$ and $\lambda = 2\pi\lambdabar$ is the de Broglie wavelength. The time dependence has been omitted. Instead of considering a preferred direction, such as two particles moving toward each other, the plane wave is represented by a superposition of incoming and outgoing spherical waves. (Outgoing plane waves can be observed emanating from the center of the disturbance when a rock is dropped into a still pond.) At a large distance r from the center, $kr \gg 1$, the spherical wave has the form $\exp(\pm ikr)/kr$ to conserve probability. The flux through a spherical shell of radius r is then independent of r. The angular dependence is determined by the Legendre polynomials, $P_l(\cos\theta)$, as appropriate for a spherical expansion. Thus for $kr \gg 1$, the incident wave can be expanded as

$$\begin{aligned} \psi_i &= \exp(ikz) \\ &= \frac{i}{2kr}\sum_l (2l+1)\left[(-1)^l \exp(-ikr)\right. \\ &\quad \left. - \exp(ikr)\right] P_l(\cos\theta) \ . \end{aligned} \tag{2.102}$$

The first term, proportional to $\exp(-ikr)$, is the incoming and the second, $\exp(ikr)$, is the outgoing wave. (If the time dependence is included, the difference between the incoming and outgoing components of the wavefunction is clear.) In scattering problems, typically $k \sim 10^{13}$ cm^{-1} or $\sim 1\ \text{fm}^{-1}$, equivalent to $k \sim 200$ MeV. The distance from the center is then $r \gg 1/k \gg 1$ fm. Since detectors sit far away from the scattering center, Eq. (2.102) should be appropriate.

The outgoing wave is more likely to be affected by scattering. On the other hand, the outgoing part will change phase and amplitude by $2\delta_l$ and η_l $(0 < \eta_l < 1)$ respectively so that

$$\begin{aligned} \psi_{\text{total}} &= \frac{i}{2kr}\sum_l (2l+1)\left[(-1)^l \exp(-ikr)\right. \\ &\quad \left. -\eta_l \exp[i(2\delta_l + kr)]\right] P_l(\cos\theta) \ . \end{aligned} \tag{2.103}$$

The scattered wave is the difference between the total and incident waves. The first term in each, representing the incoming wave, is canceled by subtraction so that

$$\begin{aligned}\psi_{\text{scatt}} &= \psi_{\text{total}} - \psi_i \\ &= \frac{-i\exp(ikr)}{2kr}\sum_l (2l+1)\left[\eta_l \exp(2i\delta_l) - 1\right] P_l(\cos\theta) \\ &= \frac{\exp(ikr)}{r} F(\theta)\ . \end{aligned} \tag{2.104}$$

Equation (2.104) represents elastic scattering since k is unchanged. This is only strictly true in the center-of-mass frame. In the laboratory frame, Eq. (2.104) is only true if the point the wave is scattered from is infinitely massive. The approximation is rather good for a low energy electron scattering off a proton since $m_p \gg m_e$.

The scattered outgoing flux through a spherical surface of solid angle $d\Omega$ is

$$v_o |F(\theta)|^2\, d\Omega = v_o \psi^*_{\text{scatt}} \psi_{\text{scatt}} r^2\, d\Omega \tag{2.105}$$

where v_o is the velocity of the outgoing wave. By definition, the right-hand side of Eq. (2.105) is the product of the elastic scattering cross section and the incident flux, $v_i \psi_i^* \psi_i d\sigma_{\text{el}}$. We have normalized ψ_i to unity so that $|\psi_i|^2 = 1$ and, for elastic scattering, $v_i = v_o$. Then

$$v_o |F(\theta)|^2\, d\Omega = v_o\, d\sigma_{\text{el}} \tag{2.106}$$

and

$$\frac{d\sigma_{\text{el}}}{d\Omega} = |F(\theta)|^2\ , \tag{2.107}$$

after canceling the v_o factors.

The total elastic scattering cross section is the integral of the scattering amplitude $F(\theta)$ over solid angle,

$$\begin{aligned}\sigma_{\text{el}} &= \int d\Omega\, |F(\theta)|^2 \qquad\qquad (2.108) \\ &= \int d\Omega \frac{1}{k^2} \sum_l \sum_{l'} (2l+1)(2l'+1) \\ &\quad \times \left[\frac{\eta_l \exp(2i\delta_l) - 1}{2i}\right]^* \left[\frac{n_{l'} \exp(2i\delta_l') - 1}{2i}\right] P_l(\cos\theta) P_{l'}(\cos\theta)\ .\end{aligned}$$

Since the Legendre polynomials are orthonormal,

$$\int d\Omega P_l(\cos\theta)P_{l'}(\cos\theta) = 4\pi\delta_{l,l'}/(2l+1) , \tag{2.109}$$

where $\delta_{l,l'}$ is the orthonormality condition. Then the elastic cross section becomes

$$\sigma_{\mathrm{el}} = 4\pi\lambda\!\!\!^{-2}\sum_l (2l+1)\left|\frac{\eta_l \exp(2i\delta_l)-1}{2i}\right|^2 . \tag{2.110}$$

The last equality comes from replacing the wave number k with the de Broglie wavelength and using the orthonormality of the Legendre polynomials. If the incoming wave is not absorbed, $\eta_l = 1$, and

$$\begin{aligned}\sigma_{\mathrm{el}} &= 4\pi\lambda\!\!\!^{-2}\sum_l (2l+1)\left|\frac{\exp(2i\delta_l)-1}{2i}\right|^2 \\ &= 4\pi\lambda\!\!\!^{-2}\sum_l (2l+1)\sin^2\delta_l . \end{aligned} \tag{2.111}$$

If the phase shift, δ_l, is zero, this corresponds to no scattering potential and $\sigma_{\mathrm{el}} = 0$.

If $\eta_l < 1$, then there is some absorption and we calculate a reaction cross section, also known as the inelastic cross section, using probability conservation. The reaction cross section is the difference between the squares of the incoming and outgoing wavefunctions weighted by r^2 and integrated over solid angle. The initial wavefunction is the first term in the incident wave, Eq. (2.102), and the outgoing wavefunction is the second term in the total wavefunction, Eq. (2.103). Carrying out this operation, we find

$$\begin{aligned}\sigma_{\mathrm{inel}} &= \int d\Omega\, r^2\left(|\psi_{\mathrm{in}}|^2 - |\psi_{\mathrm{out}}|^2\right) \\ &= \pi\lambda\!\!\!^{-2}\sum_l (2l+1)(1-\eta_l^2) . \end{aligned} \tag{2.112}$$

EXAMPLE: Show this.

Squaring the first term in Eq. (2.102) and the second term in Eq. (2.103), the r^2 factors cancel. Then

$$\begin{aligned}\sigma_{\text{inel}} &= \frac{1}{4k^2}\int d\Omega \sum_l \sum_{l'} (2l+1)(2l'+1)\left[(-1)^l(-1)^{l'}\right. \\ &- \left.\eta_l\eta_{l'}\exp[2i(\delta_l-\delta_{l'})]\right] P_l(\cos\theta)P_{l'}(\cos\theta)\ . \qquad (2.113)\end{aligned}$$

Using the orthonormality of the Legendre polynomials, the phase shifts in the exponential cancel, as does one power of $(2l+1)$, leaving us with

$$\sigma_{\text{inel}} = \frac{\pi}{k^2}\sum_l (2l+1)(1-\eta_l^2)\ . \qquad (2.114)$$

Replacing k by $1/\lambda\!\!\!^{-}$, we arrive at Eq. (2.112).

The total cross section is the sum of the elastic and inelastic cross sections, now with $\eta_l \neq 1$, as in Eq. (2.110), so that

$$\begin{aligned}\sigma_{\text{tot}} &= \sigma_{\text{el}} + \sigma_{\text{inel}} \\ &= \pi\lambda\!\!\!^{-2}\sum_l (2l+1)(1-\eta_l^2+1+\eta_l^2-2\eta_l\cos 2\delta_l) \\ &= 2\pi\lambda\!\!\!^{-2}\sum_l (2l+1)(1-\eta_l\cos 2\delta_l)\ . \qquad (2.115)\end{aligned}$$

If we compare this to the expression for $F(\theta)$, we see some similarities to the imaginary part of $F(\theta)$, so that

$$\text{Im}F(\theta) = \frac{1}{2k}\sum_l (2l+1)(1-\eta_l\cos 2\delta_l)P_l(\cos\theta)\ . \qquad (2.116)$$

If $\theta = 0$, $\cos\theta = 1$ and $P_l(1) = 1$ in the forward direction for all l. Then

$$\text{Im}F(0) = \frac{k}{4\pi}\sigma_{\text{tot}}\ . \qquad (2.117)$$

The relationship between the total cross section and the imaginary part of the forward scattering amplitude is known as the optical theorem.

The relations we have produced between the elastic, inelastic and total cross sections as functions of η_l and δ_l set bounds imposed by probability conservation. This probability conservation is also called the unitarity condition. This condition means that the intensity of the outgoing wave cannot exceed that of the incoming wave. We can see this by defining the l-dependent elastic scattering amplitude for the lth partial wave as

$$f(l) = \frac{\eta_l \exp(2i\delta_l) - 1}{2i} = \frac{i}{2}\left(1 - \eta_l \exp(2i\delta_l)\right) \tag{2.118}$$

so that

$$\sigma_{\text{el}} = 4\pi\lambda\!\!\!^{-2} \sum_l (2l+1)|f(l)|^2 \ . \tag{2.119}$$

The partial-wave amplitude can be plotted as a vector in the complex plane since $\text{Re}f(l) = (\eta_l/2)\sin 2\delta_l$ and $\text{Im}f(l) = (1/2)(1 - \eta_l \cos 2\delta_l)$. When $\eta_l = 1$, the end of the vector describes a circle of radius $1/2$ centered at $i/2$. At $\delta_l = 0$, $\text{Re}f(l) = 0$ and $\text{Im}f(l) = 1/2$ while at $\delta_l = \pi/2$, $\text{Re}f(l) = 0$ and $\text{Im}f(l) = 1$. The end of the vector lies within the unitary circle if $\eta_l < 1$, as shown in Fig. 2.8. The maximum elastic scattering cross section is at $\delta_l = \pi/2$,

$$\sigma_{\text{el}} = 4\pi\lambda\!\!\!^{-2} \sum_l (2l+1) \tag{2.120}$$

where $\sigma_{\text{tot}} = \sigma_{\text{el}}$. On the other hand, the inelastic cross section is independent of δ_l but has its maximum at $\eta_l = 0$,

$$\sigma_{\text{inel}} = \pi\lambda\!\!\!^{-2} \sum_l (2l+1) \ , \tag{2.121}$$

see Eq. (2.114), so that $\sigma_{\text{el}} = \sigma_{\text{inel}} = \sigma_{\text{tot}}/2$. When $\eta_l = 1$, the inelastic cross section vanishes.

The maximum inelastic cross section can be obtained classically if we assume that orbital angular momentum l corresponds to impact parameter $b = l\lambda\!\!\!^-$. Then particles with $l \to l+1$ are absorbed by an annular ring of cross sectional area

$$\begin{aligned} \sigma_{\text{inel}} &= \pi(b_{l+1}^2 - b_l^2) \\ &= \pi\left[(l+1)^2 - l^2\right]\lambda\!\!\!^{-2} \\ &= \pi\lambda\!\!\!^{-2}(2l+1) \ . \end{aligned} \tag{2.122}$$

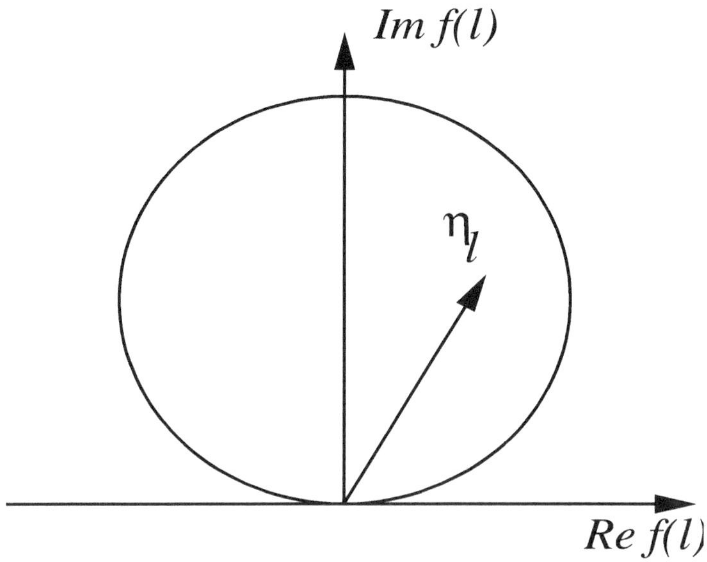

Figure 2.8: The unitary circle described by Eq. (2.118). When $\eta_l < 1$ the vector lies within the circle, as shown.

The wave-optical picture can be linked to elementary particle scattering. If $f(l)$ goes through a maximum for a particular l and $\lambda\!\!\!/$, the two particles resonate. For $f(l)$ to pass through a maximum, the phase shift passes through $\delta_l = \pi/2$. This 'resonance' has a unique total angular momentum, $J = l$, and a mass equal to the center-of-mass energy of the two particles. The resonance state has a finite width, Γ, or lifetime, τ. We drop the l subscript and let $\eta = 1$. Removing a factor of $\exp(i\delta)$ from the numerator of Eq. (2.118), we have

$$\begin{aligned} f(l) &= \frac{\exp(i\delta)[\exp(i\delta) - \exp(-i\delta)]}{2i} \\ &= \exp(i\delta)\sin\delta \\ &= \frac{\sin\delta}{\cos\delta - i\sin\delta} \\ &= \frac{1}{\cot\delta - i} \, . \end{aligned} \tag{2.123}$$

At $\delta = \pi/2$, $\cot\delta = 0$. Let E be the resonance energy and E_R its value at $\delta = \pi/2$. Expanding $\cot\delta$ around $E = E_R$,

$$\cot\delta(E) = \cot\delta(E_R) + (E - E_R)\frac{d}{dE}\cot\delta(E)\Big|_{E=E_R} + \frac{1}{2}(E - E_R)^2\frac{d^2}{dE^2}\cot\delta(E)\Big|_{E=E_R} + \ldots \quad (2.124)$$

At $E = E_R$, $\cot\delta(E_R) = 0$ and $d\cot\delta(E = E_R)/dE = -2/\Gamma$. If $|E - E_R| \simeq \Gamma \ll E_R$ the resonance is symmetric and $d^2\cot\delta(E = E_R)/dE^2 = 0$.

If the resonance is broad, the phase space varies over the width and the resonance is asymmetric. After substituting the value of the derivative with respect to E in Eq. (2.124), we have

$$\cot\delta(E) \simeq -(E - E_R)\frac{2}{\Gamma}\,. \quad (2.125)$$

Then

$$f(E) = \frac{1}{\cot\delta(E) - i} = \frac{1}{-(E - E_R)2/\Gamma - i} = \frac{\Gamma/2}{-(E - E_R) - i\Gamma/2} \quad (2.126)$$

The elastic cross section is then

$$\sigma_{\text{el}} = 4\pi\lambda\!\!\!^{-2}(2l+1)\left[\frac{\exp(2i\delta) - 1}{2i}\right]^2 = 4\pi\lambda\!\!\!^{-2}(2l+1)|f(E)|^2 = 4\pi\lambda\!\!\!^{-2}(2l+1)\frac{\Gamma^2/4}{(E - E_R)^2 + \Gamma^2/4}\,, \quad (2.127)$$

the Breit-Wigner formula. The width Γ is defined as the point where $\sigma_{\text{el}}/\sigma_{\text{el}}^{\max} = 0.5$ or $|E - E_R| = \pm\Gamma/2$. This is the "full width at half maximum". The full width at half maximum is indicated by the horizontal line in the schematic plot of $\sigma_{\text{el}}/\sigma_{\text{el}}^{\max}$ in Fig. 2.9. For definiteness, the value of E_R is taken to be the mass of the ρ^0, 0.77 GeV. To show how the elastic cross section varies with Γ, we have taken $\Gamma = 0.1$, 0.3 and 0.6 GeV respectively. The narrowest line shape is obtained with the smallest Γ. The factor of $\Gamma/4$ in the numerator also sets the maximum $\sigma/\sigma_{\max}$ at $E/E_R = 0$. The cross section ratio rises as the width increases.

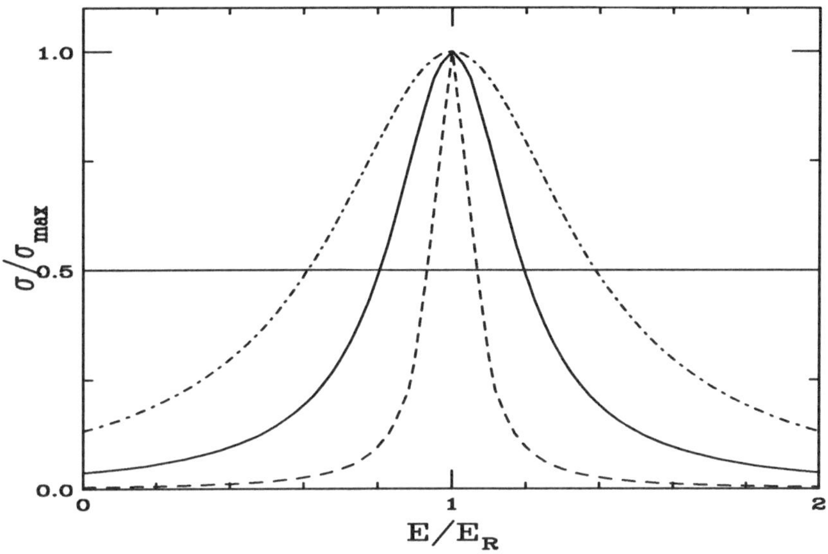

Figure 2.9: The ratio $\sigma_{\rm el}/\sigma_{\rm el}^{\max}$ as a function of E/E_R. We take $E_R =$ 0.77 GeV, the ρ mass and $\Gamma = 0.1$ (dashed), 0.3 (solid) and 0.6 (dot dashed) GeV.

The lifetime is related to the width by $\tau = \hbar/\Gamma$ so that the energy dependence of the amplitude corresponds to the radioactive decay of the resonance by Fourier transform, taking $\omega_R = E_R/\hbar$ and $\tau = \hbar/\Gamma$. The wavefunction of a decaying state is then, with $\hbar = 1$,

$$
\begin{aligned}
\psi(t) &= \psi(0)\exp(-i\omega_R t)\exp(-t/2\tau) \\
&= \psi(0)\exp[-t(iE_R - \Gamma/2)] \qquad (2.128)
\end{aligned}
$$

with intensity $I(t)$,

$$\begin{aligned} I(t) &= \psi^*(t)\psi(t) = |\psi(0)|^2 \exp(-t/\tau) \\ &= I(0)\exp(-t/\tau) \ . \end{aligned} \tag{2.129}$$

The Fourier transform is

$$\begin{aligned} g(\omega) &= \int_0^\infty dt\, \psi(t) \exp(i\omega t) \\ &= \int_0^\infty dt\, \psi(0) \exp[i(\omega - \omega_R)t] \exp(-t/2\tau) \ . \end{aligned} \tag{2.130}$$

Changing variables from ω to E with $\hbar = 1$, we have

$$\begin{aligned} g(E) &= \psi(0) \int_0^\infty dt \exp(-t[\Gamma/2 - i(E - E_R)]) \\ &= \frac{\psi(0)}{i(E - E_R) + \Gamma/2} \\ &= \frac{K}{-(E - E_R) - i\Gamma/2} \end{aligned} \tag{2.131}$$

where K is a constant proportional to $\psi(0)$.

If the resonance is elastic, such as

$$\begin{aligned} \pi^+\pi^- \to \rho^0 \to \pi^+\pi^- \ , \\ \pi^+ n \to \Delta^+ \to \pi^+ n \ , \end{aligned} \tag{2.132}$$

then the cross section is proportional to the decay probability so that $\sigma_{\rm el} \propto |g(E)|^2$. In both cases, the maximum occurs when $E = E_R$ and $\delta = \pi/2$. The maximum of $|g(E)|^2$ is $4K^2/\Gamma^2$ while we have already shown that the maximum elastic cross section is

$$\sigma_{\rm el} = 4\pi\lambda\!\!\!^{-2}(2l+1) \ . \tag{2.133}$$

Setting $|g(E)|^2$ equal to the elastic cross section at the maximum, $\sigma_{\rm el}^{\rm max} = |g(E = E_R)|^2$, gives $K^2 = (\Gamma/4)\sigma_{\rm el}^{\rm max}$.

More generally then, away from the maximum,

$$\sigma_{\rm el} = 4\pi\lambda\!\!\!^{-2}(2l+1)\frac{\Gamma^2/4}{(E - E_R)^2 + \Gamma^2/4} \ . \tag{2.134}$$

For particles with spin, $l \to J$. We also average over the spins of the initial particles, 1 and 2, so that

$$\sigma_{\rm el} = \frac{4\pi\lambda\!\!\!^{-2}(2J+1)}{(2s_1+1)(2s_2+1)} \frac{\Gamma^2/4}{(E-E_R)^2+\Gamma^2/4} \ . \tag{2.135}$$

There are, of course, more than elastic resonance decays. Instead of *e.g.* Eq. (2.132), we may have the final state $\Delta^+ \to \pi^0 p$. Thus the true width includes both elastic and inelastic channels so that

$$\Gamma = \Gamma_{\rm el} + \Gamma_{\rm inel} \ . \tag{2.136}$$

Instead of Γ^2 in the numerator of $\sigma_{\rm el}$, we should have $\Gamma_{\rm el}^2$ while in the inelastic cross section, the numerator is $\Gamma_{\rm el}\Gamma_{\rm inel}$. In the denominator, in both cases, we have the total width, Γ.

We end this section with the cross section compilations of $\sigma_{\rm tot}$ and $\sigma_{\rm el}$. These compilations exists for several beam-target combinations such as pp, $\bar{p}p$, $\pi^\pm p$ and $K^\pm p$. Only the $\bar{p}p$ cross section has been measured for $\sqrt{S} > 60$ GeV. Reactions with secondary beams such as $\pi^\pm$ and $K^\pm$ are, of course, at much lower $\sqrt{S}$.

A simple relationship between the elastic, inelastic and total cross sections can be obtained by taking $\eta_l = 0$ in Eqs. (2.110), (2.112) and (2.115). Then

$$\begin{aligned} \sigma_{\rm el} &= \pi\lambda\!\!\!^{-}\sum_l (2l+1) \ , \\ \sigma_{\rm inel} &= \pi\lambda\!\!\!^{-}\sum_l (2l+1) \ , \\ \sigma_{\rm tot} &= 2\pi\lambda\!\!\!^{-}\sum_l (2l+1) \ . \end{aligned}$$

Such a model would predict $\sigma_{\rm el}/\sigma_{\rm el}^{\rm max} = 0.5$ which does not hold asymptotically. The growing difference between the elastic and total cross sections with energy is shown in Fig. 2.10.

Some resonance structures are clearly visible in the π^+p, π^-p and K^-p cross sections. These are the $\Delta^{++}(uuu)$, $\Delta^0(udd)$ and $\Lambda(uds)$ respectively. At and below these resonances, the total and elastic cross sections are very similar. However, above $p_{\rm lab} = 2$ GeV the elastic cross

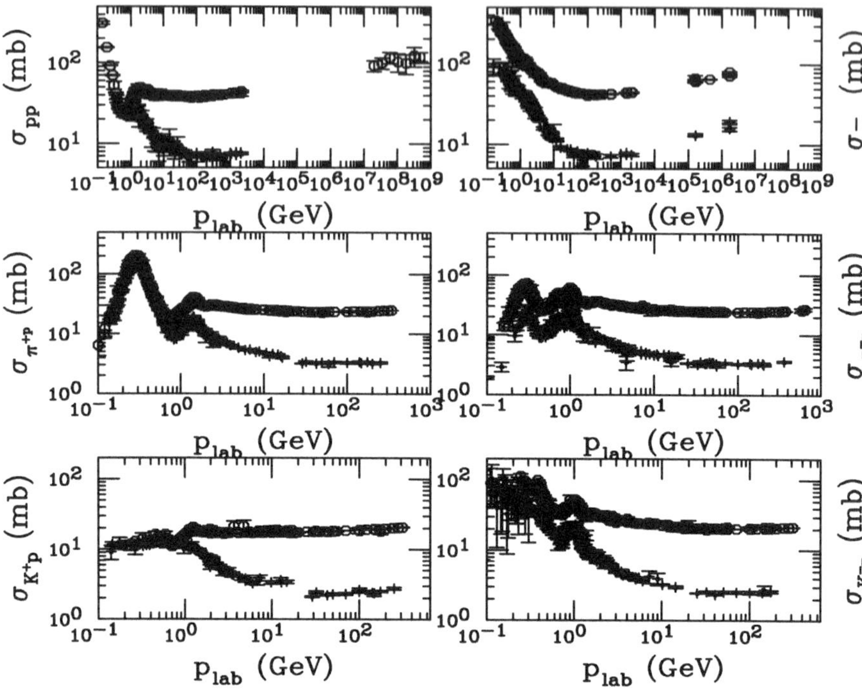

Figure 2.10: The total and elastic cross section data as a function of the laboratory momentum for h^+p (left-hand side) and h^-p (right-hand side) collisions. From top to bottom, $h = p$, π and K. The total cross sections are represented by open circles, the elastic cross sections by crosses. The data can be obtained from Ref. [5].

section drops far below that of the total. Clearly, the inelastic cross section dominates in the high $p_{\rm lab}$ region where it is possible to produce particle-antiparticle pairs without going through a resonance. This difference also points out the increasing importance of gg scattering in particle production with energy. The highest energy pp collisions come from cosmic ray data while the highest energy $\overline{p}p$ data were taken at the Fermilab Tevatron. An energy scan was taken to measure the total and elastic cross sections between $\sqrt{S} = 630$ and 1800 GeV [6]. The measurements show that both the total and elastic cross sections increase in the TeV center-of-mass energy range. A similar program is planned at the LHC [7] to measure the pp total and elastic cross sections for $\sqrt{S} \geq 2$ TeV.

The total cross sections should be independent of the projectile charge at high energy so that $\sigma(h^+p) \approx \sigma(h^-p)$. The data show this to be true at $p_{\mathrm{lab}} > 100$ GeV. Parameterizations of the total cross sections are compared to data in Fig. 2.11. The rise in the total cross

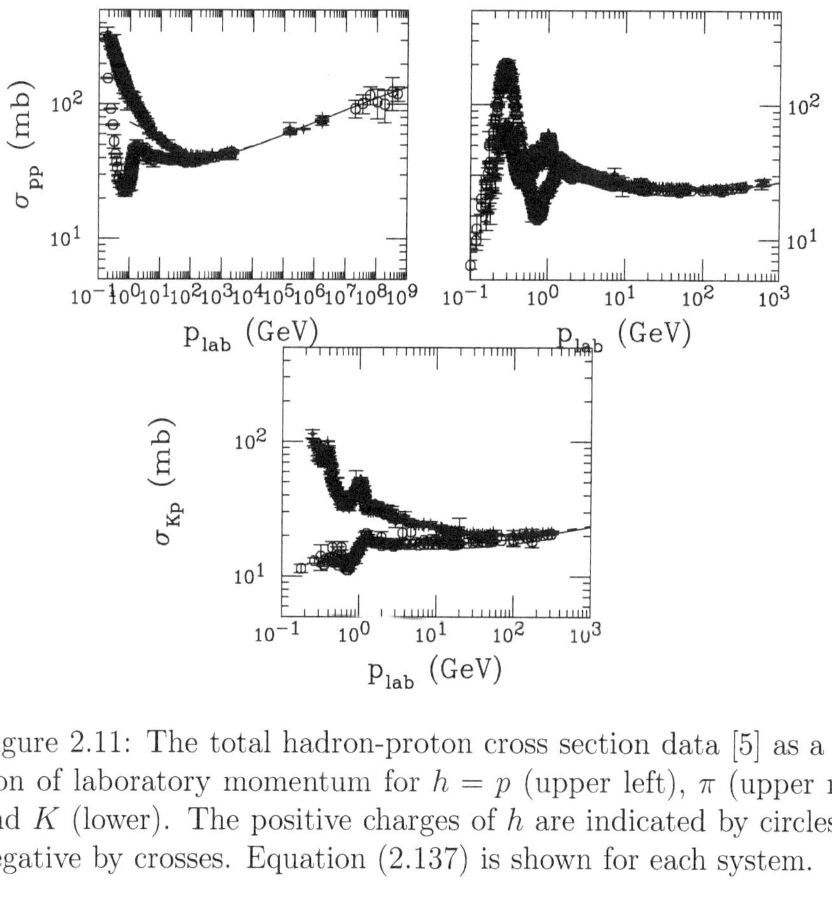

Figure 2.11: The total hadron-proton cross section data [5] as a function of laboratory momentum for $h = p$ (upper left), π (upper right) and K (lower). The positive charges of h are indicated by circles, the negative by crosses. Equation (2.137) is shown for each system.

sections with energy obtained for $h = p$ is expected to hold for all projectiles. The total cross sections above the resonance region have been parameterized as [3]

$$\sigma^{h^\pm p} = Z^{h^\pm p} + B \ln^2\left(\frac{S}{S_0}\right) + Y_1^{h^\pm p}\left(\frac{S_1}{S}\right)^{\eta_1} \mp Y_2^{h^\pm p}\left(\frac{S_1}{S}\right)^{\eta_2} . \quad (2.137)$$

The parameters B, S_0, S_1, η_1 and η_2 are assumed to be universal. The value $\sqrt{S_0} = 5.38$ GeV was chosen to avoid the resonance region.

We also note that charged particle multiplicity – the number of charged particles produced in a give hadron-hadron collision increases with energy,

$$n_{\rm ch} \approx A + B \ln S . \quad (2.138)$$

Particle production in pp collisions is dominated by pions and then kaons. Positive mesons and protons are more important at low $\sqrt{S}$. The antiparticle/particle ratio increases toward unity as $\sqrt{S}$ increases. The trend has been clearly shown in heavy-ion collisions, especially for the $\overline{p}/p$ ratio. By $\sqrt{S} = 200$ GeV, π^-/π^+, $K^-/K^+ \sim 1$ and $\overline{p}/p \sim 0.8$ [8].

2.4 The quark model, hadron-hadron interactions and parton distribution functions

So far, we have discussed real point particle scattering except for the example of $gg \to Q\overline{Q}$ where gluon beams of definite momentum cannot be produced. In this case, the gluon must be obtained from an initial hadron. Gluons, quarks and antiquarks exist within hadrons. It was known for a long time that hadrons were not point particles but had a finite size. Studies of deep-inelastic scattering of electrons on hadrons showed that the virtual photon emitted by an electron suddenly seemed to be hitting a point particle with no structure rather than a finite-size proton. These were called partons by Feynman and quarks by

Gell-Mann. The quark model postulated that baryons carried three quarks while mesons carried a quark and an antiquark. The quarks are tabulated in Table 2.1 while the quark content of some selected baryons and mesons hadrons are shown in Tables 2.2 and 2.3.

There are six types, or flavors, of quarks. All quarks carry baryon number 1/3 while antiquarks carry baryon number $-1/3$. Thus baryons have integer baryon number while mesons have zero baryon number. The up, down and strange quarks were the first to be postulated to describe the large number of particles being discovered at accelerators. The charm quark was postulated to explain the absence of strangeness-changing neutral currents. The fourth quark flavor was also required to renormalize the triangle anomaly. The J/ψ ($c\bar{c}$) – hidden charm – was the first charm hadron to be discovered through its decay to lepton pairs, followed by D mesons – open charm. The third family (bottom and top) was initially a surprise. The Υ ($b\bar{b}$) was first discovered as a peak in the lepton pair spectrum at its mass of 9.46 GeV. The top quark would round out the third family but, because of its large mass, it was not discovered until almost 20 years after the bottom quark [9]. Since the top quark is more massive than the $W^\pm$ bosons, it decays directly through $t \to Wb$ and does not form hadrons. No further quark families exist, precluded by precise measurements of the width of the Z^0 gauge boson at LEP [10]. The antiquarks with opposite electric charge and baryon number but equal masses are not listed in Table 2.1.

The quarks also carry color. The color charge was first postulated to describe the existence of the Δ^{++} (uuu) state with spin $+3/2$. Without any additional quantum numbers, this fermion wavefunction would be totally symmetric, not allowed for fermions. Color was introduced to make the Δ^{++} wavefunction antisymmetric overall. Thus the quarks listed in Table 2.1 represent a color triplet with colors red, green and blue. We discuss some of the implications of this color assignment further below.

As nuclear collisions increase in energy, more massive quarks can be produced and the effects of nuclear medium on their production and propagation can be studied. At the lowest energies, not even strange particles could be directly produced. The J/ψ was studied extensively at the CERN SPS. At RHIC, D mesons are beginning to be studied

Table 2.1: The quark types (flavors), their charges, e_f, relative to the electric charge, e, and their current masses.

Quark type	e_f	Mass (GeV)	Quark type	e_f	Mass (GeV)
up (u)	2/3	0.005	down (d)	$-1/3$	0.010
charm (c)	2/3	$1.2 - 1.8$	strange (s)	$-1/3$	0.15
top (t)	2/3	178	bottom (b)	$-1/3$	$4.5 - 5$

Table 2.2: Selected ground state baryons, including their quark content and masses.

Baryon	Quark content	Mass (GeV)
p	uud	0.938
n	ddu	0.939
Λ	uds	1.116
Σ^-	dds	1.197
Ξ^0	uss	1.315
Ω^-	sss	1.672
Λ_c^+	udc	2.285
Σ_c^{++}	uuc	2.452
Ξ_c^+	usc	2.466
Ω_c	ssc	2.698
Ξ_{cc}^+	dcc	–
Λ_b	udb	5.624

while the first hints of bottom production will also be measured. Bottom production will be copious at the LHC. Indeed, at the LHC the energy is higher than at the Tevatron $p\bar{p}$ collider where the first evidence for the top quark was gathered. Heavy quark production is an example of a 'hard' process, as we will discuss further below.

In Tables 2.2 and 2.3, it is obvious that baryons with similar quark content are more massive than their mesonic counterparts. This is most obviously seen in the difference between the nucleon and pion masses.

As more massive quarks are included, the baryon-meson mass difference decreases so that the Λ_b is only ≈ 300 MeV more massive than the $B^\pm$ and B^0. Also, more combinations of ground state mesons and baryons are available for heavier flavors: $\pi^\pm$, π^0 for light flavors, $K^\pm$, K^0, ϕ for strange quarks, D^0, $D^\pm$, D_s and J/ψ for charm and B^0, $B^\pm$, B_s, B_c and Υ for bottom, not including the excited states. The available decay channels also increase with the quark mass. Note also that the ϕ is more massive than twice the kaon mass so that $\phi \to K\overline{K}$ almost exclusively while the J/ψ and Υ are both less than twice the mass of the D and B respectively, rendering this decay channel unavailable for the lower mass charmonium and bottomonium states. These states are then very long lived since they either decay to multiple hadrons or electromagnetically. The electromagnetic decays to lepton pairs are then some of the most important. Higher mass quarkonium states exist, however, and those that lie above the $D\overline{D}$ and $B\overline{B}$ thresholds do decay almost exclusively through these channels. Their lifetimes are then correspondingly shorter. The quarkonium system will be discussed more thoroughly in Chapter 8.

The antibaryon quark composition is identical to that in Table 2.2 once the quarks are replaced with antiquarks. However, for example, the π^+ and π^- are each other's antiparticles. Therefore it is typically more difficult to produce a baryon-antibaryon pair, such as $p\overline{p}$, in a hadronic collision than a pair of mesons such as $\pi^+\pi^-$ due to the larger mass threshold that must be overcome for particle production.

The center-of-mass threshold energy for production of any given final state must be greater than the sum of the final-state masses,

$$\sqrt{S} \geq \sum_i m_i \tag{2.139}$$

At fixed-target energies, $\sqrt{S} = (m_1^2 + m_2^2 + 2E_{1\,\mathrm{lab}} m_2)^{1/2}$, where m_1 and m_2 are the masses of the projectile and target particles respectively and $E_{1\,\mathrm{lab}}$ is the total energy of the projectile. Substituting this expression for $\sqrt{S}$ into Eq. (2.139) and squaring both sides, we have

$$m_1^2 + m_2^2 + 2E_{1\,\mathrm{lab}} m_2 \geq \left(\sum_i m_i \right)^2 , \tag{2.140}$$

Table 2.3: Selected ground state mesons, including their quark content and masses.

Meson	Quark content	Mass (GeV)
π^+	$u\bar{d}$	0.140
K^0	$d\bar{s}$	0.498
K^+	$u\bar{s}$	0.494
ϕ	$s\bar{s}$	1.020
D^0	$c\bar{u}$	1.864
D^+	$c\bar{d}$	1.869
D_s^+	$c\bar{s}$	1.968
J/ψ	$c\bar{c}$	3.097
B^0	$d\bar{b}$	5.279
B^+	$u\bar{b}$	5.279
B_s^0	$s\bar{b}$	5.370
Υ	$b\bar{b}$	9.460

leading to a minimum laboratory energy of

$$E_{1\,\text{lab}} = \frac{1}{2m_2}\left(\left(\sum_i m_i\right)^2 - m_1^2 - m_2^2\right) . \tag{2.141}$$

The minimum necessary laboratory momentum for production of the final state is then $p_{1\,\text{lab}} = (E_{1\,\text{lab}}^2 - m_1^2)^{1/2}$. From Eq. (1.38) in Chapter 1, the corresponding kinetic energy of the beam is $E_{1\,\text{kin}} = E_{1\,\text{lab}} - m_1$.

EXAMPLE: Calculate the minimum laboratory energy needed to produce an antiproton in a proton-proton collision. Also calculate the beam momentum and the beam kinetic energy.
The antiproton was first discovered in 1955 at a machine, known as the Bevatron, specially designed to produce antiprotons just above threshold. At the time, the Bevatron was the world's most powerful accelerator. Since, in a *pp* collision, there are two baryons in the initial state, the baryon number in the final state must also be two. Therefore, in addition to an antiproton, there must also be three protons in the final state, two to match the baryon number in the initial state and the third to balance the negative

baryon number of the antiproton. The simplest pp interaction to produce an antiproton, $\overline{p}$, is then

$$pp \to p\overline{p}pp \ . \tag{2.142}$$

The sum of the final state masses, $\sum_i m_i$, is equal to four times the proton mass, $4m_p$. In the case of a pp collision, $m_1 = m_2 = m_p$. Thus, for this reaction, the minimum laboratory energy, Eq. (2.141), is

$$E_{1\,\mathrm{lab}} = \frac{(4m_p)^2 - 2m_p^2}{2m_p} = 7m_p = 6.573\,\mathrm{GeV} \ .$$

The laboratory momentum is

$$p_{1\,\mathrm{lab}} = (E_{1,\,\mathrm{lab}}^2 - m_p^2)^{1/2} = \sqrt{48}m_p = 6.505\,\mathrm{GeV}$$

while the kinetic energy of the beam is

$$E_{1\,\mathrm{kin}} = E_{1\,\mathrm{lab}} - m_p = 6m_p = 5.634\,\mathrm{GeV} \ .$$

The Bevatron was designed to accelerate protons to $p_{\mathrm{lab}} = 6.5$ GeV. In 1959, Emilio Segre and Owen Chamberlain won the Nobel Prize in Physics for their discovery of the antiproton [11].

The quark content given in Tables 2.2 and 2.3 that give hadrons their proper charge and baryon number are called valence quarks. The valence quarks are, however, not the only 'partons' in the hadron. There also exist quark-antiquark pairs ($q\overline{q}$) in the 'sea' as well as gluons. Neither the $q\overline{q}$ pairs nor the gluons contribute to the charge, baryon number or mass of the hadron since they exist only virtually and do not come on mass shell unless there is an interaction that disrupts the hadron wavefunction and allows them to materialize. The quark content of the hadron can be studied in depth using deep-inelastic scattering (DIS) of leptons off hadrons, particularly electrons and positrons off protons, such as the diagram shown in Fig. 2.12. The momentum transferred from the lepton to the hadron target via the emitted virtual photon, in the T channel, is $T = \mathbf{q}^2 = (\mathbf{p}_e - \mathbf{p}_{e'})^2$. Since $\mathbf{q}^2$ is negative, the usual variable for the momentum transfer squared is $Q^2 = -\mathbf{q}^2$. The parton struck by the virtual photon carries a fraction of the total

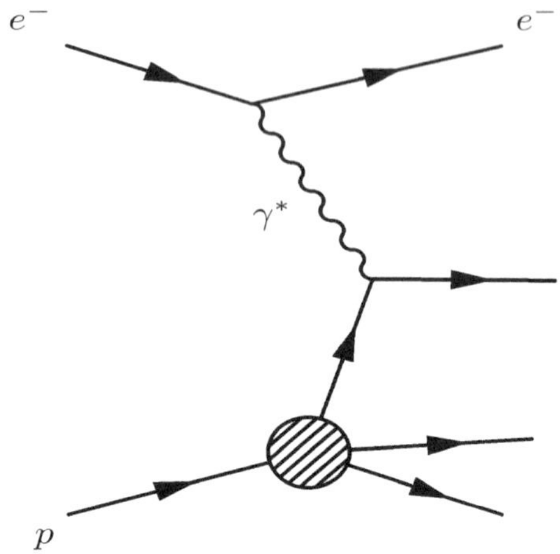

Figure 2.12: Inclusive deep-inelastic scattering of an electron off a proton. The electron emits a virtual photon which interacts with a charged parton in the proton. The interaction disrupts the proton wavefunction, breaking it up in a spray of partons.

proton momentum, $x_{\rm Bj}$ for Bjorken x, usually abbreviated as x. This momentum fraction is the ratio of the momentum transfer squared to twice the product of the hadron and virtual photon four-vectors in the rest frame of the hadron (the laboratory frame for fixed-target experiments but equivalent to the photon-hadron $\sqrt{S}$ when the masses are ignored). Thus

$$x = \frac{T}{S} = \frac{Q^2}{2(\mathbf{p}_h \cdot \mathbf{q})} = \frac{Q^2}{2m_h\nu} \tag{2.143}$$

where ν is the energy transferred to the virtual photon from the electron, $\nu = E_e - E_{e'}$.

The first DIS data was at fairly large x and moderate Q^2. They suggested that the data were independent of Q^2, called scaling. However, soon more data showed that the DIS results were a function of Q^2 and evolution equations that perturbatively described how the parton distributions changed with increasing Q^2 were developed [12].

Since the gluons do not carry electric charge, they cannot be directly measured in DIS. The earliest DIS experiments showed that the proton was a composite object. Higher energy DIS data showed the constituents were point-like. These measurements also showed that more partons than the valence partons exist in the proton since the hadron momentum is not shared equally between the three valence quarks, even giving enough momentum to gluons to produce a bound state.

In DIS, if the parton that the virtual photon interacts with carries a fraction x of the total momentum, the rest of the noninteracting partons carry away the remaining fraction, $1-x$. The fraction of the total proton momentum carried by charged partons did not, in fact, equal unity, showing that there are also charge 'neutral' partons, called gluons. Indeed, the gluons carry nearly half of the proton momentum. However, while the gluons are electric-charge neutral, they are not color-charge neutral. They exchange color between quarks and, as such, carry both a color and an anti-color. Thus the gluons are the force carriers, 'gauge bosons' of the strong interaction, quantum chromodynamics (QCD), as the photons are the gauge bosons of the electromagnetic interaction, quantum electrodynamics (QED).

There is an important difference between gluons and photons though, as we have already alluded to in our discussion of diagrams contributing to the $gg \to Q\overline{Q}$ matrix elements. In QED, when two electric charges are pulled apart, the force between them decreases due to charge screening. However, since quarks carry color while gluons carry a color and an anti-color and photons carry no charge at all, gluons interact with each other and photons do not. Thus when two color charges are pulled apart, there is, instead, color anti-screening and the color charge increases with distance. Therefore, at short distances, *e.g.* within the hadron, the partons behave like free particles with a small coupling constant, known as asymptotic freedom. At large distances, the coupling grows, confining of the partons within the hadron. Confinement is the reason why no free quarks with fractional charges have been, or will be, observed.

The property of confinement means that we can not calculate how the partons within a hadron behave as a function of x at scales Q^2 at or below the confinement scale. We can calculate how they evolve to higher scales though since this corresponds to short distance behavior, $r \sim \hbar c/Q$. We learn about the distribution of partons in the hadron as functions of x and Q^2 through comparison of the calculations to various types of data. These parton distribution functions are thus only as good as the range of the fit. When extrapolated beyond this region, the predictive power of QCD calculations is reduced. (As an aside, the extrapolation of the quark distributions outside the known region, as well as reliance on leading order rather than next-to-leading order calculations for comparison to data led to lower limits on the compositeness scale [13] than those obtained later [14]. [Compositeness refers to quarks being composite objects and not point particles.] In addition, W boson asymmetry measurements as a function of rapidity showed that improvements were needed in the light antiquark distributions [15].)

Here we will describe some of the properties of the parton distributions inside hadrons and discuss how these parton distributions are used in QCD calculations. This is not meant to be a detailed overview of the parton model. There are many excellent descriptions of this material in high energy physics textbooks such as Halzen and Martin [4] as well as Field [16] and some more recent reviews such as that in the CTEQ Handbook of Perturbative QCD [17].

We introduce the notation $f(x)$ for the parton probability density while $xf(x)$ typically refers to the parton momentum distribution. As discussed previously, the u and d quarks are already present as valence quarks in the proton. The valence quarks carry the bulk of the proton momentum at large x. The sea quarks do not contribute to the baryon number of the hadron since they are always produced in a virtual quark-antiquark pair. Therefore, for every $\overline{u}_p$ there is a u_p^s (denoted u_p^s to avoid confusion with the total u_p quark density). The u and d quark densities in the proton include both valence and sea quark densities so that

$$u_p(x) = u_p^v(x) + u_p^s(x) , \quad (2.144)$$

$$d_p(x) = d_p^v(x) + d_p^s(x) . \quad (2.145)$$

The number of valence quarks in the proton sums to three,

$$\int_0^1 dx\, u_p^v(x) = 2 \,, \tag{2.146}$$

$$\int_0^1 dx\, d_p^v(x) = 1 \,, \tag{2.147}$$

two up valence quarks and one down valence quark. Since the sea quarks are produced in pairs, $q^s\overline{q}^s = u^s\overline{u}^s, d^s\overline{d}^s, s^s\overline{s}^s \cdots$, they give a zero contribution to the total baryon number:

$$\int_0^1 dx\, [u_p^s(x) - \overline{u}_p^s(x)] = 0 \,, \tag{2.148}$$

$$\int_0^1 dx\, [d_p^s(x) - \overline{d}_p^s(x)] = 0 \,, \tag{2.149}$$

$$\int_0^1 dx\, [s_p^s(x) - \overline{s}_p^s(x)] = 0 \,, \tag{2.150}$$

and similarly for the heavy flavor sea, c_p^s, b_p^s and t_p^s. The total momentum carried by all the partons must add to the proton momentum, that is, the integral over all the parton densities weighted by the momentum fraction, the parton distributions, must be unity. When the momentum carried by all the charged partons was summed, it was only equal to about half the total proton momentum. The rest is carried by the gluons so that

$$\begin{aligned} 1 = & \int_0^1 dx\, x[u_p(x) + d_p(x) + \overline{u}_p^s(x) + \overline{d}_p^s(x) \\ & + 2\overline{s}_p^s(x) + 2\overline{c}_p^s(x) + 2\overline{b}_p^s(x) + 2\overline{t}_p^s(x) + g_p(x)] \,. \end{aligned} \tag{2.151}$$

While the heavy quark parton distributions are included in Eq. (2.151), they are not active at scales where $Q < m_Q$. Above this value, for $Q \gg m_Q$, the heavy parton is treated as massless in the evolution. We have used Eqs. (2.144) and (2.145) for $u_p(x)$ and $d_p(x)$. For the partons that appear only in the sea, since $q^s = \overline{q}^s$ from Eqs. (2.148)-(2.150), we have written $2\overline{q}^s$ in Eq. (2.151). The charge-neutral gluon appears only once but carries the largest fraction of the proton momentum.

In addition to the proton parton distributions, DIS on targets such as deuterium and other light nuclei measure the neutron distributions. One assumes that the neutron parton densities are related to those in the proton by charge symmetry so that

$$\begin{aligned} u_p(x) &= d_n(x)\,, \qquad d_p(x) = u_n(x)\,,\\ \overline{u}_p(x) &= \overline{d}_n(x)\,, \qquad \overline{d}_p(x) = \overline{u}_n(x)\,,\\ s_p(x) &= s_n(x)\,, \qquad \overline{s}_p(x) = \overline{s}_n(x) \quad \cdots\,. \end{aligned} \tag{2.152}$$

The gluon distribution is assumed to be the same for the proton and the neutron. By charge symmetry, we can also assume that

$$\int_0^1 dx\, u_n^v(x) = 1\,, \tag{2.153}$$

$$\int_0^1 dx\, d_n^v(x) = 2\,. \tag{2.154}$$

EXAMPLE: Given the rules for the proton and neutron, write down the appropriate rules for a Σ^-.
The $\Sigma^-(dds)$ is sometimes used as a secondary beam in fixed-target experiments. The mass of the strange quark is always ignored since parton distributions are calculated in the infinite momentum frame where the parton is treated as massless. Then we can assume

$$\int_0^1 dx\, s_{\Sigma^-}^v(x) = 1\,, \tag{2.155}$$

$$\int_0^1 dx\, d_{\Sigma^-}^v(x) = 2\,, \tag{2.156}$$

and

$$\begin{aligned} u_p(x) &= d_{\Sigma^-}(x)\,, \qquad d_p(x) = s_{\Sigma^-}(x)\,,\\ \overline{u}_p(x) &= \overline{d}_{\Sigma^-}(x)\,, \qquad \overline{d}_p(x) = \overline{s}_{\Sigma^-}(x)\,,\\ s_p(x) &= u_{\Sigma^-}(x)\,, \qquad \overline{s}_p(x) = \overline{u}_{\Sigma^-}(x) \quad \cdots\,. \end{aligned} \tag{2.157}$$

The gluon distribution is again assumed to be the same as in the proton.

Since secondary beams of pions are also available, some sets of parton distributions for the pion have been determined. In the case of mesons such as the pion, one of the valence partons is an antiquark. Thus for the $\pi^-(\overline{u}d)$,

$$\int_0^1 dx\, \overline{u}^v_{\pi^-}(x) = 1\,, \tag{2.158}$$

$$\int_0^1 dx\, d^v_{\pi^-}(x) = 1 \tag{2.159}$$

where now

$$\overline{u}_{\pi^-}(x) = \overline{u}^v_{\pi^-}(x) + \overline{u}^s_{\pi^-}(x)\,, \tag{2.160}$$

$$d_{\pi^-}(x) = d^v_{\pi^-}(x) + d^s_{\pi^-}(x)\,. \tag{2.161}$$

Likewise, the sea quark relations are

$$\int_0^1 dx\, [u^s_{\pi^-}(x) - \overline{u}^s_{\pi^-}(x)] = 0\,, \tag{2.162}$$

$$\int_0^1 dx\, [d^s_{\pi^-}(x) - \overline{d}^s_{\pi^-}(x)] = 0\,, \tag{2.163}$$

$$\int_0^1 dx\, [s^s_{\pi^-}(x) - \overline{s}^s_{\pi^-}(x)] = 0\,. \tag{2.164}$$

EXAMPLE: Write the rules corresponding to Eqs. (2.158)-(2.161) for the π^+.
Since the valence quarks of the π^+ are u and $\overline{d}$, the relations for the π^+ are the antiparticles of those of the π^-. Thus,

$$\int_0^1 dx\, u^v_{\pi^+}(x) = 1\,, \tag{2.165}$$

$$\int_0^1 dx\, \overline{d}^v_{\pi^+}(x) = 1 \tag{2.166}$$

and

$$u_{\pi^+}(x) = u^v_{\pi^+}(x) + u^s_{\pi^+}(x)\,, \tag{2.167}$$

$$\overline{d}_{\pi^+}(x) = \overline{d}^v_{\pi^+}(x) + \overline{d}^s_{\pi^+}(x)\,. \tag{2.168}$$

Deep-inelastic scattering alone does not distinguish between the different types of parton distributions since it is an inclusive measurement. Instead, one measures the structure function,

$$F_2^p(x) = \sum_f e_f^2 x[f_p(x) + \overline{f}_p(x)] \tag{2.169}$$

where $f = u, d, s, c, \cdots$. The charm and heavier quark parton densities are typically zero below production threshold and, when included, the heavy quarks are treated as massless as well.

The Drell-Yan process [18] has been useful in establishing the relative sea quark distributions in hadrons. Thus we now introduce the Drell-Yan process, $q\overline{q} \rightarrow \gamma^* \rightarrow l^+l^-$ to leading order, where the intermediate virtual photon has an invariant mass, M. As we discuss further a bit later, at higher orders, where at least one power of the strong coupling constant is needed, the virtual photon can be in the final state. The real next-to-leading order corrections are $q\overline{q}$ annihilation, $q\overline{q} \rightarrow \gamma^* g$, and the QCD Compton scattering diagram, $qg \rightarrow q\gamma^*$. In both cases, the intermediate state is a quark. In the latter process, the q can be replaced by a $\overline{q}$. If the photon is not virtual but real, these processes are the leading order processes for real photon production in QCD.

Drell-Yan dilepton production is another example of a 'hard' probe. Hard probes are so named because their large scale such as mass (M in the case of Drell-Yan and quarkonium production and m_Q in the case of heavy quarks) or transverse momentum (jets originating from massless partons or real high p_T photons) makes perturbative QCD calculations possible. The QCD properties of asymptotic freedom and confinement both play a role in determining when perturbative QCD can be used. The large momentum or mass scale means that the interaction takes over a short time interval or distance scale, $r = \hbar c/Q$. If $Q \geq 2$ GeV, $r \leq 0.1$ fm. On such short distance scales, much smaller than the hadronic size scale, the strong coupling constant is relatively small and we are in the asymptotically free regime. Thus we can talk about two single partons interacting with each other. However, the two partons are still confined to separate hadrons. That is why we also need the global analysis of the parton distribution functions: they make the connection between the real hadronic world and the virtual world of partons in the hadron. The hard scale separates these two regimes.

The separation of the perturbatively-calculable partonic cross section from the nonperturbative, short-distance parton distribution functions was proved possible through the factorization theorem [19]. Once the parton distribution functions have been determined by global fits to data for a given projectile hadron, they can be used for the calculation of any final state. Thus the proton parton distribution functions are universal. Although we do not have to worry about hadronization in the Drell-Yan case, the factorization theorem further states that the final state should be independent of the initial state. This means that when partons produced in an interaction go on shell and become the hadrons finally observed in the detector, it should not matter whether the final state partons were produced by photons, electrons or hadrons. The nonperturbative process that produces the final state hadron from the initial parton is parameterized by fragmentation functions. We will briefly discuss some aspects of hadronization (or fragmentation) in a later chapter.

In Drell-Yan production to leading order, a quark from the projectile annihilates with its corresponding antiquark in the target, producing a virtual photon which then decays to a lepton pair with invariant mass M, $q\overline{q} \rightarrow \gamma^\star \rightarrow l^+l^-$, as shown in Fig. 2.13. The hadronic Drell-

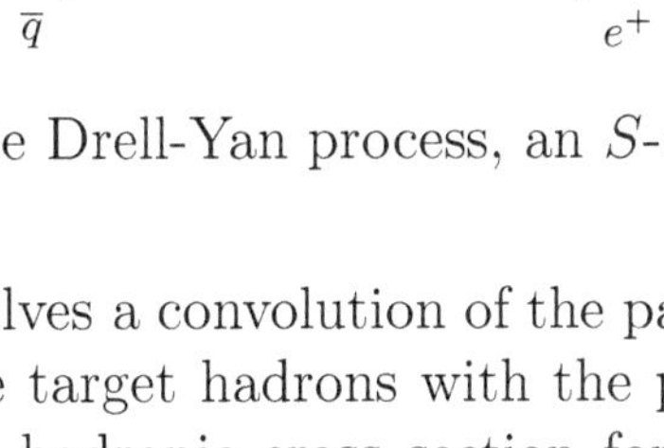

Figure 2.13: The Drell-Yan process, an S-channel process.

Yan cross section involves a convolution of the parton densities of both the projectile and the target hadrons with the partonic cross section. At leading order, the hadronic cross section for any given final state can be expressed as

$$\sigma_{h_1 h_2} = \sum_{f_1,f_2=q,\overline{q},g} \int_0^1 dx_1 dx_2\, f_{1/p}(x_1,Q^2) f_{2/p}(x_2,Q^2) \widehat{\sigma}_{f_1 f_2} \quad (2.170)$$

where $\widehat{\sigma}_{f_1 f_2}$ is the partonic cross section for production of the final state through initial $f_1 f_2$ scattering. The scale of the interaction is Q^2. The partonic cross section for Drell-Yan production as a function of M, with $Q^2 = M^2$, is

$$\frac{d\widehat{\sigma}_{f\overline{f}}}{dM} = \frac{8\pi\alpha^2}{9M} e_f^2 \delta(s - M^2) \quad (2.171)$$

where s is the square of the parton-parton center-of-mass energy, $s = x_1 x_2 S$, x_1 is the fractional momentum carried by the projectile parton, x_2 is the fractional momentum of the target parton, and S is the square of the hadron-hadron center-of-mass four-momentum.

Note that in Eq. (2.171) we have replaced the equivalent four-dimensional delta function, as in Eq. (2.88) for the $2 \rightarrow 2$ scattering case, by the delta function $\delta(s - M^2)$. We can replace the energy denominator, $2E$, by an integral over the energy, insisting on a positive energy by a theta function and that the virtual photon be on mass shell with the delta function, $\delta(\mathbf{p}_3^2 - M^2)$, so that

$$\frac{1}{2E_3} = \int dE_3 \delta(\mathbf{p}_3^2 - M^2)\theta(E_3^2) \; . \quad (2.172)$$

The full phase-space integral for the $2 \rightarrow 1$ process where the final-state virtual photon is labeled 3, is then

$$\int \frac{d^3 p_3}{2E_3} \delta^4(\mathbf{p}_1' + \mathbf{p}_2' - \mathbf{p}_3) = \int d^4 p_3\, \delta^4(\mathbf{p}_1' + \mathbf{p}_2' - \mathbf{p}_3) \delta(\mathbf{p}_3^2 - M^2) \quad (2.173)$$

where $\mathbf{p}_1'$ and $\mathbf{p}_2'$ are the four-momenta of the incoming partons. On the parton level, the Mandelstam invariants for $2 \rightarrow 2$ scattering are

$$s = (\mathbf{p}_1' + \mathbf{p}_2')^2 = (\mathbf{p}_3 + \mathbf{p}_4)^2 \; , \quad (2.174)$$

$$t = (\mathbf{p}_1' - \mathbf{p}_3)^2 = (\mathbf{p}_2' - \mathbf{p}_4)^2 \; , \quad (2.175)$$

$$u = (\mathbf{p}_1' - \mathbf{p}_4)^2 = (\mathbf{p}_2' - \mathbf{p}_3)^2 \; , \quad (2.176)$$

analogous to the hadronic invariants defined in Eqs. (1.28)-(1.30) in Chapter 1. We do not prime the final-state four-vectors. Note that in this specific case of $2 \to 1$ annihilation in the Drell-Yan process, $\mathbf{p}_4 = 0$ and $\mathbf{p}_3^2 = M^2$. We do the integration over d^4p_3 in Eq. (2.173), leaving us with $2\pi\delta((\mathbf{p}_1' + \mathbf{p}_2')^2 - M^2)$ and since $s = (\mathbf{p}_1' + \mathbf{p}_2')^2$ for the partonic variables, the delta function finally is $\delta(s - M^2)$. (Similar manipulation can be done to obtain, for massless initial and final-state partons, $\delta(s + t + u)$, as seen in some papers. If the parton masses are not all zero, as in the case for heavy quark-antiquark pair production (with quark mass m), then the delta function may be written as either $\delta(s + t + u - 2m^2)$ or $\delta(s + t_1 + u_1)$ with $t_1 = t - m^2$ and $u_1 = u - m^2$.)

To obtain the hadroproduction cross section as a function of pair mass, the partonic cross section is convoluted with the quark and antiquark densities evaluated at scale M,

$$\begin{aligned}\frac{d\sigma^{\rm DY}}{dM} &= \frac{8\pi\alpha^2}{9M}\int_0^1 dx_1 dx_2\,\delta(s - M^2)\sum_f e_f^2[f_p(x_1, M^2)\overline{f}_p(x_2, M^2) \\ &\quad + \overline{f}_p(x_1, M^2) f_p(x_2, M^2)]\ . \end{aligned} \tag{2.177}$$

Typically one is interested in the Drell-Yan cross section in some mass and rapidity (or $x_F = x_1 - x_2$) range since experiments have only a finite acceptance in rapidity space. To access this combined differential cross section, another delta function, either $\delta[y - (1/2)\ln(x_1/x_2)]$ or $\delta(x_F - x_1 + x_2)$ is introduced.

EXAMPLE: Derive the expression for the cross section in terms of x_F and M.

The x_F delta function is introduced into Eq. (2.177) and we replace s with x_1x_2S. Then

$$\begin{aligned}\frac{d\sigma^{\rm DY}}{dM dx_F} &= \frac{8\pi\alpha^2}{9M}\int_0^1 dx_1 dx_2\,\delta(x_F - x_1 + x_2)\delta(x_1x_2S - M^2) \\ &\quad \times \sum_f e_f^2[f_p(x_1, M^2)\overline{f}_p(x_2, M^2) + \overline{f}_p(x_1, M^2) f_p(x_2, M^2)]\ . \end{aligned} \tag{2.178}$$

We integrate over x_1 using the x_F delta function to fix $x_{1_0} = x_F + x_2$ so that

$$\frac{d\sigma^{\rm DY}}{dM dx_F} = \frac{8\pi\alpha^2}{9M} \int_0^1 dx_2 \, \delta[(x_F + x_2)x_2 S - M^2] \tag{2.179}$$
$$\times \sum_f e_f^2 [f_p(x_F + x_2, M^2)\overline{f}_p(x_2, M^2) + \overline{f}_p(x_F + x_2, M^2) f_p(x_2, M^2)] \, .$$

A change of variable in the second delta function is then used to integrate over x_2,

$$\delta[(x_F + x_2)x_2 S - M^2] \longrightarrow \frac{\delta(x_2 - x_{2_0})}{(2x_2 + x_F)S} = \frac{\delta(x_2 - x_{2_0})}{(x_2 + x_{1_0})S} \tag{2.180}$$

where the second relation makes use of the result from the first delta function. Inserting this into Eq. (2.179), we find

$$\frac{d\sigma^{\rm DY}}{dM dx_F} = \frac{8\pi\alpha^2}{9M} \frac{1}{S(x_{1_0} + x_{2_0})} \sum_f e_f^2 [f_p(x_F + x_{2_0}, M^2)\overline{f}_p(x_{2_0}, M^2)$$
$$+ \overline{f}_p(x_F + x_{2_0}, M^2) f_p(x_{2_0}, M^2)] \, . \tag{2.181}$$

The change of variable to y and M is left as an exercise. In this case, we will find

$$x_1 = (M/\sqrt{S}) \exp(y) \, , \quad x_2 = (M/\sqrt{S}) \exp(-y) \, . \tag{2.182}$$

The values of x_1 and x_2 for Drell-Yan production given in Eq. (2.182) can also be obtained by using four-momentum conservation in $2 \to 1$ scattering where the initial quark-antiquark pair annihilate to a single particle, the virtual photon. Recall in Chapter 1 that we wrote down the four-momenta for the incoming hadrons in Eqs. (1.49) and (1.50). To derive Eq. (2.182), we need to use the four-momenta of the partons in the hadrons. Thus, we assume that the parton four-momentum is the momentum fraction x multiplied by the hadron four-momentum. In that case, $\mathbf{p}_1' = x_1 \mathbf{p}_1$ where $\mathbf{p}_1'$ is the parton four-momentum and $\mathbf{p}_1$ is the hadron four-momentum. Likewise $\mathbf{p}_2' = x_2 \mathbf{p}_2$. Then

$$\mathbf{p}_1' = (x_1\sqrt{S}/2, 0, x_1\sqrt{S}/2) \, , \tag{2.183}$$
$$\mathbf{p}_2' = (x_2\sqrt{S}/2, 0, -x_2\sqrt{S}/2) \, . \tag{2.184}$$

EXAMPLE: Show that $s = x_1 x_2 S$.
Since the incoming partons are assumed to be massless, $(\mathbf{p}_1')^2 = 0$ and $(\mathbf{p}_2')^2 = 0$. Thus $s = 2\mathbf{p}_1' \cdot \mathbf{p}_2' = x_1 x_2 S$.

Instead of the $2 \to 2$ kinematics discussed in Chapter 1, here we have only a single final-state particle, the virtual photon, at leading order in QCD. In this case, at leading order the interaction is purely electromagnetic. The cross section is thus proportional to a power of the electromagnetic coupling constant, α, and not of the strong coupling constant, α_s. The virtual photon has four-momentum

$$\mathbf{p}_{\gamma^*} = (m_T \cosh y, p_T, m_T \sinh y) \tag{2.185}$$

where $m_T = \sqrt{M^2 + p_T^2}$. Four-momentum conservation tells us

$$\mathbf{p}_1' + \mathbf{p}_2' = \mathbf{p}_{\gamma^*} \ . \tag{2.186}$$

Since, in center-of-mass kinematics, the initial partons are not carrying any transverse momentum, then $p_T = 0$ in Eq. (2.185) and $m_T = M$. We obtain x_1 and x_2 using energy and longitudinal momentum conservation,

$$\frac{x_1\sqrt{S}}{2} + \frac{x_2\sqrt{S}}{2} = M \cosh y \ , \tag{2.187}$$

$$\frac{x_1\sqrt{S}}{2} - \frac{x_2\sqrt{S}}{2} = M \sinh y \ . \tag{2.188}$$

Adding Eqs. (2.187) and (2.188) gives

$$x_1\sqrt{S} = M(\cosh y + \sinh y) = M \exp(y) \tag{2.189}$$

while subtracting them gives

$$x_2\sqrt{S} = M(\cosh y - \sinh y) = M \exp(-y) \ . \tag{2.190}$$

Dividing these equations by $\sqrt{S}$, we obtain Eq. (2.182).

In Fig. 2.14, we show the values of x_1 (left-hand side) and x_2 (right-hand side) for a virtual photon with $M = 4$ GeV for several values of $\sqrt{S}$ spanning the available range of energies relevant for ion-ion collisions. The maximum rapidity for any given M and $\sqrt{S}$ can be found by setting x_1 or x_2 to unity since neither x_1 nor x_2 can exceed unity. Then $|y_{\rm max}| = \ln(\sqrt{S}/M)$. The available rapidity space thus widens with $\sqrt{S}$ and narrows with M. The narrowest rapidity range corresponds to $\sqrt{S} = 20$ GeV while the widest is for $\sqrt{S} = 14$ TeV. Note that while x_1 increases with y, x_2 decreases with y. One can obtain x_2 from x_1 by changing the sign of the rapidity, as expected by inspection of Eq. (2.182).

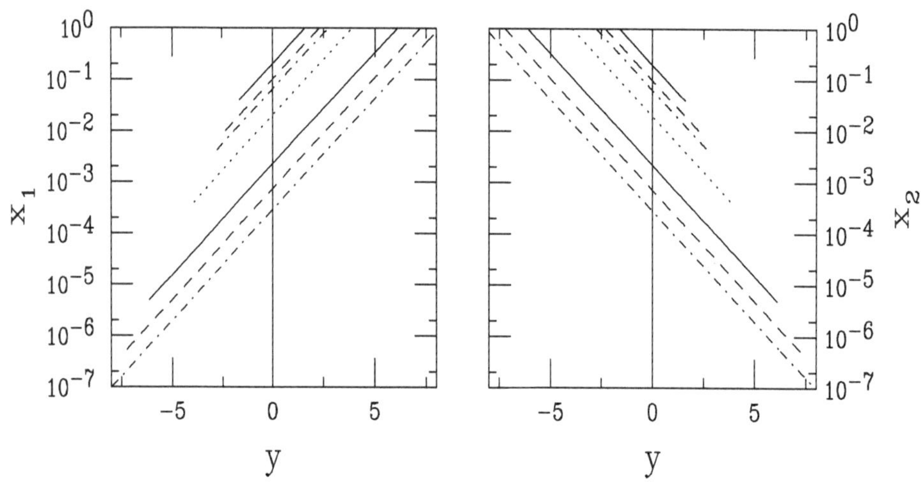

Figure 2.14: The values of x_1 (left-hand side) and x_2 (right-hand side) for $M = 4$ GeV and, from top to bottom, $\sqrt{S} = 20$ (solid), 40 (dashed), 60 (dot-dashed), 200 (dotted), 1800 (solid), 5500 (dashed) and 14000 (dot-dashed) GeV.

To conclude our discussion on the Drell-Yan process, we recall the corrections needed to go to the next order in accuracy for the cross section. The two additional processes, real corrections to the cross section, were $q\overline{q} \to \gamma^* g$ and $qg \to q\gamma^*$ (or $\overline{q}g \to \overline{q}\gamma^*$), shown in Fig. 2.15. The virtual photon emitted then produces the final state lepton pair.

In Fig. 2.13, the quark and antiquark, as well as the final-state lepton pair, couple to the virtual photon. Thus there is an electromagnetic coupling at each vertex, leading to the two powers of the electromag-

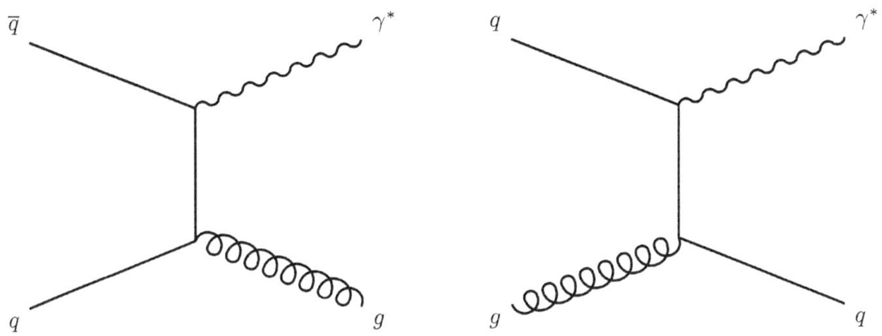

Figure 2.15: The next-to-leading order diagrams for Drell-Yan production. On the left-hand side we show $q\overline{q} \rightarrow \gamma^* g$ while the right-hand side is $qg \rightarrow q\gamma^*$.

netic coupling constant, α, in Eq. (2.171). However, Fig. 2.15 shows that, at next-to-leading order, in addition to a quark-photon vertex, there is another vertex where a quark couples to a gluon instead. Thus the diagrams in Fig. 2.15 will be of order $\alpha\alpha_s$ where α_s is the strong coupling constant. (When the splitting of the virtual photon to a lepton pair is taken into account, the full Drell-Yan cross section is of order $\alpha^2\alpha_s$.) Note that since the kinematics of the lepton pair are taken to be those of the virtual photon, the momenta of the individual leptons does not enter the calculation of the Drell-Yan cross section.

The importance of these next-to-leading order corrections to the Drell-Yan cross section depend on the pair mass since α_s, although referred to as the strong coupling *constant*, is not really constant but a function of scale and number of quark flavors. Since α_s depends on the scale, it is said to be a 'running' coupling. The leading order expression for α_s is

$$\alpha_s = \frac{12\pi}{(33 - 2n_f)\ln(Q^2/\Lambda^2)} \qquad (2.191)$$

where Λ is a cutoff scale that may be thought of as marking the boundary between the partonic and hadronic worlds. It is typically on the order of 200 MeV or so but depends on the value of n_f. To keep the dependence of α_s smooth as a function of Q^2, the value of Λ is adjusted at the values of Q^2 equal to the square of the heavy quark masses. When $1.5 < Q < 4.75$ GeV, above the charm threshold and below the bottom threshold, $n_f = 4$, referred to as four 'active' flavors. Below the charm mass, $n_f = 3$, while above the bottom mass, $n_f = 5$. Above the top mass, 178 GeV, $n_f = 6$. As Q approaches Λ, α_s blows up. The scale dependence of α_s to leading order with $\Lambda = 200$ MeV for $n_f = 4$ is shown in Fig. 2.16. At next-to-leading order, another term with a negative sign is included, reducing the value of α_s relative to the leading order evaluation.

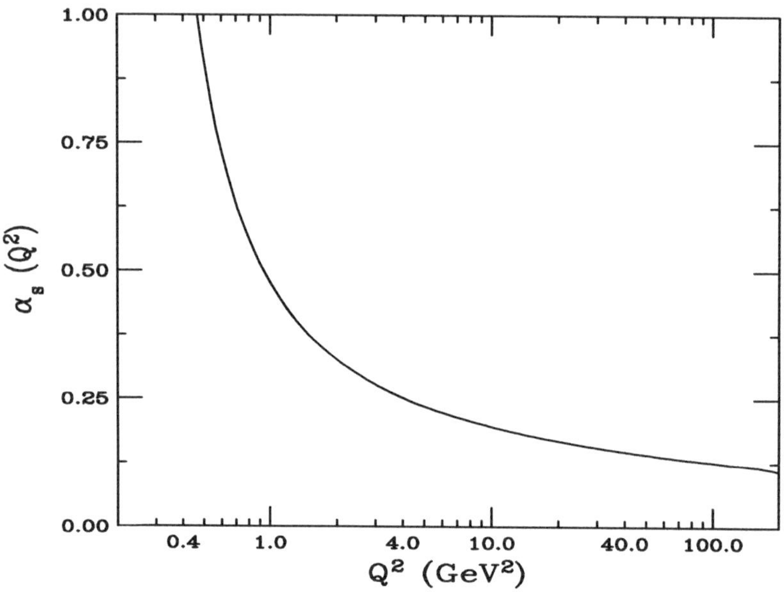

Figure 2.16: The leading order strong coupling as a function of scale with $\Lambda = 200$ MeV for $n_f = 4$.

In the Drell-Yan process, we take $Q \propto M$. While usually one takes $Q = M$, the scale dependence of the cross section can be investigated by choosing $M/2 < Q < 2M$. If the perturbative expansion is good, the cross section should be independent of scale. When $Q^2 > 10$ GeV^2, $\alpha_s \ll 1$ and α_s is not changing very rapidly with Q^2. Here the scale dependence of the cross section should be rather small. However, for $Q^2 < 4$ GeV^2, the scale dependence can be strong. Indeed, for $Q^2 \leq 1$

GeV^2, α_s is no longer really small compared to unity. Indeed, perturbative QCD is generally no longer applied to $Q^2 \leq 1$ GeV^2 since the expansion parameter for going to higher orders of the cross section, α_s, is not small.

An additional test of the reliability of the cross section calculation to a particular order in α_s, besides the scale dependence, is the calculation of what is known as the K factor. The K factor was first applied experimentally as a correction factor to raise the leading order theoretical result to the level of the measured cross section. After the next-to-leading order Drell-Yan cross section was calculated, this experimental K factor was nearly unity. Theoretically, the K factor can be computed by comparing two fixed-order calculations as a function of some variable such as mass, rapidity, p_T (if available) or $\sqrt{S}$ (if the total cross section is finite). The first caveat exists because some leading order cross sections, such as the Drell-Yan process, are only good at $p_T = 0$ since the pair has no transverse momentum if the projectile parton momentum is defined as in Eqs. (2.183) and (2.184). We make the second caveat since a total cross section can only be calculated for finite mass partons and leptons where one can take the parton p_T to zero since the mass scale regulates the denominator of the cross section, as in Eq. (2.171). (Note that we cannot calculate the Drell-Yan cross section to zero pair mass for this reason but can only refer to a cross section at a given pair mass.) The Drell-Yan K factor then is largest for the lowest values of pair mass, $M \sim 2 - 3$ GeV, but decreases as M increases. The Drell-Yan K factor is typically less than 1.5. Larger K factors are found for processes where the strong coupling enters in the leading order diagrams. Some of the largest K factors, $K \sim 2 - 3$ between the next-to-leading and leading order cross sections, are found for charm production since it is already order α_s^2 in the leading order diagrams, see Figs. 2.5 and 2.6. Thus there are possibly large uncertainties in the calculations of quantities of interest in heavy-ion collisions which should be taken into account when drawing conclusions.

EXAMPLE: Use four-momentum conservation to solve for x_1 and x_2 in the case of $2 \to 2$ scattering, similar to next-to-leading order Drell-Yan production but where the final-state particles have the same mass.

We use Eqs. (2.183) and (2.184) above with Eqs. (1.51) and (1.52) in Chapter 1. We begin again with transverse momentum, finding

$$0 = \vec{p}_{T_3} + \vec{p}_{T_4}$$

so that the final state particles have equal and opposite transverse momentum and $p_T = |\vec{p}_{T3}| = |\vec{p}_{T4}|$. Then $m_{T_3} = m_{T_4} = m_T$ and Eqs. (1.51) and (1.52) become

$$\begin{aligned} \mathbf{p}_3 &= (m_T \cosh y_3, p_T, m_T \sinh y_3) , \\ \mathbf{p}_4 &= (m_T \cosh y_4, -p_T, m_T \sinh y_4) . \end{aligned}$$

Four-momentum conservation dictates

$$\mathbf{p}_1' + \mathbf{p}_2' = \mathbf{p}_3 + \mathbf{p}_4 \tag{2.192}$$

so that, using energy and longitudinal momentum conservation,

$$\begin{aligned} \frac{x_1\sqrt{S}}{2} + \frac{x_2\sqrt{S}}{2} &= m_T(\cosh y_3 + \cosh y_4) , \\ \frac{x_1\sqrt{S}}{2} - \frac{x_2\sqrt{S}}{2} &= m_T(\sinh y_3 + \sinh y_4) . \end{aligned} \tag{2.193}$$

We can solve these two equations for x_1 and x_2, obtaining

$$x_1 = \frac{m_T}{\sqrt{S}}[\exp(y_3) + \exp(y_4)] , \tag{2.194}$$

$$x_2 = \frac{m_T}{\sqrt{S}}[\exp(-y_3) + \exp(-y_4)] . \tag{2.195}$$

We have discussed some assumptions concerning the parton distributions in hadrons. However, not all assumptions prove correct, as experiments show. An example of how experiment can prove assumptions wrong was a study of the Gottfried sum rule using the Drell-Yan process. The difference between the proton and neutron structure functions can then be written as

$$\frac{1}{x}\left(F_2^{\gamma p}(x) - F_2^{\gamma n}(x)\right) = \frac{1}{3}\left[u^v(x) - d^v(x)\right] + \frac{2}{3}\left[\overline{u}(x) - \overline{d}(x)\right] . \tag{2.196}$$

Using Eqs. (2.146) and (2.147) for the proton valence quark distributions, after integrating both sides of Eq. (2.196) over x, one obtains the Gottfried sum rule,

$$\int_0^1 \frac{dx}{x}\,[F_2^{\gamma p}(x) - F_2^{\gamma n}(x)] = \frac{1}{3} + \frac{2}{3}\int_0^1 dx\,[\overline{u}(x) - \overline{d}(x)] = \frac{1}{3}\ . \quad (2.197)$$

If the sea is flavor symmetric, $\overline{u} = \overline{d}$. This was a dearly held assumption. However, measurements of perturbative lepton pair production through the Drell-Yan process on proton and deuteron targets [20] showed that, in fact, $\overline{u} < \overline{d}$, see Ref. [21] for a more detailed discussion. Henceforward new sets of parton density fits have included these Drell-Yan data to separate the $\overline{u}$ and $\overline{d}$ distributions.

There are several groups that regularly use new data to update and extend the range of the parton distributions. More recently, these have included more realistic assessments of the uncertainties from both the theoretical (choice of data sets, treatment of the deuteron, assumptions inherent in the initial parameterizations, low x behavior) and experimental (statistical and systematic errors on data, choice of cuts, analysis techniques) standpoints.

An example of a recent set evolved to next-to-leading order in the couplings is shown in Fig. 2.17 for a Q^2 scale of 10 GeV^2 a typical scale for *e.g.* J/ψ production. The distributions are part of a global analysis performed by CTEQ (Collaboration of Theorists and Experimentalists in QCD). This particular set is called CTEQ6M since it is the 6$^{\rm th}$ global analysis by CTEQ and done in the $\overline{\rm MS}$ subtraction scheme [22]. (See a textbook on field theory in perturbative QCD, such as that by Field [16] for an explanation of the $\overline{\rm MS}$ scheme.) Other parton distribution function evaluations for protons by Martin and collaborators [23] and Glück and collaborators [24] have been performed with more or less regular updates, as more data becomes available. Since all evaluations use basically the same data sets, the resulting proton parton distribution functions are similar in the region where data is available and all sets must satisfy the same rules for baryon number and total momentum conservation.

The valence distributions, shown as the solid and dashed curves for u_v and d_v respectively, peak at $\langle x \rangle \approx 0.2$. The difference between the u_p^s

and d_p^s distributions determined through studies of the Gottfried sum rule, Eq. (2.197), and further detailed in Drell-Yan studies, is obvious for $x > 0.002$. The strange sea distribution is considerably less than those of light sea quarks. While the gluon distribution is similar to that of the sea quarks, it is much higher at low x and has to be reduced by an order of magnitude to appear on the same plot with the quarks.

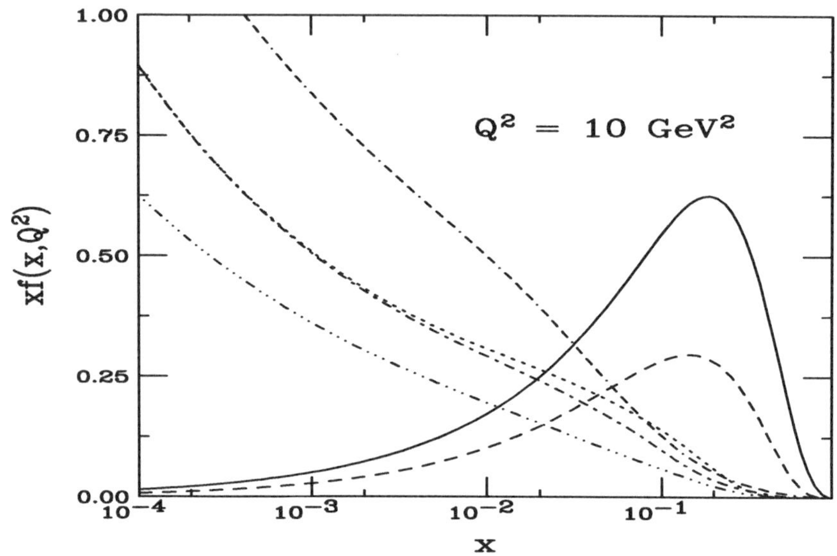

Figure 2.17: The CTEQ6M parameterization of the proton parton distribution functions as a function of momentum fraction x at $Q^2 = 10$ GeV^2. The curves correspond to u_p^v (solid), d_p^v (dashed), u_p^s (dot-dashed), d_p^s (dotted), s_p^s (dot-dot-dot-dashed) and g_p (dash-dash-dash-dotted). The gluon distribution is multiplied by 0.1.

The valence distributions behave as $x^a(1-x)^b$ where $a \sim 0.5$ so that they drop to zero as $x \to 0$ and peak at $\langle x \rangle \approx 0.2$. The up valence distribution is about twice that of the down, as expected from Eq. (2.146). The behavior at large x can be approximated by quark counting rules [25], postulated by Gunion in the late 1970's. He predicted that the power b is $b \approx 2n_s - 1$ where n_s is the number of 'spectator quarks'. In the case of valence quarks in the proton, uud, there are two spectators so that $b \approx 3$. For a pair of sea quarks, $q\overline{q}$, with $uudq\overline{q}$, $n_s = 4$

and $b \approx 7$. On the other hand, with a gluon, $uudg$, there are 3 spectators and $b \approx 5$. The earliest sets of parton distribution functions assumed that the sea quark and gluon distributions would saturate at the minimum scale, Q_0, so that these distributions were just $(1-x)^b$. However, the first data from the electron-proton collider (HERA) at DESY (Deutches Electronisches Synchrotron in Hamburg, Germany) showed that already at moderate values of Q, near Q_0, these distributions increased with decreasing x. Some of the most recent data can be found in Ref. [26]. Thus, most modern parameterizations of the parton distribution functions assume that the sea quark and gluon distributions grow as $x^{-d}(1-x)^b$ where $d \sim 0.2 - 0.3$. They then grow large as $x \longrightarrow 0$ and drop off more steeply at large x than the valence quark distributions.

The more limited quality of the pion data does not allow such sophisticated refinement of the pion parton distribution functions. The most recent set of pion parton distribution functions, by Glück and collaborators [27], shown in Fig. 2.18, is from 1999 and includes the assumption that, at low x, the sea quark and gluon distributions of the pion also grow with decreasing x. They assume that, for a π^-, the valence parton distributions for the $\overline{u}$ and d are equal, as are those of the light quark sea. The strange sea is small and almost negligible in their evaluation. The sea quark and gluon distributions persist to larger values of x than in the proton, as may be expected for quark counting rules. The valence distributions also peak at a larger value of x, $\langle x \rangle = 0.3$ for $Q^2 = 10$ GeV^2.

EXAMPLE: Determine the exponent b for the valence, sea and gluon distributions in the pion.

Recall that $b = 2n_s - 1$. With a pion, $n_s = 1$ for valence quarks, 3 for sea quarks and 2 for gluons. Thus $b = 1$, 5 and 3 respectively, relative to 3, 7 and 5 for the proton.

The hadronic cross section in Eq. (2.170) consists of two parts, the parton-parton interaction cross section, calculated in perturbative QCD, and the convolution of the parton densities, known as the parton luminosity. The expression for the parton luminosity for colliding

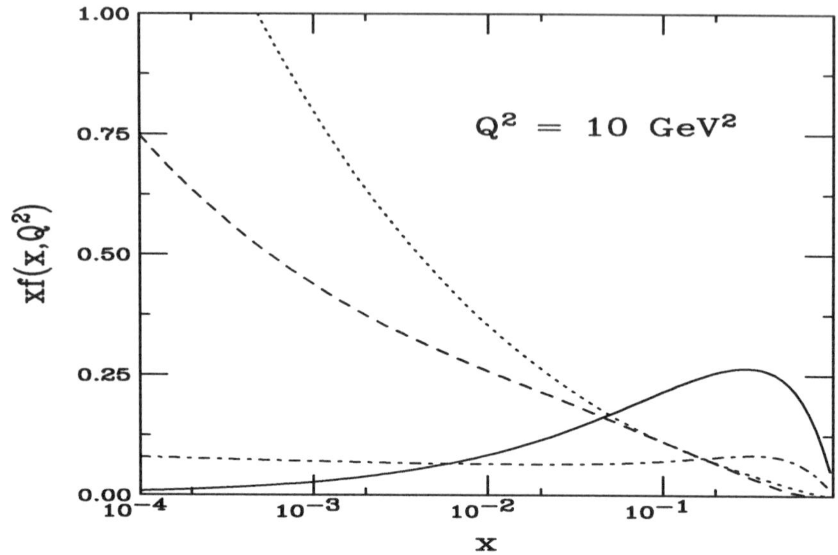

Figure 2.18: The GRS99 parameterization of the pion parton distribution functions as a function of momentum fraction x at $Q^2 = 10$ GeV^2. The curves correspond to $\overline{u}^v_{\pi^-} = d^v_{\pi^-}$ (solid), the light quark sea (*e.g.* $\overline{d}^s_{\pi^-}$) (dashed), $s^s_{\pi^-}$ (dot-dashed) and g_{π^-} (dotted). The gluon distribution is multiplied by 0.1 while the strange sea is multiplied by 10.

partons i and j in hadrons h_1 and h_2 at fixed Q^2 is

$$\begin{aligned}\Phi_{ij}(\tau) &= \tau \int_0^1 dx_1 \int_0^1 dx_2 \delta(\tau - x_1 x_2) \\ &\quad \times \phi_{i/h_1}(x_1, Q^2)\phi_{j/h_2}(x_2, Q^2) \ . \qquad (2.198)\end{aligned}$$

In the case of $q\overline{q}$ annihilation, as in the Drell-Yan process, Eq. (2.177), $\phi_{i/p} = f_p^i$ and $\phi_{j/p} = \overline{f}_p^j$ and f is summed over the active quark flavors. A quark flavor is said to be active if the scale of the produced final state is larger than the quark mass so that the quark can be considered massless. For example, in the process $q\overline{q} \to c\overline{c}$, only u, d and s quarks are active flavors and included in the sum. In the case of $b\overline{b}$ production, $f = u$, d, s and c. For the $gg \to Q\overline{Q}$ channel, $i = j = g$. The delta function in Eq. (2.198) can easily be done to integrate over either x_1 or x_2, giving

$$\Phi_{ij}(\tau) = \tau \int_\tau^1 \frac{dx_2}{x_2} \phi_{i/h_1}\left(\frac{\tau}{x_2}, Q^2\right) \phi_{j/h_2}(x_2, Q^2) \ . \tag{2.199}$$

As an illustration, the $q\overline{q}$ and gg luminosities in pp collisions are shown on the left and right-hand sides of Fig. 2.19 for the CTEQ61L parton densities [22]. Instead of plotting the luminosity as a function of τ, it is shown as a function of the variable $\eta = \tau S/M^2 - 1$ with $Q = M = 4$ GeV for $\sqrt{S} = 20$, 200 and 5500 GeV. A weight factor of $\ln(10)\eta/(1+\eta)$ has been included to produce a peak at the most likely values of τ for the given M and $\sqrt{S}$. The $q\overline{q}$ convolution in a pp collision is

$$\begin{aligned} \phi_{q/p}(x_1, M^2)\phi_{\overline{q}/p}(x_2, M^2) = \\ & u_p(x_1, M^2)\overline{u}_p^s(x_2, M^2) + \overline{u}_p^s(x_1, M^2)u_p(x_2, M^2) \\ + \ & d_p(x_1, M^2)\overline{d}_p^s(x_2, M^2) + \overline{d}_p^s(x_1, M^2)d_p(x_2, M^2) \\ + \ & 2s_p^s(x_1, M^2)\overline{s}_p^s(x_2, M^2) \ . \end{aligned} \tag{2.200}$$

Recall that $u_p(x) = u_p^v(x) + \overline{u}_p^s(x)$ since $\overline{u}_p^s(x) = u_p^s(x)$ and $d_p(x) = d_p^v(x) + \overline{d}_p^s(x)$. Both the $u_p^s(x_1)\overline{u}_p(x_2)$ and $\overline{u}_p^s(x_1)u_p(x_2)$ terms are given because $\overline{u}_p(x) \neq u_p(x)$ and $d_p(x) \neq \overline{d}_p(x)$ while the s and $\overline{s}$ distributions in the proton are assumed to be identical. On the other hand, the gg convolution is

$$\phi_{g/p}(x_1, M^2)\phi_{g/p}(x_2, M^2) = g_p(x_1, M^2)g_p(x_2, M^2) \tag{2.201}$$

Note that while the $q\overline{q}$ and gg luminosities are similar for the lowest energy, at the higher energies, the gg luminosities dominate. In all cases, the peak of the luminosity is close to $\eta = 1$ or $\tau = (M^2/S)(\eta+1)$, corresponding to $\tau = 8 \times 10^{-2}$, 8×10^{-4} and 1.06×10^{-6} respectively. As the energy increases, the maximum possible η also increases. Note that $\eta \to 0$ corresponds to production threshold.

To show the application of these parton luminosities to real calculations, Figs. 2.20 and 2.21 present the next-to-leading order Drell-Yan rapidity and mass distributions in $\sqrt{S} = 200$ and 500 GeV pp collisions

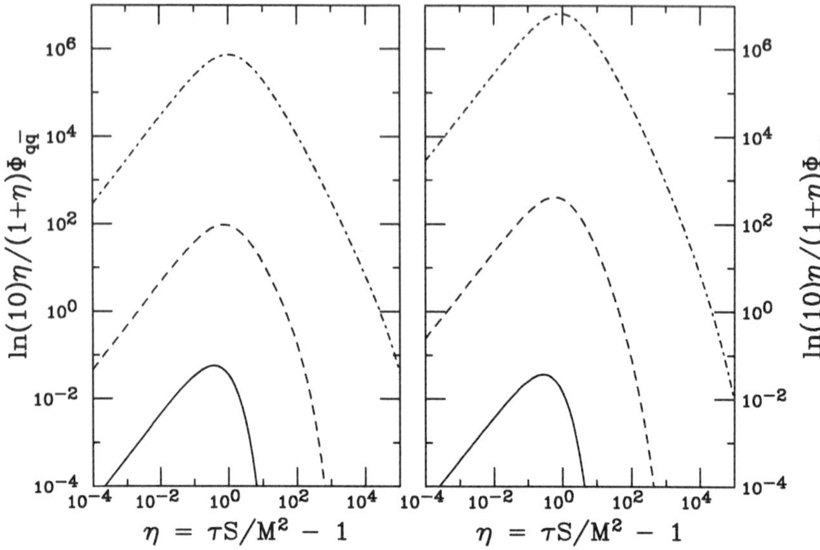

Figure 2.19: The $q\bar{q}$ and gg parton luminosities as a function of $\eta = \tau S/M^2 - 1$ for $M = 4$ GeV and $\sqrt{S} = 20$, 200 and 5500 GeV, indicated by the solid, dashed and dot-dashed curves respectively. The luminosities are calculated for the CTEQ61L parton densities.

using the CTEQ6M parton densities, as in Fig. 2.17. The rapidity distributions are shown in three relevant mass bins: $2 < M < 3$ GeV, the region just below the J/ψ mass; $4 < M < 9$ GeV, the mass region between the J/ψ and Υ masses; and $11 < M < 20$ GeV, above the Υ family masses. The energies used are the maximum Au+Au and pp collision energies, respectively. The maximum accessible rapidity depends on the mass,

$$y_{\rm max} = \ln\left(\frac{\sqrt{S}}{M}\right) \tag{2.202}$$

Thus, for $\sqrt{S} = 200$ GeV, the maximum rapidity accessible at $M = 20$ GeV is $y = 2.3$. The largest y value shown for the highest mass bin is greater than 2.3 since the cross section is dominated by the lowest mass contribution.

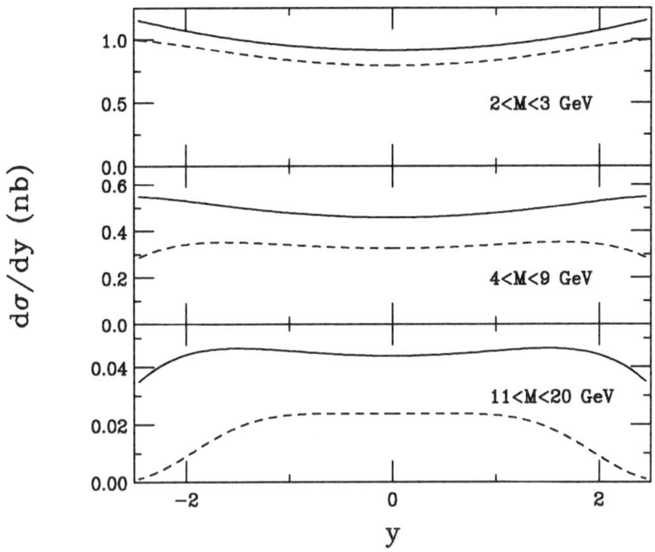

Figure 2.20: The Drell-Yan dilepton rapidity distributions in pp collisions at 200 (dashed) and 500 (solid) GeV in the mass windows: $2 < M < 3$ GeV (top); $4 < M < 9$ GeV (center); and $11 < M < 20$ GeV (bottom).

Since the results are shown for pp collisions, the rapidity distributions are symmetric around $y = 0$. The distributions at both energies tend to increase with y until the x value corresponds to the average x of the valence quark distribution at the scale $Q = M$. After this maximum, the rapidity distributions begin to drop rather rapidly, $d\sigma/dy \to 0$. At $\sqrt{S} = 200$ GeV, this turnover has occurred in the highest mass bin and the entire rapidity range is shown. This is not the case for the higher energy where the distribution is still relatively flat over $|y| \leq 2$ for $11 < M < 20$ GeV.

The changing width of the rapidity distribution is reflected in the mass distributions shown for three rapidity ranges: $|y| < 1$, the central region; and $1.2 < |y| < 2.2$, forward and backward of midrapidity. Note that the fall of the cross sections with M is steeper for 200 GeV than 500 GeV. This is particularly true at forward rapidities where the rapidity distribution becomes narrower at lower masses, see Fig. 2.20.

On the other hand, the shape of the mass distribution is similar for all rapidity ranges at 500 GeV.

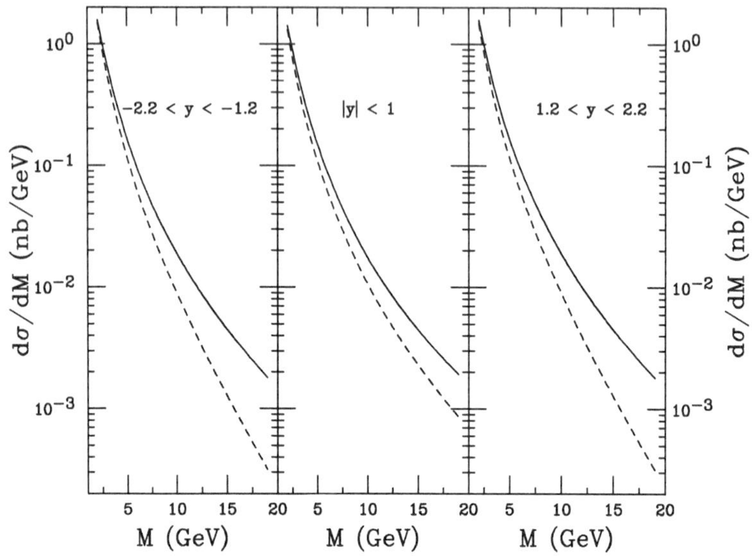

Figure 2.21: The Drell-Yan dilepton mass distributions in pp collisions at 200 (dashed) and 500 GeV (solid) in the rapidity windows: $-2.2 < y < -1.2$ (left); $|y| < 1$ (center); and $1.2 < y < 2.2$ (right).

In the pp case the rapidity distributions are symmetric and the mass distributions at forward and backward rapidities are identical. This would not be the case, however, for asymmetric collisions such as the d+Au runs at $\sqrt{S_{NN}} = 200$ GeV since the neutron and proton numbers are not equal and $\sigma_{pp}^{\rm DY} \neq \sigma_{nn}^{\rm DY} \neq \sigma_{pn}^{\rm DY} \neq \sigma_{np}^{\rm DY}$ for real parton densities. These cross sections are not equal to each other because the u and d valence and sea distributions are not equal.

The growth of the sea quark and gluon distributions with decreasing x in Fig. 2.17 cannot continue indefinitely. If it did, the densities would become infinite, violating unitarity, a concept discussed in the next section. Thus, in addition to the normal evolution of the parton distributions, a number of ideas have been proposed to cure this growth. Nonlinear evolution terms have been introduced which slow the evolution rate of the hadron parton densities [28]. When included in global fits to the parton distribution functions, they can, somewhat counter-intuitively lead to enhanced gluon distributions at low x and Q^2 due to the need to also fit the DIS data [29]. The color-glass condensate has also been proposed as a way to describe the low x behavior [30]. In both pictures, the effects of both nonlinear evolution and the color-glass condensate should set in at larger values of x than in free protons.

Deep-inelastic scattering experiments have also been carried out on nuclear targets, albeit over a more limited range of x and Q^2, see Ref. [31] for a review. If the protons and neutrons in the nucleus behave as free particles, the ratio of the nuclear results for *e.g.* F_2^A to those on a deuterium target, per nucleon, $(F_2^A/A)/(F_2^d/2)$ should be equal to unity. However, when heavy nuclear results are compared to $F_2^d/2$, a definite, A-dependent change in shape is observed. Since a depletion relative to deuterium was first seen at relatively large x, $x > 0.2$, by the European Muon Collaboration, this region is referred to as the EMC region. Also, a depletion is observed for $x < 0.01$, known as the shadowing region. At intermediate x values, $0.01 < x < 0.2$, there is an enhancement, called antishadowing. Since the data is rather sparse in the perturbative region, there can be large differences in the model parameterizations, especially in regions where no data yet exist. Thus global analyses of the nuclear parton densities are far from as accurate and sophisticated as those of the free proton.

Figure 2.22 shows a parameterization by Eskola and collaborators [32] for $A = 40$ and 208, the atomic numbers for calcium and lead, typical targets in nuclear DIS, also at $Q^2 = 10$ GeV^2. The curves are the ratios of the nuclear parton distribution functions relative to those of the proton. The modification increases with A so that the effect is stronger in lead than in calcium. The ratios are shown for $Q^2 = 10$ GeV^2. The modification is strongest at low Q^2 and, especially for the gluons, decreases rapidly with increasing Q^2.

EXAMPLE: What values of x are accessible at rapidities $y = -2, 0$ and 2 for $\sqrt{S} = 200$ GeV and $M = 3$ and 10 GeV and which shadowing regime?

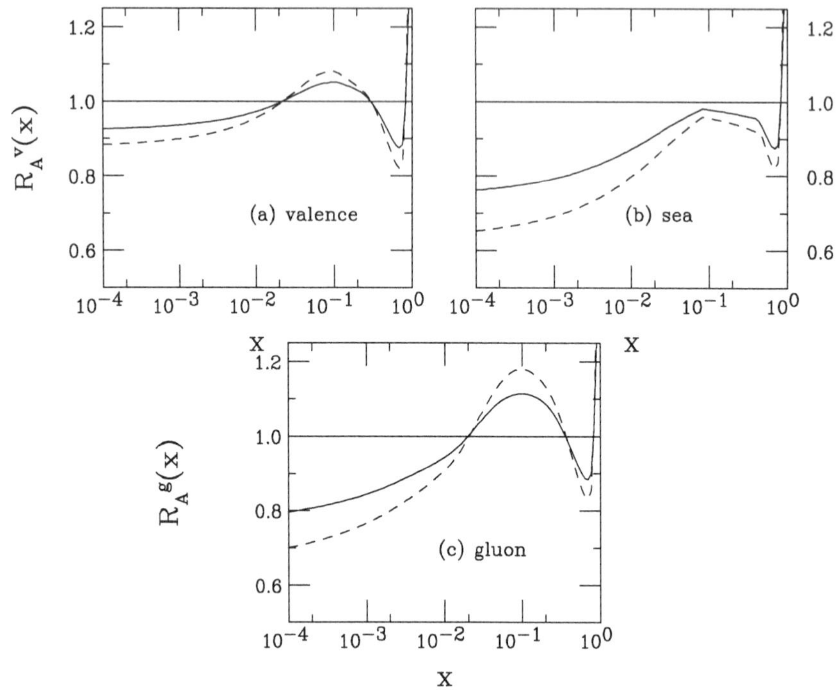

Figure 2.22: The EKS98 parameterization of the modification of the parton distribution functions in the nucleus relative to the proton as a function of momentum fraction x at $Q^2 = 10$ GeV^2. The up valence quark ratios (a), up sea quark ratios (b) and the gluon ratios (c) are shown for calcium, $A = 40$, (solid) and lead, $A = 208$, (dashed) nuclei.

Recall from Eq. (2.182) that $x = (M/\sqrt{S})\exp(y)$. For $y = 0$, $M/\sqrt{S} = 0.015$ and 0.05 for $M = 3$ and 10 respectively. Then $\exp(y = 2) = 7.4$ and $\exp(y = -2) = 0.135$. The values of x are tabulated below. The x values at negative rapidity correspond to the shadowing region while that at $y = 2$ and $M = 10$ is in the EMC region. The other values of x lie in the antishadowing region.

The effects of shadowing on Drell-Yan production are shown in Figs. 2.23 and 2.24 in $\sqrt{S} = 200$ GeV d+Au collisions using the EKS98 parameterizations from Fig. 2.22. Since the proton and neutron num-

Table 2.4: Tabulated values of x for the given values of M and y at $\sqrt{S} = 200$ GeV.

y	$M = 3$ GeV	$M = 10$ GeV
-2	0.002	0.0068
0	0.015	0.05
2	0.11	0.37

bers of the nuclei (isospin) need to be taken into account for quark-dominated processes, the d+Au/pp ratios are also shown without any additional shadowing effects. A comparison of the pp and d+Au mass and rapidity distributions shows that the shapes are different due to isospin effects, even before shadowing is taken into account. In the low mass and mid-to-forward rapidity regions, shadowing dominates over isospin. However, for masses greater than 10 GeV, isospin effects tend to dominate. The combined effect of shadowing and isospin, d+Au with shadowing relative to pp, is the product of the shadowing ratio and the ratio of d+Au/pp without antishadowing.

In the x region accessible at RHIC, effects of gluon shadowing on Drell-Yan production are negligible even though there is a gluon-quark contribution to Drell-Yan production at next-to-leading order. Since Drell-Yan production is dominated by $q\overline{q}$ initial states, it is not surprising that the shadowing results for the mass and rapidity distributions reflect only quark shadowing.

Figure 2.23 shows the ratios of d+Au with shadowing relative to d+Au without shadowing as a function of mass in the three rapidity bins: backward (left), midrapidity (center) and forward (right). The effects of shadowing are smallest in the backward region where x_2 is largest. This region shows some antishadowing at $M < 10$ GeV due to the large valence quark contribution. At higher masses, the larger x_2 values are in the EMC region. At leading order, when $M = 4$ GeV, $-2.2 < y < -1.2$ corresponds to $0.18 > x_2 > 0.066$ while for $M = 10$ GeV, $0.45 > x_2 > 0.17$. At midrapidity, some low x shadowing is exhibited for $M < 12$ GeV. In the most forward rapidity bin, low x

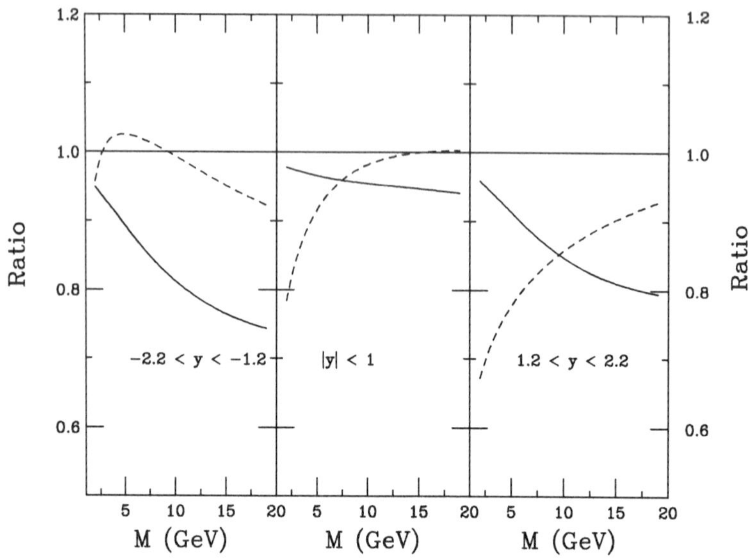

Figure 2.23: The Drell-Yan dilepton shadowing and isospin ratios at 200 GeV in the rapidity windows: $-2.2 < y < -1.2$ (left); $|y| < 1$ (center); and $1.2 < y < 2.2$ (right). The shadowing results, the ratios with to without shadowing are given by the dashed curves while the solid curve is the d+Au/pp ratio without shadowing.

shadowing is present over the full mass range. Here for $M = 4$ GeV, $1.2 < y < 2.2$ corresponds to $0.006 > x_2 > 0.0022$ while $M = 10$ GeV corresponds to $0.015 > x_2 > 0.0055$.

The solid curves in Fig. 2.23 are the d+Au/pp ratios without shadowing. The isospin effect is due to the valence quark content of the collision partners. Figure 2.17 shows that the u valence distribution is larger than the d valence distribution and the average x is larger for the u valence distribution. Thus the isospin effect is strongest where valence quarks dominate the PDFs. It is largest at forward and backward rapidities where x_1 is large for the deuteron and x_1 is large for the gold nucleus, respectively. While the ratios in these two rapidity regions are rather similar, the overall effect is greatest in the backward region where x_1 is large for gold with the biggest difference between neutron and proton numbers. Since the Z/A ratio is different for deuterons and

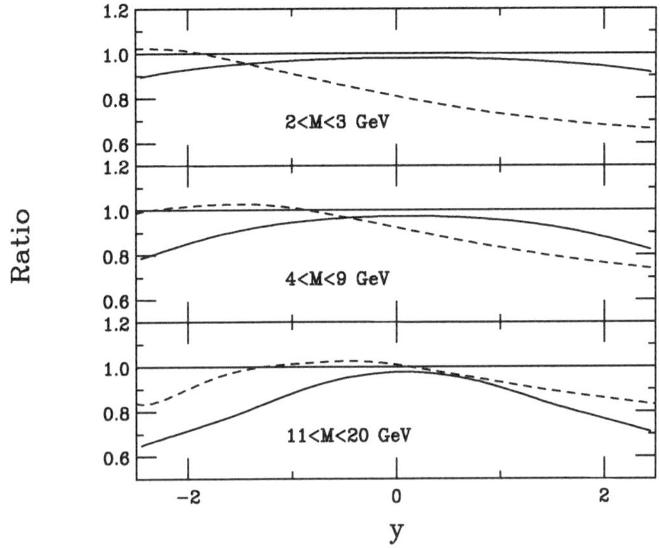

Figure 2.24: The Drell-Yan dilepton shadowing and isospin ratios at 200 GeV in the mass windows: $2 < M < 3$ GeV (top); $4 < M < 9$ GeV (center); and $11 < M < 20$ GeV (bottom). The shadowing results, the ratios with to without shadowing are given by the dashed curves while the solid curves are the d+Au/pp ratios without shadowing.

gold, the isospin effect is asymmetric around $y = 0$. At midrapidity both x_1 and x_2 are relatively small and similar, reducing the overall effect.

The calculated ratios are shown as a function of rapidity in the three mass bins in Fig. 2.24. In the smallest mass bin, the shadowing effects are largest, increasing from a slight antishadowing effect at large negative rapidity to a small x shadowing effect at large positive rapidity. By the largest mass bin, there is evidence of an EMC region at large negative rapidity and only a small shadowing effect at positive rapidity. Again the isospin effect is seen to increase with mass. The rapidity asymmetry, also seen in Fig. 2.23, is obvious here.

Note that the shadowing effects shown in these figures are strongly dependent on the shadowing parameterization. While the quark and antiquark parameterizations are based on more data than are the gluon parameterizations, considerable uncertainty remains, especially extrapolating the rather low Q^2 results to higher values. The nuclear parton densities are now at the level of accuracy of the proton parton densities before the advent of HERA, the *ep* collider. Hopefully more data will reduce the uncertainties.

2.5 Photoproduction and two-photon physics in heavy-ion collisions

In addition to studies of strong interactions of dense matter at ion colliders, electromagnetic interactions may be studied by examining collisions where the two nuclei do not physically touch each other. Instead they pass by each other at distances greater than $2R_A$ in symmetric AA collisions where R_A is the radius of nucleus A. These interactions are referred to as ultra-peripheral collisions or UPCs [33, 34]. It is possible to study electromagnetic interactions through photoproduction (a photon from one nucleus interacts with a quark or gluon from the opposite nucleus) and two-photon processes (two photons, one from the field of each nucleus) using UPCs because highly accelerated ions are surrounded by photon fields. These photons are almost real since their virtuality is less than $(\hbar c/R_A)^2$. For $A > 16$ the virtuality is less than $(60\,\mathrm{MeV})^2$, safely negligible.

Since the electromagnetic interaction is long range, photons can interact with partons in the opposite nucleus even when the nuclei themselves do not interpenetrate. Because the photon energies are less than those of the nucleons, these photo-nuclear interactions have a smaller average center-of-mass energy than hadronic parton-parton collisions. However, even though the energy is smaller, the coherent photon beams have a flux proportional to the square of the nuclear charge, Z, enhancing the rates relative to those of photoproduction in *pp* collisions. Thus the photoproduction rates can be high.

Photoproduction occurs two ways in UPCs: 'direct' and 'resolved' production. 'Direct' photoproduction occurs when a photon emitted from one nucleus fuses with a parton from the other nucleus, forming

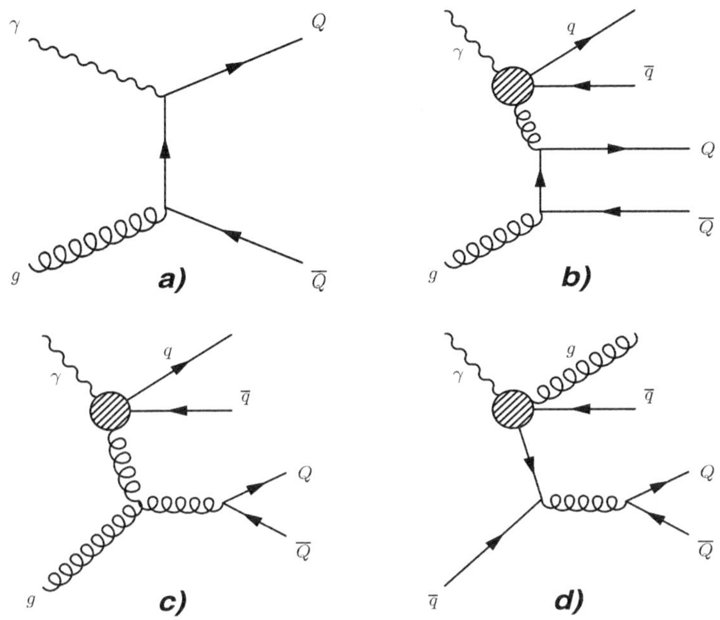

Figure 2.25: Feynman diagrams for heavy quark photoproduction from (a) direct and (b)-(d) resolved photons. The crossed diagrams for (a) and (b) are not shown.

the final state. The diagrams for $Q\overline{Q}$ production are shown in Fig. 2.25. The $\gamma g \to q\overline{q}$ process, the most direct probe of the nuclear gluon distribution, is the same for heavy (Q) and light (q) quarks. In the latter case, the light quarks appear as streams of particles, known as jets.

A generic direct photoproduction cross section for ultra-peripheral AA collisions is obtained by convoluting the partonic photoproduction cross section, $d^2\sigma_{\gamma i}/dtdu$, with the photon flux from one nucleus, d^3N_γ/dkd^2b, and the parton distribution in the opposite, $F_i^A(x_2, Q^2)$,

$$S^2 \frac{d^2\sigma_{\gamma A}^{\rm dir}}{dTdUd^2b} = 2\int dz \int_{k_{\rm min}}^{\infty} dk \frac{d^3N_\gamma}{dkd^2b} \int_{x_{2_{\rm min}}}^{1} \frac{dx_2}{x_2} \times \left[\sum_{i=q,\overline{q},g} F_i^A(x_2, Q^2) s^2 \frac{d^2\hat{\sigma}_{\gamma i}}{dtdu} \right] . \quad (2.203)$$

When the final state is a $Q\overline{Q}$ pair, $i = g$. The fractional momentum of the nucleon carried by the gluon is x_2 while the photon momentum is denoted by k. The minimum photon momentum needed to produce the final state is $k_{\rm min}$. The spatial coordinates are b and z, the transverse and longitudinal coordinates. The factor of two in Eq. (2.203) arises because both nuclei emit photons and thus serve as targets. If pA collisions are considered, this factor is not included.

The photon can also fluctuate into states with multiple $q\overline{q}$ pairs and gluons, *i.e.* $|n(q\overline{q})m(g)\rangle$, n $q\overline{q}$ pairs and m gluons, the combination of which remains a color singlet with zero flavor and baryon number. One of these photon components can interact with a quark or gluon from the target nucleus ('resolved' production) [35]. The quark and gluon constituents of the photon open up more production channels and could, in principle, lead to larger rates for resolved production in certain regions of phase space. For example, resolved photoproduction of heavy quarks also includes the $q\overline{q} \rightarrow Q\overline{Q}$ channel, not available for direct photoproduction. The total photoproduction cross section is the sum of the direct and resolved contributions.

The photon flux is calculated using the Weiszacker-Williams method. The flux is typically greater for large Z nuclei since a distance b away from the nucleus, the flux is

$$\frac{d^3N_\gamma}{dkd^2b} = \frac{Z^2\alpha w^2}{\pi^2 kb^2}\left[K_1^2(w) + \frac{1}{\gamma_L^2}K_0^2(w)\right] \tag{2.204}$$

where $w = kb/\gamma_L$, k is the photon energy, γ_L is the Lorentz boost in the collision center-of-mass frame and $K_0(w)$, $K_1(w)$ are modified Bessel functions. The photon flux decreases exponentially above a cutoff energy determined by the size of the nucleus. In the center-of-mass frame, the cutoff is $k_{\rm max} \approx \gamma_L \hbar c/R_A$. In the rest frame of the target nucleus, the cutoff is boosted to $E_{\rm max} = (2\gamma_L^2 - 1)\hbar c/R_A$. The corresponding center-of-mass energy of a γN collision is $\sqrt{S_{\gamma N}} \approx \sqrt{2E_{\rm max}m_p}$.

The total photon flux striking the target nucleus is the integral of Eq. (2.204) over the transverse area of the target for all impact parameters subject to the constraint that the two nuclei do not interact hadronically. A reasonable analytic approximation for AA collisions is given by the photon flux integrated over radii $b > 2R_A$. The analytic photon flux is

$$\frac{dN_\gamma}{dk} = \frac{2Z^2\alpha}{\pi k}\left[w_R K_0(w_R)K_1(w_R) - \frac{(w_R)^2}{2}\left(K_1^2(w_R) - K_0^2(w_R)\right)\right] \tag{2.205}$$

where $w_R = 2kR_A/\gamma_L$ while for proton-nucleus interactions $w_R = k(r_p + R_A)/\gamma_L$.

EXAMPLE: Show this.
We integrate over the transverse spatial coordinate, $d^2b = 2\pi b db$ from the minimum value, $b_{\rm min} = 2R_A$ to infinity. We replace w by kb/γ_L in Eq. (2.204) to obtain

$$\frac{dN_\gamma}{dk} = \frac{2Z^2\alpha}{\pi}\frac{k}{\gamma_L^2}\int_{b_{\rm min}}^\infty db\, b\left[K_1^2\left(\frac{kb}{\gamma_L}\right) + \frac{1}{\gamma_L^2}K_0^2\left(\frac{kb}{\gamma_L}\right)\right] . \tag{2.206}$$

Next, we change variables to $x = kb/\gamma_L$ so that $b = x\gamma_L/k$ and $x_{\rm min} = 2kR_A/\gamma_L$. The final form of the integral is

$$\frac{dN_\gamma}{dk} = \frac{2Z^2\alpha}{\pi k}\int_{x_{\rm min}}^\infty dx\, x\left[K_1^2(x) + \frac{1}{\gamma_L^2}K_0^2(x)\right] . \tag{2.207}$$

We use the Bessel function integral

$$\int dx\, xK_n^2(x) = \frac{x^2}{2}[K_n^2(x) - K_{n-1}(x)K_{n+1}(x)] \tag{2.208}$$

to simplify the results. Note that $K_{-n}(x) = K_n(x)$ if n is an integer, useful for the K_0^2 integration, and also $K_2 = K_0 + 2K_1/x$ for the K_1^2 integration. Putting all this together gives

$$\int_{x_{\rm min}}^\infty dx\, x[K_1^2(x) + \frac{1}{\gamma_L^2}K_0^2] = \frac{x^2}{2}[K_1^2(x) - K_0^2(x) - \frac{2}{x}K_0(x)K_1(x) + \frac{1}{\gamma_L^2}(K_0^2(x) - K_1^2(x))]|_{x_{\rm min}}^\infty . \tag{2.209}$$

Note that $K_0(x), K_1(x) \to 0$ as $x \to \infty$. Regrouping and making use of the identity $1 - 1/\gamma_L^2 = \beta^2 \approx 1$, we finally obtain Eq. (2.205) for $w_R \equiv x_{\rm min}$.

Calculations of processes such as heavy quark production by photon-gluon fusion, $\gamma g \to Q\overline{Q}$, can be treated similarly to parton-parton interactions in the initial nucleon-nucleon interactions in central collisions since k is a continuum up to $E_{\rm beam} = \gamma_L m_p$ where $E_{\rm beam}$ is the beam energy. The photon flux in Eq. (2.203) thus takes the part of the parton densities. In the case of resolved production, in addition to the photon flux, there is also the density of partons in the photon.

The flux in Eq. (2.205) is used if only one photon is involved in the interaction. However, a photon from the field around one nucleus may interact with the field around the opposite nucleus, leading to $\gamma\gamma$ or two-photon interactions. The rate of two-photon interactions is determined by the the $\gamma\gamma$ luminosity, $L_{\gamma\gamma}$,

$$\frac{dL_{\gamma\gamma}}{dk_1 dk_2} = \int_{b>R_A}\int_{r>R_A} d^2b d^2r \frac{d^3N_\gamma}{dk_1 d^2b}\frac{d^3N_\gamma}{dk_2 d^2r} P(|\vec{b}-\vec{r}|) \qquad (2.210)$$

where $P(|\vec{b}-\vec{r}|)$ is the probability of a hadronic interaction.

The square of the two-photon center-of-mass energy, $S_{\gamma\gamma}$, is depends on the photon energies, $S_{\gamma\gamma} = 4k_1k_2$. The kinematic limit for two-photon interactions, $4k_{\rm max}^2 \approx S_{\gamma\gamma}$, is significantly reduced relative to $\sqrt{S}$ since $k_{\rm max} \ll p_{\rm beam}$.

The effective luminosities for photon-nucleus and photon-photon interactions at the LHC are shown in Fig. 2.26. The left-hand side compares the effective γA luminosity in pPb and Pb+Pb interactions. The two cases shown for pPb are when the lead nucleus emits the photon (γp) and when the proton emits the photon (γPb). The luminosity is higher when the photon comes from the nucleus.

Two factors contribute to the difference. The first, and largest, is the factor of Z^2 in the nucleus relative to the proton. The second is the relatively small size of the proton. The luminosity of the AB collision, L_{AB}, is the determined by the specifics of the machine, see Ref. [36] for a discussion of luminosities in nucleus-nucleus collisions at the LHC. The AB luminosity is multiplied by the photon flux where $n(k)$ on the y-axis of the γA luminosity is the flux calculated in Eq. (2.205). The expected pPb luminosity, $L_{p{\rm Pb}}$, could be up to a factor of 1000 higher than $L_{\rm PbPb}$. Since, as discussed in Chapter 1, the LHC will provide higher velocity protons than ions, γ_L is higher in pPb. This difference is seen in the photon energy range of the luminosity on the left-hand side of Fig. 2.26.

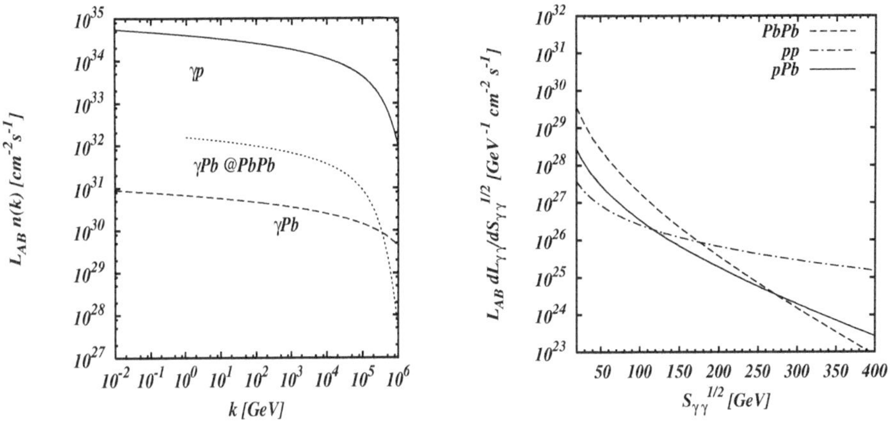

Figure 2.26: Left-hand side: The effective γAluminosity, $L_{AB}n(k)$ with $n(k) \equiv dN_\gamma/dk$, is shown for the cases where the photon is emitted from the proton (γPb) and the ion (γp) as well as when the proton is emitted from the ion in a Pb+Pb collision (γPb@Pb+Pb). (b) The photon-photon luminosities, $L_{AB}dL_{\gamma\gamma}dS_{\gamma\gamma}$, are compared for pp, pPb and Pb+Pb collisions at the LHC. From [37].

The right-hand side of Fig. 2.26 shows the effective two-photon luminosities as a function of the photon-photon center-of-mass energy in pp, pPb and Pb+Pb collisions at the LHC. The effective luminosities are the product of the AB luminosity and the $\gamma\gamma$ luminosity calculated in Eq. (2.210). At high $S_{\gamma\gamma}$, only the pp luminosity is large since the top pp energy more than a factor of two larger than the Pb+Pb center-of-mass energy, thus allowing higher energy photons from both beams. Since the pPb center-of-mass energy is also larger than the Pb+Pb energy, at $\sqrt{S_{\gamma\gamma}} > 250$ GeV the effective pPb luminosity becomes larger than the effective Pb+Pb luminosity.

Predictions for heavy quark and dijet production in ultra-peripheral nucleus-nucleus and proton-nucleus collisions can be found in Refs. [38, 39]. Photoproduction events are relatively clean since they are typically low multiplicity. Only one nucleus breaks up in γA interactions, leaving a rapidity gap between the produced system and the nucleus that emitted the photon. In two-photon interactions, the multiplicity is even lower and there should be rapidity gaps between the produced system and both collision partners.

Exercises

1. Calculate the threshold laboratory energy, the laboratory momentum and the kinetic energy for the following reactions:

   - $pp \to p\Lambda K$
   - $pn \to pnK^+K^-$
   - $pp \to pp\pi^+\pi^-\pi^0$
   - $\pi p \to K\Lambda$
   - $pd \to pn\Delta^+\pi^0$
   - $pp \to J/\psi pp$
   - $\pi J/\psi \to D\overline{D}$
   - $p\overline{p} \to t\overline{t}$

   Assume that the 'beam' particle is the leftmost of the two interacting particles. If the charge state is not given, use the one that conserves all relevant quantum numbers. Work in units of GeV/c^2 and keep 3 significant figures in the particle masses.

2. Given the parton rules for pions, write down the appropriate rules for K^+ and K^-.

3. Derive the Drell-Yan cross section as a function of y and M.

4. For $2 \to 2$ scattering, derive the partonic t and u invariants as a function of x_1 and x_2.

5. Assuming the valence parton distribution can be written as $xq_p^v(x) = Ax(1-x)^3$ while the sea quark and gluon distributions are $xq^s(x) = C(1-x)^7$ and $xg(x) = B(1-x)^5$ respectively. Normalize the valence distribution to three quarks to find A. Then, assuming that the gluons carry 30% of the proton momentum, find C by total momentum conservation.

6. Use the values of A, B and C obtained in Problem 5 to calculate the $q\overline{q}$ and gg luminosities in pp and $\overline{p}p$ collisions for $M = 10$ GeV and $\sqrt{S} = 40$, 500 and 14000 GeV.

Table 2.5: Nuclear beams, mass numbers, A, nuclear charge Z, and beam energies, $E_{\rm beam}$ for AA collisions at RHIC and the LHC.

Nucleus	A	Z	$E_{\rm beam}$ (GeV)
		RHIC	
O	16	8	125
Si	28	14	125
I	127	53	104
Au	197	79	100
		LHC	
O	16	8	3500
Ar	40	18	3150
Kr	84	36	3070
Sn	119	50	2920
Pb	208	82	2750

7. Write the $q\overline{q}$ parton density convolution for asymmetric AB collisions, taking into account the different numbers of protons, Z_A and Z_B, and neutrons, N_A and N_B in nuclei A and B.

8. Using the parton densities obtained in Problem 5, calculate the Drell-Yan cross section as a function of rapidity, over the entire rapidity range, for $M = 10$ GeV at $\sqrt{S} = 62$, 200, and 5500 GeV pp collisions as well as for $M = 4$ and 20 GeV at $\sqrt{S} = 200$ GeV. Then calculate the Drell-Yan cross section as a function of M at $y = 0$ for $\sqrt{S} = 20$, 200 and 5500 GeV.

9. From the beam energies at RHIC and LHC tabulated in Table 2.5, calculate the Lorentz factors, γ_L, $k_{\rm max}$, and $E_{\rm max}$ in the nuclear rest frame, as well as the corresponding maximum center-of-mass energy, $\sqrt{S_{\rm max}}$, in this frame for single photon interactions with protons. Calculate the center-of-mass energy for two-photon interactions, $\sqrt{S_{\gamma\gamma}}$, for the same nuclei and energies.

Chapter 3

Geometry

3.1 Introduction

So far, we have concentrated on the individual nucleon-nucleon collisions and have addressed neither the fact that these nucleons are bound in a nucleus nor the effect of the finite nuclear size upon observables. We address both these issues in this chapter. We begin with the determination of the nuclear density distributions and go on to address the links between the 'centrality' of the collision and the observables.

3.2 Nuclear density distributions

We hinted at some of the finite nuclear size effects in Chapter 2 when we discussed the possibility of electromagnetic interactions in AA collisions. We only calculated the photon flux at distances greater than twice the nuclear radius. However, we have not discussed how that radius was determined.

There are generally two types of measurements that can fix the nuclear size: those that probe the distribution of electric charge in the nucleus and those that probe the nuclear matter distribution. Experiments that measure the charge density distributions study the Coulomb interactions of a charged particle with the nucleus. Electron scattering is one way to measure the charge distribution. While, as we discussed in the previous chapter, at high energies the quark substructure is probed, at lower energies only the protons are probes since the charge neutral neutrons are not 'seen' by electrons. Matter distributions measure all the nucleons, not just the protons. Rutherford scattering (low energy hadron-nucleus interactions) is one example of matter distribution measurements. Rutherford directed an α-particle (^{4}He) source on a gold foil and watched the deflections of the incident α to test J. J. Thomson's proposal that electrons were spread evenly in a lump of positively-charged matter: the 'plum pudding' model of the atom. If the atom was 'pudding'-like, the α-particles would not be significantly deflected passing through the foil. Most were not but some were scattered more than 90°, back toward the source, repelled by a much larger charge. Thus it was determined that the positive charge of the atom was localized in a small area of the atom, the atomic nucleus. It is usually assumed that the charge and matter distributions are identical so that fits of the nuclear charge density distributions can be applied equally well to the nuclear matter distributions.

We now describe the calculation of the scattering cross section and the determination of the nuclear density distributions. Recall that the transition probability involves the initial and final-state wavefunction, $\psi_i = \exp(i\mathbf{p}_i \cdot \mathbf{r})$ and $\psi_f = \exp(i\mathbf{p}_f \cdot \mathbf{r})$ respectively, of the scattered electron and the interaction potential, $V(\mathbf{r})$, so that

$$T_{fi} = \int d^4\mathbf{r}\, \psi_f^* V(\mathbf{r}) \psi_i \; .$$

If the source is static, the potential is independent of time so that

$$\begin{aligned} T_{fi} &= \int dt \exp[-i(E_i - E_f)t] \int d^3r \, \exp(-i(\vec{p}_f - \vec{p}_i) \cdot \vec{r}) V(\vec{r}) \\ &= 2\pi\delta(E_i - E_f) \int d^3r \, \exp(i\vec{q} \cdot \vec{r}) V(r) \end{aligned} \quad (3.1)$$

where we have defined $\vec{q} = \vec{p}_i - \vec{p}_f$. The nuclear charge density distribution, $Ze\rho(\vec{r}')$, enters the interaction potential where $\vec{r}'$ is a point in the nuclear volume. Thus the electron, at position $\vec{r}$ during the scattering, feels a potential energy over the distance $|\vec{r} - \vec{r}'|$ due to the charge difference dQ at $\vec{r}'$, as shown in Fig. 3.1, so that

$$dV = -\frac{edQ}{4\pi|\vec{r} - \vec{r}'|} \ . \tag{3.2}$$

The charge dQ is related to the charge density distribution over the

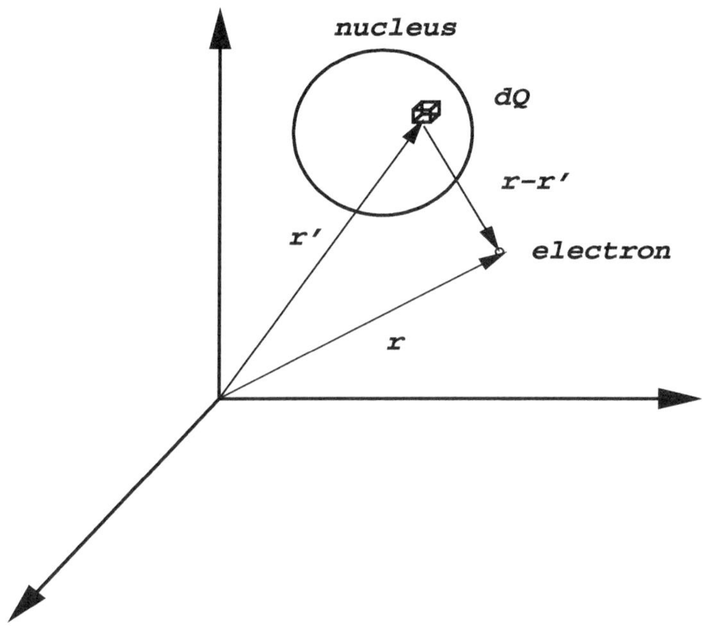

Figure 3.1: The charge density distribution is obtained from the potential energy between the charge element dQ in the nucleus and the electron probe where the distance between the two charges is $\vec{r} - \vec{r}'$.

nuclear volume,

$$dQ = Ze\rho(\vec{r}')d^3r' \tag{3.3}$$

where d^3r' is the nuclear volume element. The total interaction potential is obtained by summing the charge over the volume,

$$V(r) = -\frac{Ze^2}{4\pi}\int d^3r' \frac{\rho(\vec{r}')}{|\vec{r} - \vec{r}'|} \ . \tag{3.4}$$

We define the matrix element as

$$\mathcal{M} = \mathcal{A}\int d^3r' \int d^3r \exp(i\vec{q}\cdot\vec{r})\frac{Ze^2}{4\pi}\frac{\rho(\vec{r'})}{|\vec{r}-\vec{r'}|} \tag{3.5}$$

where $\mathcal{A}$ is the amplitude for an electron scattering from a point particle. It is calculated using the appropriate Feynman rules for fermions. The derivation can be found in *e.g.* Ref. [40]. If we define $\vec{s} = \vec{r} - \vec{r'}$, then $\exp(i\vec{q}\cdot\vec{r}) = \exp(i\vec{q}\cdot\vec{s})\exp(i\vec{q}\cdot\vec{r'}) = \exp(iqs\cos\theta)\exp(i\vec{q}\cdot\vec{r'})$. Then we have

$$\mathcal{M} = \mathcal{A}\frac{Ze^2}{4\pi}\int d^3s\frac{\exp(iqs\cos\theta)}{s}\int d^3r' \exp(i\vec{q}\cdot\vec{r'})\rho(\vec{r'}) \ . \tag{3.6}$$

The first integral in Eq. (3.6) becomes, after squaring the matrix element, proportional to the cross section for an electron scattering off a point-like heavy target,

$$\frac{d\sigma}{dq^2} = Z^2\frac{d\sigma_{ep}}{dq^2}|F(q)|^2 \tag{3.7}$$

where we have written the cross section as a function of the Mandelstam invariant T with $T = -q^2$. We have a factor of Z^2 in front of the electron-proton scattering cross section to account for the nuclear charge in the matrix element. The additional factor due to the nuclear charge density is the square of the nuclear form factor, $F(\vec{q})$, where

$$F(\vec{q}) = \int d^3r \exp(i\vec{q}\cdot\vec{r'})\rho(\vec{r'}) \ . \tag{3.8}$$

The form factor is found by measuring the scattering cross section as a function of the scattering angle θ' since, for elastic scattering, $q = 2p\sin(\theta'/2)$ where p is the electron momentum ($p = |\vec{p}_i| = |\vec{p}_f|$ for elastic scattering). Taking the inverse Fourier transform of the form factor gives the nuclear charge density distribution.

EXAMPLE: Show that $q = 2p\sin(\theta'/2)$ *when* $p = |\vec{p}_i| = |\vec{p}_f|$.
We defined $\vec{q} = \vec{p}_i - \vec{p}_f$ before. Thus

$$q = [(\vec{p}_i - \vec{p}_f)^2]^{1/2} \ .$$

After setting $p = |\vec{p}_i| = |\vec{p}_f|$, we have

$$q = [2p^2(1 - \cos\theta')]^{1/2} .$$

The half-angle formula tells us that $1 - \cos\theta' = 2\sin^2(\theta'/2)$. Then

$$q = [4p^2 \sin^2(\theta'/2)]^{1/2} = 2p\sin(\theta'/2) . \tag{3.9}$$

EXAMPLE: Determine the matrix element $|\mathcal{M}|^2$ and thus the electron-proton cross section,

$$\frac{d\sigma_{ep}}{dq^2} = \frac{4\pi\alpha^2}{q^4}\cos^2(\theta'/2) , \tag{3.10}$$

starting from Eq. (3.6).
We first calculate the matrix element for ep scattering in the lab frame assuming the proton is point-like. In this frame, for a target with a mass much greater than the electron mass, the target can be considered to be at rest both before and after the scattering, known as recoilless scattering,

$$\mathcal{M}_{ep} = \mathcal{A}\frac{e^2}{4\pi}\int d^3s \frac{\exp(iqs\cos\theta)}{s} . \tag{3.11}$$

Writing $d^3s = 2\pi s^2 ds d(\cos\theta)$,

$$\mathcal{M}_{ep} = \mathcal{A}\frac{e^2}{2}\int ds\, d(\cos\theta) s\, \exp(iqs\cos\theta) . \tag{3.12}$$

The integral over $d(\cos\theta)$ leaves us with an integral only over s,

$$\begin{aligned} \mathcal{M}_{ep} &= \mathcal{A}\frac{e^2}{2}\int \frac{ds\, s}{iqs}\exp(iqs\cos\theta)\Big|_{-1}^{1} \\ &= \mathcal{A}\frac{e^2}{2}\int \frac{ds}{iq}[\exp(iqs) - \exp(-iqs)] . \end{aligned} \tag{3.13}$$

The integral of $\exp(iqs)$ over s will diverge for the upper limit of the s integral, $s \to \infty$. To keep the integral finite, we introduce a screening term in the potential, $\exp(-r/a) \sim \exp(-s/a)$ where $a \sim R_{\rm atom} \sim 10^4 R_A$ since $R_{\rm atom} \sim 10^{-8}$ cm while the nuclear radius is on the order of a few times

10^{-13} cm. Eventually we let $a \to \infty$. The integration over s is then finite, giving

$$\begin{aligned}
\mathcal{M}_{ep} &= \mathcal{A}\frac{e^2}{2}\int \frac{ds}{iq}(\exp[-s(1/a-iq)] - \exp[-s(1/a+iq)]) \\
&= \frac{e^2}{2iq}\left[-\frac{\exp[-s(1/a-iq)]}{1/a-iq} + \frac{\exp[-s(1/a+iq)]}{1/a+iq}\right]\Bigg|_0^\infty \\
&= \frac{e^2}{2iq}\left[\frac{1}{1/a-iq} - \frac{1}{1/a+iq}\right] . \qquad (3.14)
\end{aligned}$$

After adding the two contributions, the factors of iq cancel, along with the factor of two in the denominator, so that

$$\mathcal{M}_{ep} = \mathcal{A}\frac{e^2}{(1/a)^2 + q^2} = \mathcal{A}\frac{e^2}{q^2} \qquad (3.15)$$

where, in the last equality, we have taken $a \to \infty$. The square of the matrix element, needed for the cross section, is

$$|\mathcal{M}_{ep}|^2 = |\mathcal{A}|^2\frac{(4\pi)^2\alpha^2}{q^4} \qquad (3.16)$$

where we have replaced the electric charge by the electromagnetic coupling constant, $\alpha = e^2/4\pi$. The derivation of $|\mathcal{A}|^2$ can be found in Ref. [40] for $2 \to 2$ scattering. It is

$$\begin{aligned}
|\mathcal{A}|^2 &= 8[(\mathbf{p}_1 \cdot \mathbf{p}_2)(\mathbf{p}_3 \cdot \mathbf{p}_4) + (\mathbf{p}_1 \cdot \mathbf{p}_4)(\mathbf{p}_2 \cdot \mathbf{p}_3) \\
&\quad - (\mathbf{p}_1 \cdot \mathbf{p}_3)m_p^2 - (\mathbf{p}_2 \cdot \mathbf{p}_4)m_e^2 + 2(m_e m_p)^2] . \qquad (3.17)
\end{aligned}$$

We will evaluate this amplitude a bit later. First, we determine the expression for the cross section in our particular case. We begin with Eq. (2.89) in Chapter 2. The expression for the Lorentz-invariant phase space, Eq. (2.90), when one particle is at rest, is

$$d\text{Lips} = (2\pi)^4\delta^4(\mathbf{p}_1 + \mathbf{p}_2 - \mathbf{p}_3 - \mathbf{p}_4)\frac{d^3p_3}{(2\pi)^3 2E_3}\frac{1}{(2\pi)^3 2E_4} .$$

We let $\mathbf{p}_1$ and $\mathbf{p}_3$ be the electron four-momenta before and after the interaction, respectively, while the proton four-momenta are $\mathbf{p}_2$ and $\mathbf{p}_4$. Here, we assume that the proton remains stationary both before and after the scattering so that there is no recoil and $\mathbf{p}_2 = \mathbf{p}_4 = (m_p, 0)$. We have also thus only included a factor of $1/(2\pi)^3 E_4$ in the phase space for the final-state proton. The four-dimensional delta function is then $\delta^4(\mathbf{p}_1 + \mathbf{p}_2 - \mathbf{p}_3 - \mathbf{p}_4) = \delta(E_1 - E_3)\delta^3(\vec{p}_1 - \vec{p}_3)$ since the m_p terms in the energy conservation delta function cancel. The flux is $F = 4[(\mathbf{p}_1 \cdot \mathbf{p}_2)^2 - m_e^2 m_p^2]^{1/2} = 4p_1 m_p$ since $\mathbf{p}_1 \cdot \mathbf{p}_2 = E_1 m_p$. The expression for the cross section is then

$$d\sigma_{ep} = \frac{|\mathcal{M}_{ep}|^2}{64\pi^2 p_1 m_p E_1 E_2} d^3p_3 \delta(E_1 - E_3) \ . \tag{3.18}$$

We can write the differential element as $d^3p_3 = p_3^2 dp_3 d\Omega = E_3^2 dE_3 d\Omega$ since, for an electron, $m_e \sim 0$ and $E_1 \sim p_1$, $E_3 \sim p_3$. Then the electron four-momenta are $\mathbf{p}_1 = (p, \vec{p})$ and $\mathbf{p}_3 = (p, \vec{p})$ since $|\vec{p}_1| = |\vec{p}_3| = |\vec{p}|$. Setting $p_1 = E_1$ and $E_2 = m_p$ and integrating over dE_3 using the energy conservation delta function gives us

$$\frac{d\sigma_{ep}}{d\Omega} = \frac{|\mathcal{M}_{ep}|^2}{64\pi^2 m_p^2} \tag{3.19}$$

Given the values of the four-momenta $\mathbf{p}_1 \cdots \mathbf{p}_4$ we have deduced from energy-momentum conservation, we can calculate the amplitude $|\mathcal{A}|^2$ in Eq. (3.17). Letting $m_e \to 0$, we drop the last two terms in Eq. (3.17). We also have $\mathbf{p}_1 \cdot \mathbf{p}_2 = \mathbf{p}_3 \cdot \mathbf{p}_4 = \mathbf{p}_2 \cdot \mathbf{p}_3 = \mathbf{p}_1 \cdot \mathbf{p}_4 = Em_p = pm_p$. The remaining 4-vector product is $\mathbf{p}_1 \cdot \mathbf{p}_3 = p^2 - p^2 \cos\theta' = q^2/2$. Then

$$|\mathcal{A}|^2 = 8[2p^2 m_p^2 - m_p^2(p^2 - p^2\cos\theta')] = 16p^2 m_p^2 \cos^2(\theta'/2) \ . \tag{3.20}$$

The complete squared matrix element is then

$$|\mathcal{M}_{ep}|^2 = \frac{256\pi^2\alpha^2 m_p^2}{q^4} p^2 \cos^2(\theta'/2) \tag{3.21}$$

so that the cross section becomes

$$\frac{d\sigma_{ep}}{d\Omega} = \frac{4\alpha^2}{q^4} p^2 \cos^2(\theta'/2) \ . \tag{3.22}$$

Instead of the cross section in a certain angular interval, as in Eq. (3.22), the cross section in Eq. (3.10) is a function of q^2. We can change from a cross section differential in $d\Omega$ to one differential in dq^2 in two steps. We first change $d\Omega$ to dT, the momentum transfer squared and then, by relating dT to dq^2, we can replace $d\Omega$ by dq^2. Recall from Eq. (2.99) in Chapter 2 that $dT = (p_1 p_3/\pi) d\Omega = (p^2/\pi) d\Omega$. We can replace dT by dq^2

since $T = (\mathbf{p}_1 - \mathbf{p}_3)^2 = -(\vec{p}_1 - \vec{p}_3)^2 = -\vec{q}^2 \equiv q^2$. Then $dq^2 = (p^2/\pi)d\Omega$. Finally,

$$\frac{d\sigma_{ep}}{dq^2} = \frac{4\pi\alpha^2}{q^4}\cos^2(\theta'/2) . \tag{3.23}$$

Experiments have shown that the nuclear charge density at the center of the nucleus is very similar for all nuclei. The nucleons do not form clumps either at the center or near the surface. The measured distributions, both of charge and matter, are relatively independent of distance until almost at the nuclear surface. Therefore, the number of nucleons per volume is approximately constant,

$$\frac{A}{4\pi R_A^3/3} \sim \text{constant} \tag{3.24}$$

where R_A is the nuclear radius. To satisfy this constancy, we need

$$R_A \propto A^{1/3} . \tag{3.25}$$

The proportionality constant, r_0, has been determined to be ~ 1.2 fm.

One simple way to satisfy the approximate constancy of the density distribution as a function of distance is to assume that the nucleus is like a hard sphere with

$$\rho_A(\vec{r}) = \rho_0 \theta_R \tag{3.26}$$

where $\theta_R = 1$ for $r < R_A$ and zero for $r > R_A$. In this case, ρ_0 is the constant in Eq. (3.24) and

$$\rho_0 = \frac{3}{4\pi r_0^3} . \tag{3.27}$$

This is known as the 'hard sphere' or 'sharp surface' approximation. Of course it is not a good approximation at distances close to and greater than the nuclear radius since interactions can still occur when the tails of the density distributions overlap — impossible if the nucleus indeed had a sharp surface.

Another function that has characteristics similar to the hard sphere approximation but with a long tail at large distance is the Fermi-Dirac distribution, discussed more fully in Chapter 4. In the case of nuclear density distributions, this functional form is known as the Woods-Saxon parameterization. The nuclear density distribution is fit using either two or three parameters,

$$\rho_A(r) = \rho_0 \frac{1 + \omega(r/c)^2}{\exp[(r-c)/z] + 1} \tag{3.28}$$

for the three parameter case. In the two parameter Woods-Saxon distribution, $\omega = 0$. The parameter c is $\sim \langle r^2 \rangle^{1/2}$. For the hard sphere approximation,

$$\langle r^2 \rangle = \frac{4\pi\rho_0 \int dr r^2 r^2}{4\pi\rho_0 \int dr r^2} = \frac{3}{5} R_A^2 \, . \tag{3.29}$$

In a real nucleus, c is a bit larger than $\langle r^2 \rangle^{1/2}$, $c \sim 1.07A^{1/3}$. The parameter z is typically ~ 0.545 fm [41].

The density decreases rapidly near the surface, falling from 90% of its central value to 10% of that value within about 2.3 fm. This distance, the skin thickness, is related to the parameter z in Eq. (3.28).

EXAMPLE: Show that the skin thickness is ~ 2.3 *fm.*
Let $\rho_A(r_1) = 0.9\rho_0$ and $\rho_A(r_2) = 0.1\rho_0$. Then, using a two-parameter Woods-Saxon distribution for the nuclear density, we have, for r_1,

$$0.9 = \frac{1}{\exp[(r_1 - c)/z] + 1} \, .$$

Solving for the exponential, we obtain

$$\exp[(r_1 - c)/z] = \frac{1}{9} \, .$$

Likewise, for r_2,

$$\exp[(r_2 - c)/z] = 9 \, .$$

Taking the logarithm of the ratio of the two, we have

$$r_2 - r_1 = z \ln 81 \sim 4.4z = 2.4 \text{ fm}$$

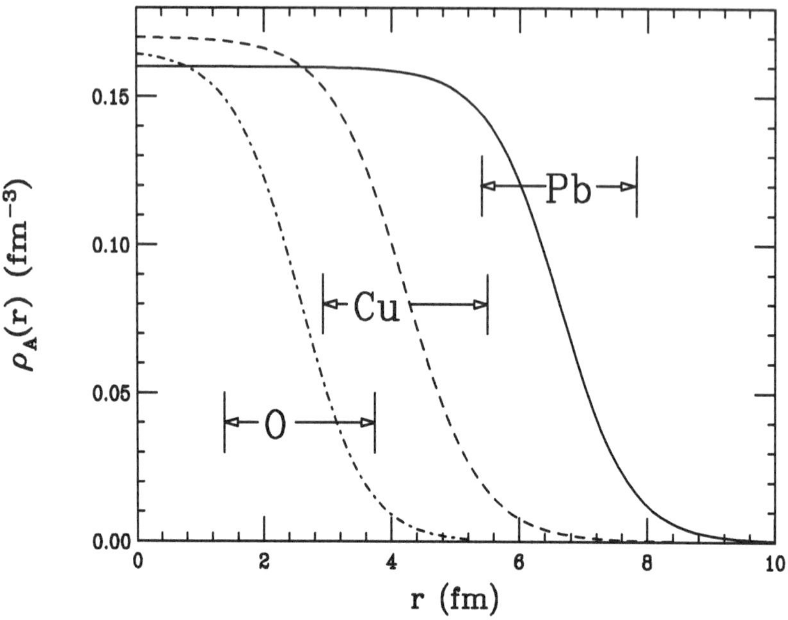

Figure 3.2: The nuclear density distributions for oxygen (dot-dashed), copper (dashed) and lead (solid) nuclei.

for $z = 0.545$ fm. The individual values of r_1 and r_2 are

$$
\begin{aligned}
r_1 &= c - z \ln 9 \\
r_2 &= c + z \ln 9 \ .
\end{aligned}
$$

Figure 3.2 shows the charge density distributions as a function of distance r for oxygen, O, copper, Cu, and lead, Pb, nuclei. The skin thickness of each nucleus is indicated. Other forms of density distributions may be used, particularly for smaller nuclei, such as Gaussian, harmonic oscillator or Yukawa-type distributions, but the Woods-Saxon form has been used for $A > 16$ with success. The nuclear matter distributions are normalized so that the integral of ρ over the nuclear volume is equal to the nuclear mass number,

$$\int d^3 r \rho_A(r) = A \ . \tag{3.30}$$

Table 3.1: The parameters for Woods-Saxon density distributions for several nuclei, fitted from charge density data.

A	c (fm)	z (fm)	ω	ρ_0 (fm^{-3})
O(16)	2.608	0.513	-0.051	0.1654
Si(27)	3.07	0.519	0	0.1739
Ca(40)	3.766	0.586	-0.161	0.1699
Cu(63)	4.214	0.586	0	0.1701
I(110)	5.33	0.535	0	0.1577
Au(197)	6.38	0.535	0	0.1693
Pb(208)	6.624	0.549	0	0.1600

Table 3.1 presents the Woods-Saxon parameterizations for several nuclei. The value of ρ_0 is fixed from the normalization in Eq. (3.30) while the other parameters, obtained from measurements of the charge density distributions, are tabulated in Ref. [42]. Not all nuclei of interest for nuclear collisions are found parameterized in Ref. [42]. For such nuclei, one may take the averages given above, $c \sim 1.07A^{1/3}$ fm, $z \sim 0.545$ fm, and $\omega = 0$ with ρ_0 calculated from Eq. (3.30).

Before finishing this section, one last nucleus should be discussed. The deuteron, $A = 2$ and $Z = 1$, has been accelerated at RHIC in d+Au collisions at $\sqrt{S_{NN}} = 200$ GeV. These dA collisions may be preferred over pA collisions for technical reasons: the deuteron beam and the heavy-ion beam have similar Z/A ratios, simplifying the magnetic optics.

While the proton can be treated as a point particle for the purposes of geometrical studies, the deuteron cannot. It is a relatively loosely bound proton-neutron state with a consequently large spatial extent. We use the Hulthén wave function [43] to calculate the deuteron density distribution,

$$\rho_{\rm d}(r) = \left|\frac{u(r)}{r}\right|^2 + \left|\frac{w(r)}{r}\right|^2 , \tag{3.31}$$

where

$$\begin{aligned} u(r) &= N\sqrt{1-\epsilon^2}(1-\exp[-\beta(\alpha r - x_c)])\exp(-\alpha r) \\ w(r) &= N\epsilon(1-\exp[-\gamma(\alpha r - x_c)])^2\left[1+\frac{3}{\alpha r}(1-\exp[-\gamma\alpha r])\right. \\ &\quad \left. + \frac{3}{\alpha^2 r^2}(1-\exp[-\gamma\alpha r])^2\right]\exp(-\alpha r)\ . \end{aligned}$$

Two sets of parameters are usually given. The first is $\epsilon = 0.02438$, $\beta = 9.045$, $\gamma = 4.799$ and $x_c = 0.13$ fm while the second is $\epsilon = 0.03232$, $\beta = 4.68$, $\gamma = 2.494$ and $x_c = 0$. In both cases, $\alpha = 0.2317$/fm. The density is also normalized so that $\int d^2r dz \rho_A(s) = A$.

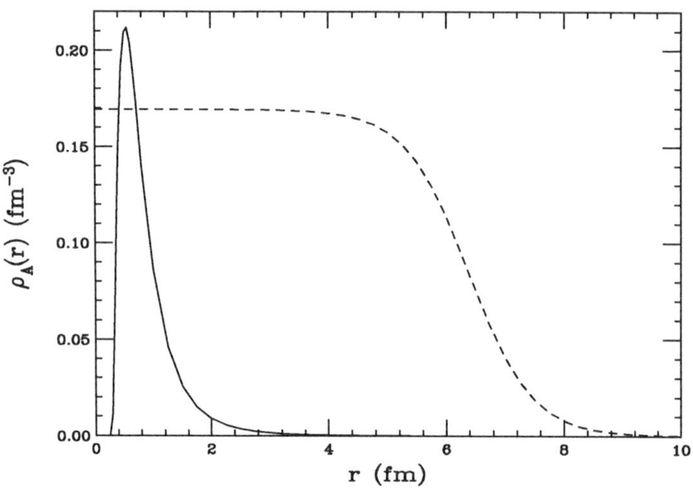

Figure 3.3: The nuclear density distribution for deuterium (solid) compared to gold (dashed).

Figure 3.3 compares the deuteron density distribution using the first set of parameters with that of gold. The deuteron distribution has a slightly higher peak density but a longer tail for its relative size. In fact, for very large r, more than 20 fm, the deuteron density is higher than the gold density. Thus the finite deuteron size is important.

3.3 Geometry of nucleus-nucleus collisions

The system produced in a nucleus-nucleus collision is expected to be closer to the conditions necessary for quark-gluon plasma production if more nucleons are involved in the initial collision. Thus the closer the collision is to 'head-on', or the more 'central' a collision is, the more likely plasma production will be. More glancing or 'peripheral' collisions are less likely to lead to ideal conditions for plasma production. This is referred to as 'centrality'.

Several variables are used to quantify collision centrality including the number of participants, $N_{\rm part}$, and the number of collisions, $N_{\rm coll}$. Both are given in terms of the impact parameter, b, the distance between the centers of two colliding nuclei. Low b corresponds to more central collisions, high b to peripheral collisions. Of course the impact parameter is not directly measurable but experience at the BNL AGS and CERN SPS has shown that the number of produced particles, the multiplicity, and the energy of the produced particles in the transverse direction, the transverse energy, E_T, are both proportional to the number of nucleon participants. Thus since the dependence of $N_{\rm part}$ on impact parameter can be calculated rather precisely for a given nuclear density distribution, we can get a good handle on the collision centrality.

The Glauber model of multiple collisions [44] is used to relate the impact parameter to the number of participants and the number of collisions. It is a geometrical model based on the assumption of constant inelastic cross section, $\sigma_{\rm inel}$, for each subsequent collision.

Figure 3.4 shows the colliding system in the rest frame of one of the nuclei, the nucleus on the right – the 'target' in a fixed target collision. The 'projectile' nucleus is contracted in the direction of motion. The transverse distance between the centers of the two nuclei is referred to as the 'impact parameter', usually labeled $\vec{b}$. The variable $\vec{s}$ is the distance between the center of one nucleus and a point inside the other nucleus.

It is perhaps instructive to start by considering a pA collision. For

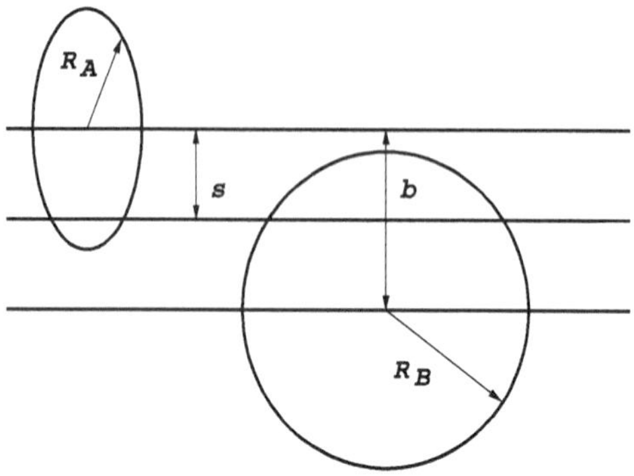

Figure 3.4: The picture of a nucleus-nucleus collision in the rest frame of the target nucleus.

any given impact parameter, as long as the proton is impinging on the nucleus at some b, the proton drills a cylindrically-shaped tube through the nucleus. The length of this tube depends on the distance from the center or the 'thickness' of the nucleus at that point. The nuclear thickness function, $T_A(b)$, is just the integral of the nuclear density over longitudinal dimension,

$$T_A(b) = \int dz \rho_A(b, z) \ , \tag{3.32}$$

and depends only on b. For example, if

$$\rho_A = \rho_0 \theta_R = \rho_0 \theta(z^2 - R_A^2)\theta(b^2 - R_A^2) \ , \tag{3.33}$$

then, for a given b, the longitudinal distance to the surface from the center is $z = \pm(R_A^2 - b^2)^{1/2}$. If the hard-sphere approximation is used,

$$T_A(b) = \rho_0 \int_{-\sqrt{R_A^2-b^2}}^{\sqrt{R_A^2-b^2}} dz = 2\rho_0 \sqrt{R_A^2 - b^2} \ . \tag{3.34}$$

Note that $T_A(b)$ has units of inverse area. Thus, according to Eq. (3.34), the maximum thickness at the center of the nucleus is $2\rho_0 R_A$ while at the edge, $b \sim R_A$, the nucleus has zero thickness. We normalize so that

$$\int d^2b T_A(b) = 4\pi\rho_0 \int_0^{R_A} db\, b \sqrt{R_A^2 - b^2} = \frac{4\pi\rho_0}{3} R_A^3 \ .$$

Since $\rho_0 = 3/4\pi r_0^3$, we have

$$\int d^2b T_A(b) = A \; , \tag{3.35}$$

independent of the parameterization of the nuclear density distribution since ρ_0 can be fixed from the normalization condition.

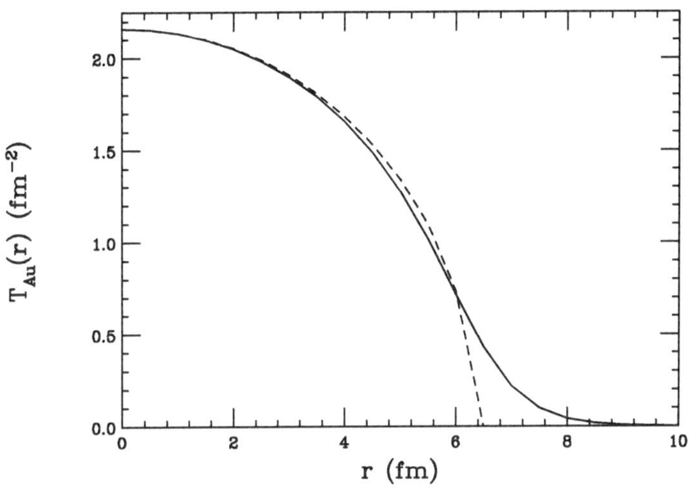

Figure 3.5: The nuclear thickness function, $T_A(b)$, for a gold nucleus. The Woods-Saxon parameterization (solid curve) is compared to the sharp surface result (dashed).

Figure 3.5 compares the nuclear thickness function, $T_A(b)$, calculated with the 2-parameter Woods-Saxon density distribution to that assuming a sharp surface, as in Eq. (3.34). The two results are essentially identical for $r < 3$ fm and differ only by a few percent for $3 < r < 6$ fm but differ significantly for $r > R_A$ where the sharp surface result is zero. The difference between the two assumptions about the density are thus most important for peripheral collisions where, although the impact parameter is large, the tails of the two distributions still overlap. As we will see, these tails are important for accounting for fluctuations in the number of participant nucleons and the number of nucleon-nucleon collisions.

When two nuclei collide, the nucleons in the row at transverse distance $\vec{s}$ in nucleus A collide with the nucleons in the row at $\vec{b}-\vec{s}$ in nucleus B as in Fig. 3.4. The interaction of these two rows of nucleons is sometimes referred to as row-on-row collisions. We can calculate the probability of a nucleon-nucleon collision at impact parameter b when nucleus A is at position $(\vec{b}_A, z_A)$ and nucleus B is at position $(\vec{b}_B, z_B)$. The probability of a NN collision at impact parameter $\vec{b}$ is the product of three components: the probability for finding a nucleon from nucleus A in the volume element $d^2b_A dz_A$; the probability of finding a nucleon from nucleus B in the volume element $d^2b_B dz_B$; and the probability of an inelastic nucleon-nucleon collision. The probability for finding a nucleon in the volume element $d^2b_A dz_A$ at position $(\vec{b}_A, z_A)$ in nucleus A is

$$\frac{\rho_A(\vec{b}_A, z_A)}{A} d^2b_A dz_A \ , \tag{3.36}$$

proportional to the density distribution in the nuclear volume divided by the mass number A. The probability is normalized so that integration over the nuclear volume gives unity since we have defined $\int d^3r \rho_A(r) = A$. The simplest inelastic nucleon-nucleon collision probability is a delta function times the NN inelastic cross section, $\sigma_{\rm inel}\delta(\vec{b}-\vec{b}_A-\vec{b}_B)$. Then

$$dP_{AB}(b) = \frac{\rho_A(\vec{b}_A, z_A)}{A} d^2b_A \, dz_A \frac{\rho_B(\vec{b}_B, z_B)}{B} d^2b_B \, dz_B \delta(\vec{b}-\vec{b}_A-\vec{b}_B) \ . \tag{3.37}$$

The total probability as a function of b is the integral over the volume elements of both nuclei, normalized per AB,

$$\begin{aligned} P_{AB}(b) &= \frac{\sigma_{\rm inel} T_{AB}(b)}{AB} \\ &= \frac{\sigma_{\rm inel}}{AB} \int d^2b_A dz_A d^2b_B dz_B \rho_A(\vec{b}_A, z_A) \rho_B(\vec{b}_B, z_B) \delta(\vec{b}-\vec{b}_A-\vec{b}_B) \\ &= \frac{\sigma_{\rm inel}}{AB} \int d^2s dz_A dz_B \rho_A(\vec{s}, z_A) \rho_B(|\vec{b}-\vec{s}|, z_B) \end{aligned} \tag{3.38}$$

where we have integrated over b_B using the delta function and identified b_A with s. Dropping the common factors of AB and replacing $\int \rho_A(\vec{r}, z)dz$ with $T_A(r)$, we have

$$T_{AB}(b) = \int d^2s T_A(s) T_B(|\vec{b} - \vec{s}|) \ , \tag{3.39}$$

the nuclear overlap function. Note that $T_{AB}(b)$, like the nuclear profile function, $T_A(b)$, has dimensions of inverse area, fm^{-2}. The integral of $T_{AB}(b)$ over all impact parameters is a number equal to the product of the nuclear mass numbers,

$$\int d^2b T_{AB}(b) = AB \ . \tag{3.40}$$

In Fig. 3.6, we show the overlap integral as a function of impact parameter for several symmetric collision systems, from oxygen and silicon to gold and lead. Since $T_{AB}(b)$ is purely a function of geome-

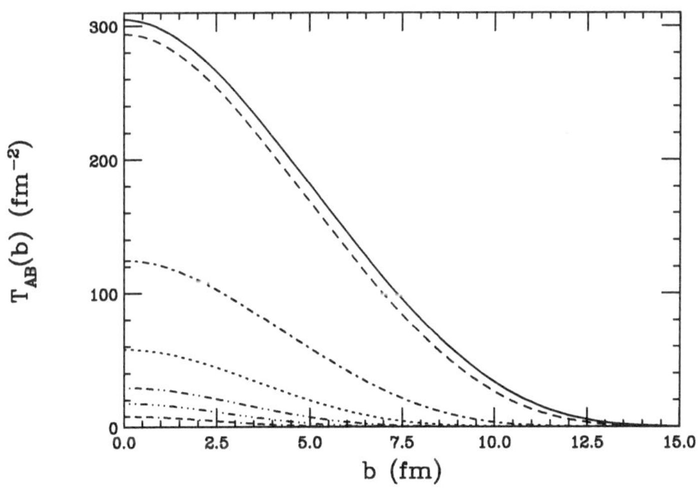

Figure 3.6: The nuclear overlap integral as a function of impact parameter b. for O+O (dash-dash-dash-dotted), Si+Si (dot-dot-dot-dashed), Ca+Ca (dot-dot-dash-dashed), Cu+Cu (dotted), I+I (dot-dashed), Au+Au (dashed) and Pb+Pb (solid) collisions.

try, it is independent of the collision energy. The product $\sigma_{\rm inel}T_{AB}(b)$ gives the number of nucleon-nucleon collisions as a function of impact parameter,

$$N_{\rm coll}(b) = \sigma_{\rm inel}T_{AB}(\vec{b}) \ . \tag{3.41}$$

Note that $N_{\rm coll}(b)$ does depend on energy since the inelastic cross section grows with $\sqrt{S}$. The inelastic cross section is 32 mb at $\sqrt{S} = 20$ GeV, 42 mb at $\sqrt{S} = 200$ GeV and 60 mb at $\sqrt{S} = 5.5$ TeV. Thus the number of collisions can increase by nearly a factor of two for the same system at 5.5 TeV relative to 20 GeV, a large increase. Since the units of $T_{AB}(b)$ are fm^{-2}, after multiplying by the inelastic NN cross section, the number of collisions is dimensionless.

The probability for n inelastic NN collisions at impact parameter $\vec{b}$ out of a possible AB collisions is

$$P_{AB}(n,\vec{b}) = \left(\begin{array}{c} AB \\ n \end{array} \right) \left[\frac{\sigma_{\rm inel}T_{AB}(\vec{b})}{AB}\right]^n \left[1 - \frac{\sigma_{\rm inel}T_{AB}(\vec{b})}{AB}\right]^{AB-n} \tag{3.42}$$

where

$$\left(\begin{array}{c} AB \\ n \end{array} \right) = \frac{(AB)!}{n!(AB-n)!} \ . \tag{3.43}$$

The two factors in Eq. (3.42) are the probability for n hits with $AB-n$ misses. The total cross section, $\sigma_{\rm inel}^{AB}$, is obtained from the sum over n from 1 to AB and the integral over impact parameter is

$$\sigma_{\rm in}^{AB} = \int d^2b \sum_n P_{AB}(n,\vec{b}) = \int d^2b \left(1 - \left[1 - \frac{\sigma_{\rm inel}T_{AB}(b)}{AB}\right]\right) \ . \tag{3.44}$$

EXAMPLE: Derive Eq. (3.44).

Let $\sigma_{\rm inel}T_{AB}(b)/AB = n_{\rm coll}(b)$. The lower case n is used to distinguish this ratio from the definition in Eq. (3.41) above. Then we rewrite Eq. (3.42) as

$$P_{AB}(n,\vec{b}) = \left(\begin{array}{c} AB \\ n \end{array} \right) n_{\rm coll}^n(b)(1-n_{\rm coll}(b))^{AB-n} \ .$$

From the properties of binomial expansions, we know that

$$(a+x)^m = \sum_{n=0}^{\infty} \binom{m}{n} x^n a^{m-n} = a^m + \sum_{n=1}^{\infty} \binom{m}{n} x^n a^{m-n}$$

so that

$$\sum_{n=1}^{\infty} \binom{m}{n} x^n a^{m-n} = (a+x)^m - a^m \,. \tag{3.45}$$

Substituting $a = 1 - n_{\rm coll}(b)$ and $x = n_{\rm coll}(b)$, we obtain

$$\begin{aligned} P_{AB}(\vec{b}) &= \sum_{n=1}^{AB} P_{AB}(n, \vec{b}) \qquad (3.46) \\ &= [n_{\rm coll}(b) + (1 - n_{\rm coll}(b))]^{AB} - [1 - n_{\rm coll}(b)]^{AB} \\ &= 1 - [1 - n_{\rm coll}(b)]^{AB} \,, \end{aligned}$$

giving us Eq. (3.44).

EXAMPLE: Determine $\langle n \rangle$, $\langle n^2 \rangle$ and the standard deviation, $\langle n^2 \rangle - \langle n \rangle^2$ as a function of impact parameter.
The average number of nucleon-nucleon collisions is

$$\begin{aligned} \langle n \rangle &= \frac{1}{P_{AB}(b)} \sum_{n=1}^{AB} n P(n, \vec{b}) \\ &= \frac{1}{P_{AB}(b)} \sum_{n=1}^{AB} n \binom{AB}{n} n_{\rm coll}^n(b)(1 - n_{\rm coll}(b))^{AB-n} \qquad (3.47) \end{aligned}$$

We can manipulate the binomial factor so that

$$\begin{aligned} \sum_{n=1}^{AB} n \frac{(AB)!}{n!(AB-n)!} &= \sum_{n=1}^{AB} \frac{(AB)!}{(n-1)!(AB-n)!} \\ &= AB \sum_{n=1}^{AB} \frac{(AB-1)!}{(n-1)!(AB-n)!} \qquad (3.48) \end{aligned}$$

since $(AB - n)! = ((AB - n) - (n - 1))!$. Then

$$\langle n \rangle = \frac{AB n_{\rm coll}(b)}{P_{AB}(b)} \sum_{n=1}^{AB} \binom{AB-1}{n-1} n_{\rm coll}^{n-1}(b)(1 - n_{\rm coll}(b))^{AB-n} \,. \tag{3.49}$$

We make the substitution $m = n - 1$ so that

$$\begin{aligned} \langle n \rangle &= \frac{ABn_{\rm coll}(b)}{P_{AB}(b)} \sum_{m=1}^{AB} \left(\begin{array}{c} AB \\ m \end{array} \right) n_{\rm coll}^m(b)(1-n_{\rm coll}(b))^{AB-m} \\ &= ABn_{\rm coll}(b) = \sigma_{\rm inel} T_{AB}(b) \ . \end{aligned} \tag{3.50}$$

Doing the same manipulations as above twice gives $\langle n^2 \rangle$. We first have

$$\begin{aligned} \langle n^2 \rangle &= \frac{1}{P_{AB}(b)} \sum_{n=1}^{AB} n^2 P(n,\vec{b}) \\ &= \frac{1}{P_{AB}(b)} \sum_{n=1}^{AB} n^2 \left(\begin{array}{c} AB \\ n \end{array} \right) n_{\rm coll}^n(b)(1-n_{\rm coll}(b))^{AB-n} \end{aligned} \tag{3.51}$$

After making the changes in Eq. (3.48) for one power of n and substituting $m = n - 1$ as in Eq. (3.50), we obtain

$$\begin{aligned} \langle n^2 \rangle &= \frac{ABn_{\rm coll}(b)}{P_{AB}(b)} \\ & \times \sum_{m=1}^{AB} (1+m) \left(\begin{array}{c} AB \\ m \end{array} \right) n_{\rm coll}^m(b)(1-n_{\rm coll}(b))^{AB-m} \end{aligned} \tag{3.52}$$

The first term in $1 + m$ is equivalent to $\langle n \rangle$, calculated in Eq. (3.50). The second term can be obtained by replacing m by $n - 1$ so that

$$\begin{aligned} \sum_{m=1}^{AB} m \frac{(AB)!}{m!(AB-m)!} &= \sum_{n=1}^{AB} (n-1) \frac{(AB-1)!}{(n-1)!(AB-n)!} \\ &= (AB-1) \sum_{n=1}^{AB} \frac{(AB-2)!}{(n-2)!(AB-n)!} \ . \end{aligned} \tag{3.53}$$

Then

$$\begin{aligned} \langle n^2 \rangle &= ABn_{\rm coll}(b) \left[1 + \frac{(AB-1)n_{\rm coll}(b)}{P_{AB}(b)} \right. \\ & \left. \times \sum_{n=1}^{AB} \left(\begin{array}{c} AB-2 \\ n-2 \end{array} \right) n_{\rm coll}^{n-2}(b)(1-n_{\rm coll}(b))^{AB-n} \right] \ . \end{aligned}$$

Shifting the index in the second term to $p = n - 2$, then we finally find

$$\langle n^2 \rangle = ABn_{\rm coll}(b)[1 + (AB-1)n_{\rm coll}(b)] \ . \tag{3.54}$$

The standard deviation is

$$\langle n^2 \rangle - \langle n \rangle^2 = ABn_{\rm coll}(b)(1 - n_{\rm coll}(b)) \ . \tag{3.55}$$

The number of participants in an AB collision can be obtained starting from the probability for a nucleon-nucleon collision in the interaction of a hadron with a nucleus. The expressions are somewhat simpler for a hadron-nucleus collision since $B = 1$ and $T_{AB}(b) \to T_A(b)$. Thus the probability becomes

$$P_{hA}(n, \vec{b}) = \begin{pmatrix} A \\ n \end{pmatrix} \left[\sigma_{\rm inel} \frac{T_A(b)}{A}\right]^n \left[1 - \sigma_{\rm inel} \frac{T_A(b)}{A}\right]^{A-n} \ . \tag{3.56}$$

The sum over probabilities is then

$$\sum_{n=1}^{A} P_{hA}(n, \vec{b}) = 1 - \left[1 - \sigma_{\rm inel} \frac{T_A(b)}{A}\right]^A \ . \tag{3.57}$$

If $\sigma_{\rm inel} T_A(b)/A \ll 1$, then the second term in Eq. (3.57) can be approximated as an exponential so that

$$\sum_{n=1}^{A} P_{hA}(n, \vec{b}) = 1 - \exp(-\sigma_{\rm inel} T_A(b)) \tag{3.58}$$

since the A's cancel in the exponent.

We can then use Eq. (3.58) to find the number of participant nucleons in a nucleus-nucleus collision. The number of participants in nucleus A is proportional to the nuclear profile function at transverse position $\vec{s}$, $T_A(s)$, weighted by the sum over the probability for a nucleon-nucleus collision at transverse position $|\vec{b} - \vec{s}|$ in nucleus B from Eq. (3.58). The number of participants in nucleus B is obtained the same way. Thus at a given b, the number of participants is

$$\begin{aligned} N_{\rm part}(b) \ &= \ \int d^2s \, \Big[T_A(s)(1 - \exp[-\sigma_{\rm inel} T_B(|\vec{b} - \vec{s}|)]) \\ &\quad + T_B(|\vec{b} - \vec{s}|)(1 - \exp[-\sigma_{\rm inel} T_A(s)])\Big] \ . \end{aligned} \tag{3.59}$$

If $\sigma_{\rm inel} T_A(s)$ and $\sigma_{\rm inel} T_B(|\vec{b} - \vec{s}|)$ are large, as for a central collision, the exponentials can be neglected and we can approximate the number of participants as

$$N_{\rm part}(b) = \int d^2s \left[T_A(s) + T_B(|\vec{b} - \vec{s}|) \right] \; . \tag{3.60}$$

In Fig. 3.7, we show the number of participant nucleons as a function of impact parameter for the same systems shown in Fig. 3.6. Typically

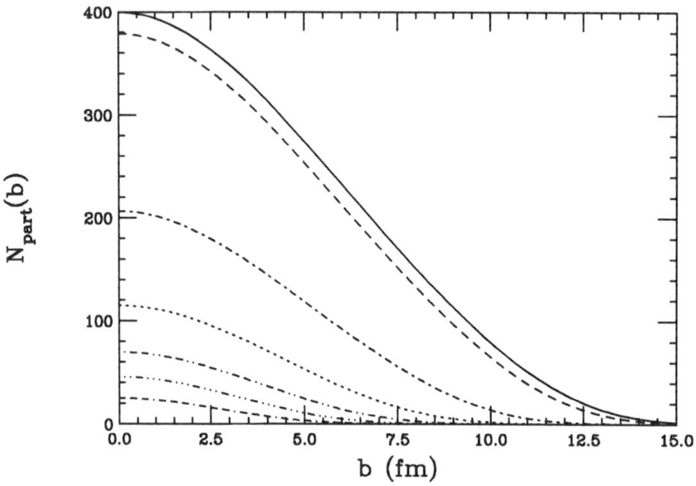

Figure 3.7: The number of nucleon participants as a function of impact parameter b. for O+O (dash-dash-dash-dotted), Si+Si (dot-dot-dot-dashed), Ca+Ca (dot-dot-dash-dashed), Cu+Cu (dotted), I+I (dot-dashed), Au+Au (dashed) and Pb+Pb (solid) collisions.

$N_{\rm part}(b)$ is larger than $T_{AB}(b)$ for the same values of impact parameter. The Woods-Saxon parameterization of the nuclear density gives a somewhat smaller number of participants at $b = 0$ than does the sharp surface density. There is some variation in the number of participants as a function of energy, albeit less than in the number of collisions, since $\sigma_{\rm inel}$ only appears in the exponential factors in $N_{\rm part}(b)$ while $N_{\rm coll}(b)$ is directly proportional to $\sigma_{\rm inel}$.

As instructive examples, we can calculate the maximum number of participants for the sharp surface parameterization in two simple cases.

The first, symmetric collisions where $A = B$, is relevant for heavy-ion colliders. The result is rather trivial since, at $b = 0$,

$$N_{\rm part}(0) = 2 \int d^2s T_A(s) = 2A \ . \tag{3.61}$$

In the second case, we consider asymmetric collisions with $B \gg A$. These types of collisions were done in the initial BNL and CERN fixed-target programs. The BNL AGS collided silicon beams on gold targets while the CERN SPS collided oxygen and sulfur beams on heavy targets. At $b = 0$, nucleus A drills a cylindrically-shaped hole through the center of B. Here all the nucleons in A are participants while the number of participants in B are those in the volume of the cylinder swept out by nucleus A so that

$$N_{\rm part}(0) = A + B \left(\frac{V_{A/B}}{V_B} \right)$$

where $V_{A/B}$ is the volume of B removed by A and $V_{A/B} = 2\pi R_A^2 R_B$. Then

$$\begin{aligned} N_{\rm part}(0) &= A + B \left(\frac{2\pi R_A^2 R_B}{4\pi R_B^3/3} \right) \\ &= A + \frac{3}{2} B \cdot \frac{A^{2/3} B^{1/3}}{B} \\ &= A + \frac{3}{2} A^{2/3} B^{1/3} \ . \end{aligned} \tag{3.62}$$

We can also estimate the nuclear overlap function at $b = 0$, $T_{AB}(0)$, with the sharp surface approximation for the same two cases. When $A = B$,

$$\begin{aligned} T_{AB}(0) &= 4\rho_0^2 \int d^2s (R_A^2 - s^2) \\ &= 8\pi\rho_0^2 (\frac{1}{2} s^2 R_A^2 - \frac{1}{4} s^4) \Big|_0^{R_A} = 2\pi \rho_0^2 R_A^4 \\ &= \frac{9}{8\pi r_0^2} A^{4/3} \ . \end{aligned} \tag{3.63}$$

In the last step we have used the sharp surface values of the central density, ρ_0, and nuclear radius, R_A, $\rho_0 = 3/4\pi r_0^3$ and $R_A = r_0 A^{1/3}$. In the sharp surface approximation when $A \neq B$,

$$T_{AB}(0) = \frac{9}{32\pi r_0^2}\left\{2(AB^{1/3} + BA^{1/3}) - (B^{2/3} - A^{2/3})^2 \ln\left(\frac{B^{1/3} + A^{1/3}}{B^{1/3} - A^{1/3}}\right)\right\} . \quad (3.64)$$

EXAMPLE: Show this.
We start with the integral

$$T_{AB}(0) = 4\rho_0^2 \int d^2s \sqrt{R_A^2 - s^2}\sqrt{R_B^2 - s^2} = 4\pi\rho_0^2 \int_0^{R_A^2} ds^2 (R_A^2 R_B^2 - s^2(R_A^2 + R_B^2) + s^4)^{1/2} .$$

With the substitutions $x = s^2$, $c = 1$, $a = R_A^2 R_B^2$, $b = -(R_A^2 + R_B^2)$ in $R = a + bx + cx^2$, we have an integral of $\sqrt{R}$ over x. This is a standard form where

$$\int_{x_{\rm min}}^{x_{\rm max}} dx\sqrt{R} = \left[\frac{2cx + b}{4c}\sqrt{R} + \frac{4ac - b^2}{8c^{3/2}} \ln(2\sqrt{cR} + 2cx + b)\right]_{x_{\rm min}}^{x_{\rm max}} .$$

Substituting the values of R, a, b and c as well as inserting the upper and lower limits of integration, we obtain Eq. (3.64).

3.4 Probes of centrality

There are several different measurements that can be performed to determine the centrality of a collision. These include nuclear 'stopping' or rapidity loss, the comparison of 'transverse' energy production to energy going forward, and particle production or 'multiplicity'. Since transverse energy and multiplicity are rather closely connected, we discuss only stopping and transverse energy in the remainder of this section.

Nuclear 'stopping power' refers to the amount of rapidity lost per nucleon-nucleon collision and was first inferred by Busza and Goldhaber

[45] in hadron-nucleus interactions. If the path length through matter, proportional to R_A, is long enough and the energy is not too high, the initial nucleons may be slowed enough to remain at midrapidity ($y \sim 0$). Measurements at the CERN SPS indicated a rather high degree of stopping.

In a hadron-hadron collision, some of the forward hadron momentum is converted to particle production. These produced particles are typically less energetic than the initial hadrons. The produced particles also generally have some energy directed transverse to the colliding hadrons. This energy is called 'transverse energy', E_T. In nucleus-nucleus collisions, each individual nucleon-nucleon collision contributes to the total transverse energy production. A comparison of the collision E_T to the energy of the 'spectator nucleons', those nucleons in the initial nuclei that did not take part in the collision but went down the beam pipe, is a good measure of centrality. The spectator energy is measured by 'zero degree calorimeters', (ZDCs) so named because they sit at zero degrees relative to the transverse direction. Thus E_T and $E_{\rm ZDC}$ are highly correlated.

3.4.1 Stopping

The Glauber model used to calculate the number of participant nucleons and the number of nucleon-nucleon collisions cannot be fully correct because it assumes the same cross section for each intermediate collision. However, after each collision, the nucleon that interacts loses some of its rapidity so that each subsequent collision it suffers is at lower effective $\sqrt{S}$. While the pp total and inelastic cross sections do not change much over a broad range of energies, at lower energies, especially for fixed-target interactions, there may be an effect. This loss of rapidity is referred to as stopping [45].

The amount of stopping in a collision can be related to the light-cone momentum fraction x. This is not the same fraction as that entering the parton distribution functions. Here it is the fraction of hadron momentum lost by the entire hadron, not the fraction of the hadron momentum carried by a single parton in the hadron. The initial fraction is unity, $x_0 = 1$. The momentum lost in the first collision reduces the fraction to x_1. The probability of finding a nucleon with fraction x_1 after one collision is [46, 47]

$$\omega(x_0, x_1) = \alpha x_1^{\alpha-1} \tag{3.65}$$

where α is a parameter related to the rapidity lost by the nucleons in the collision, as we will see. Equation (3.65) can be generalized to n collisions by

$$\omega(x_{n-1}, x_n) = N\alpha x_n^{\alpha-1} \tag{3.66}$$

where $x_n \leq x_{n-1}$, normalized so that

$$\int_0^{x_{n-1}} dx_n \omega(x_{n-1}, x_n) = 1 \ . \tag{3.67}$$

EXAMPLE: Calculate N in Eq. (3.66).
From Eq. (3.67), we have

$$\begin{aligned} 1 &= N\alpha \int_0^{x_{n-1}} dx_n x_n^{\alpha-1} \\ &= N x_n^{\alpha} \Big|_0^{x_{n-1}} = N x_{n-1}^{\alpha} \end{aligned}$$

so that

$$N = x_{n-1}^{-\alpha} \tag{3.68}$$

and

$$\omega(x_{n-1}, x_n) = \alpha \left(\frac{x_n}{x_{n-1}} \right)^{\alpha-1} \frac{1}{x_{n-1}} \ . \tag{3.69}$$

The stopping will redistribute the nucleon momentum distribution since a reduction in the rapidity is also a reduction in the longitudinal momentum. If the nucleon momentum distribution is $D^{(n-1)}(x_{n-1})$ after $n-1$ collisions, after n collisions the momentum distribution is further reduced, becoming

$$D^{(n)}(x_n) = \int_{x_n}^1 dx_{n-1} D^{(n-1)}(x_{n-1}) \omega(x_{n-1}, x_n) \ , \tag{3.70}$$

normalized so that

$$\int dx_n D^{(n)}(x_n) = 1 \ . \tag{3.71}$$

The distribution with no momentum loss, $n = 0$, is a delta function defining $x_0 = 1$

$$D^{(0)}(x_0) = \delta(x_0 - 1) \ . \tag{3.72}$$

After n subsequent collisions,

$$D^{(n)}(x_n) = \alpha x_n^{\alpha-1} \frac{(-\alpha \ln x_n)^{n-1}}{(n-1)!} \ . \tag{3.73}$$

EXAMPLE: Calculate $D^{(n)}(x_n)$ for $n = 1$, 2 and 3.
When $n = 1$,

$$\begin{aligned} D^{(1)}(x_1) &= \int_{x_1}^{1} dx_0 D^{(0)}(x_0)\omega(x_0, x_1) \\ &= \int_{x_1}^{1} dx_0 \delta(x_0 - 1)\alpha x_1^{\alpha-1} \\ &= \alpha x_1^{\alpha-1} \ . \end{aligned} \tag{3.74}$$

For $n = 2$,

$$\begin{aligned} D^{(2)}(x_2) &= \int_{x_2}^{1} dx_1 D^{(1)}(x_1)\omega(x_1, x_2) \\ &= \int_{x_2}^{1} dx_1 \, \alpha x_1^{\alpha-1} \, \alpha \frac{x_2^{\alpha-1}}{x_1^{\alpha}} = \alpha^2 x_2^{\alpha-1} \int_{x_2}^{1} \frac{dx_1}{x_1} \\ &= \alpha^2 x_2^{\alpha-1} \ln x_1 \Big|_{x_2}^{1} \\ &= \alpha x_2^{\alpha-1}(-\alpha \ln x_2) \ . \end{aligned} \tag{3.75}$$

Finally, when $n = 3$,

$$\begin{aligned}
D^{(3)}(x_3) &= \int_{x_3}^{1} dx_2 D^{(2)}(x_2)\omega(x_2, x_3) \\
&= \int_{x_3}^{1} dx_2\, \alpha x_2^{\alpha-1}(-\alpha \ln x_2)\, \alpha \frac{x_3^{\alpha-1}}{x_2^{\alpha}} \\
&= -\alpha^3 x_3^{\alpha-1} \int_{x_3}^{1} \frac{dx_2}{x_2} \ln x_2 = -\frac{\alpha^3}{2} x_3^{\alpha-1} \frac{1}{2} \ln^2 x_2 \Big|_{x_3}^{1} \\
&= \frac{\alpha}{2} x_3^{\alpha-1}(-\alpha \ln x_3)^2 \, . \qquad (3.76)
\end{aligned}$$

The average value of the nucleon momentum fraction after n collisions, $\langle x_n \rangle$, is

$$\begin{aligned}
\langle x_n \rangle &= \frac{\int_0^1 dx_n\, x_n\, D^{(n)}(x_n)}{\int_0^1 dx_n\, D^{(n)}(x_n)} \\
&= \int_0^1 dx_n\, x_n \int_{x_n}^1 dx_{n-1}\, \alpha x_{n-1}^{\alpha-1} \frac{(-\alpha \ln x_{n-1})^{n-2}}{(n-2)!} \alpha \frac{x_n^{\alpha-1}}{x_{n-1}^{\alpha}} \qquad (3.77)
\end{aligned}$$

due to the normalization condition in Eq. (3.71). The integral is then

$$\begin{aligned}
\langle x_n \rangle &= (-1)^{n-2}\alpha^n \int_0^1 dx_n\, x_n^{\alpha} \int_{x_n}^1 \frac{dx_{n-1}}{x_{n-1}} \frac{\ln^{n-2} x_{n-1}}{(n-2)!} \\
&= (-1)^{n-2}\alpha^n \int_0^1 dx_n\, x_n^{\alpha} \frac{\ln^{n-1} x_{n-1}}{(n-1)!} \Big|_{x_n}^{1} \\
&= \alpha^n \int_0^1 dx_n\, x_n^{\alpha} \frac{\ln^{n-1} x_n}{(n-1)!} \\
&= \left(\frac{\alpha}{\alpha+1} \right)^n \, . \qquad (3.78)
\end{aligned}$$

The integral over x_n in Eq. (3.78) can be shown to be $(\alpha + 1)^n$ by iteration. From Eq. (3.78), it is clear that

$$\langle x_n \rangle = \frac{\alpha}{\alpha+1} \langle x_{n-1} \rangle \, . \qquad (3.79)$$

Thus, after each collision, the nucleon momentum fraction decreases by $\alpha/(\alpha+1)$. We can relate the momentum fraction to the rapidity by making the change of variable,

$$x_n = \frac{m_T}{m}\exp(y_n - y_N) \tag{3.80}$$

where y_N is beam rapidity. The normalization of $D^{(n)}$ changes so that

$$\begin{aligned} \int_0^1 dx_n\, D^{(n)}(x_n) &= \int_{y_{n,\,\mathrm{min}}}^{y_{n,\,\mathrm{max}}} dy_n x_n D^{(n)}(x_n) \\ &= \int_{y_{n,\,\mathrm{min}}}^{y_{n,\,\mathrm{max}}} dy_n D^{(n)}(y_n) = 1 \end{aligned} \tag{3.81}$$

where $dx_n = x_n dy_n$. The limits of the integration over y are $y_{n,\,\mathrm{min}} = -\infty$ and $y_{n,\,\mathrm{max}} = y_N - \ln(m_T/m)$. We then have

$$\begin{aligned} D^{(n)}(y_n) &= x_n D^{(n)}(x_n) \\ &= \alpha x_n^\alpha \frac{(-\alpha \ln x_n)^{n-1}}{(n-1)!} \\ &= \alpha \left(\frac{m_T}{m}\right)^\alpha \exp[\alpha(y_n - y_N)] \\ &\quad \times \frac{(-\alpha[\ln(m_T/m) + y_n - y_N])^{n-1}}{(n-1)!}\,. \end{aligned} \tag{3.82}$$

To calculate the average rapidity loss, we write x_n as $\exp[\alpha(\ln(m_T/m) + y_n - y_N)] = \exp(\alpha z_n)$ where $z_n = \ln(m_T/m) + y_n - y_N$. Thus

$$D^{(n)}(y_n) = D^{(n)}(z_n) = \alpha \exp(\alpha z_n)\frac{(-\alpha z_n)^{n-1}}{(n-1)!} \tag{3.83}$$

The average y_n is then

$$\begin{aligned} \langle y_n \rangle &= \frac{\int_{z_{n,\,\mathrm{min}}}^{z_{n,\,\mathrm{max}}} dz_n\, y_n\, D^{(n)}(z_n)}{\int_{z_{n,\,\mathrm{min}}}^{z_{n,\,\mathrm{max}}} dz_n\, D^{(n)}(z_n)} \\ &= \int_{z_{n,\,\mathrm{min}}}^{z_{n,\,\mathrm{max}}} dz_n\,(z + y_N - \ln(m_T/m))D^{(n)}(z_n) \\ &= y_N - \ln(m_T/m) + \int_{z_{n,\,\mathrm{min}}}^{z_{n,\,\mathrm{max}}} dz_n\, z_n D^{(n)}(z_n) \end{aligned} \tag{3.84}$$

since $\int dz_n D^{(n)}(z_n) = 1$ by definition. The limits of the z_n integration are $z_{n,\,\rm min} = -\infty$ and $z_{n,\,\rm max} = 0$. After iteration,

$$\begin{aligned} \langle y_n \rangle - y_N + \ln(m_T/m) &= \alpha^n (-1)^{n-1} \int_{-\infty}^{0} dz_n \exp(\alpha z_n) \frac{z_n^{n-1}}{(n-1)!} \\ &= -\frac{\alpha^n}{\alpha^{n-1}} \frac{n!}{(n-1)!} \\ &= -\frac{n}{\alpha} \end{aligned} \tag{3.85}$$

If we calculate the rapidity shift for the $n-1$ collision, we similarly find

$$\langle y_{n-1} \rangle - y_N + \ln(m_T/m) = -\frac{n-1}{\alpha} \; . \tag{3.86}$$

Thus the rapidity loss between collision $n-1$ and the $n^{\rm th}$ collision is

$$\langle y_{n-1} \rangle - \langle y_n \rangle = \frac{1}{\alpha} \; . \tag{3.87}$$

Busza and Goldhaber [45] inferred a loss of ~ 0.5 units of rapidity per collision or $\alpha \sim 2$. A proton passing through a lead nucleus can collide an average of 6 times, a loss of about 3 units of rapidity. How completely the nuclei can be stopped depends on the initial rapidity and the nuclear size. Lower energy (lower rapidity) collisions can be more completely stopped than high energy collisions where the rapidity is large. In addition, the larger the nuclear radius, the more times an incoming nucleon can collide during the passage through the nucleus. More rapidity is lost with larger targets. The NA49 collaboration at the CERN SPS observed more stopping in Pb+Pb collisions at $p_{\rm lab} = 158$ GeV/nucleon than in S+S collisions at $p_{\rm lab} = 200$ GeV/nucleon at the CERN SPS [48].

3.4.2 Energy production

Another important measure of centrality is the transverse energy produced in the collision. When used together with the energy of the nucleons that do not interact, the collision centrality can be pinpointed rather accurately.

The noninteracting nucleons are called "spectators". Their energy can be measured either in a beam dump (large absorber) far away from the target in fixed-target interactions and in beam-beam counters placed far away from the rest of the detector in a collider which do not stop the beam. At RHIC, each detector was given a set of beam-beam counters to assure a common measure of centrality even when the detectors cover different regions of phase space. Since the spectators are not deflected away from the beam, they are traveling collinear to the beam with transverse angle $\theta = 0$. Thus the spectators are at zero degrees in the transverse direction. The detectors that measure the spectators are usually calorimeters which record energy deposition. Thus they are often called zero-degree calorimeters or ZDCs. If the zero degree energy, $E_{\rm ZDC}$, is large, the nucleus-nucleus collision was rather peripheral and a large portion of the beam energy continued without deflection. (Note that $E_{\rm ZDC}^{\rm max} = AE_{\rm lab}/{\rm nucleon}$.) Thus large $E_{\rm ZDC}$ corresponds to a peripheral collision. On the other hand, if $E_{\rm ZDC} \sim 0$, then a large portion of the beam energy was converted to particle production and generation of transverse energy. These collisions are central, nearly head on.

The zero degree energy for one beam in an AB collision depends on the number of participants in the interacting nucleus in that beam. The dominant contribution to the zero degree energy is the spectator, noninteracting nucleons with a smaller contribution from very forward participants and produced particles [49] so that

$$\langle E_{\rm ZDC}(b)\rangle = E_{\rm ZDC}^{\rm spect}(b) + E_{\rm ZDC}^{\rm part}(b) \ . \tag{3.88}$$

The first term is proportional to the number of spectators in *e.g.* nucleus A, $A - N_{A\,{\rm part}}$, while the contribution from the participants in the second term is proportional to the total $N_{\rm part}$. Thus the zero degree energy for nucleus A is

$$\langle E_{\rm ZDC}(b)\rangle = E_A(A - N_{A\,{\rm part}}(b)) + aN_{\rm part}(b) \ . \tag{3.89}$$

In an AA collision, $N_{A\,{\rm part}} = N_{\rm part}/2$. In collider mode, $E_A = \sqrt{S_{NN}}/2$ and the zero degree energy of both beams would be considered. If $A \neq B$ in the collider kinematics, as was the case for d+Au collisions at RHIC, the spectator energy of B would be $\sqrt{S_{NN}}/2(B - N_{B\,\rm part})$. In fixed-target mode, $E_A = E_{\rm lab}/$nucleon.

On the other hand, transverse energy, E_T, is a measure of particle production. The E_T of a given collision is found by summing the energy transverse to the beam direction for all particles,

$$E_T = \sum_i E_i \sin\theta_i \tag{3.90}$$

where i is a cell in a detector.

The CERN SPS experiment NA50 took data on J/ψ and Drell-Yan production in fixed-target Pb+Pb collisions at $E_{\rm lab} = 158$ GeV/nucleon. They followed the NA38 experiment which took fixed-target data with 200 GeV/nucleon oxygen and sulfur beams on uranium targets. While NA38 studied J/ψ production with transverse energy measurements alone, NA50 also had a zero degree calorimeter to better fix the centrality by correlating E_T with $E_{\rm ZDC}$. The collaboration measured E_T with an electromagnetic calorimeter, collecting only neutral particle decays and some of the charged. Thus their measured E_T values were not as large as those of other SPS experiments which measured charged particle energies more effectively. The Pb+Pb correlation is shown in Fig. 3.8.

Of course, only a fraction of the energy lost by the spectators to particle production will be observed as transverse energy. There are several reasons for this. First, the detector acceptance is finite, usually some rapidity slice on either side of $y = 0$, midrapidity. (Collider detectors are usually symmetric around midrapidity.) Particles that are rather collinear to the beam will typically not be detected. Next, these detectors often have some magnetic field associated with them. Very low momentum particles are more affected by these fields and may curl up without being detected at all. There may also not be enough types of detectors to record all produced particles. For example, not all detectors measure neutral particles. Specialized detectors are needed to measure muons. Neutrinos, which interact only extremely rarely, are typically not observed in colliders but carry away a significant amount of energy. In high-energy hadron-hadron collisions, the decay $W^- \rightarrow e^-\overline{\nu}_e$ is characterized by a high energy electron with no other observed

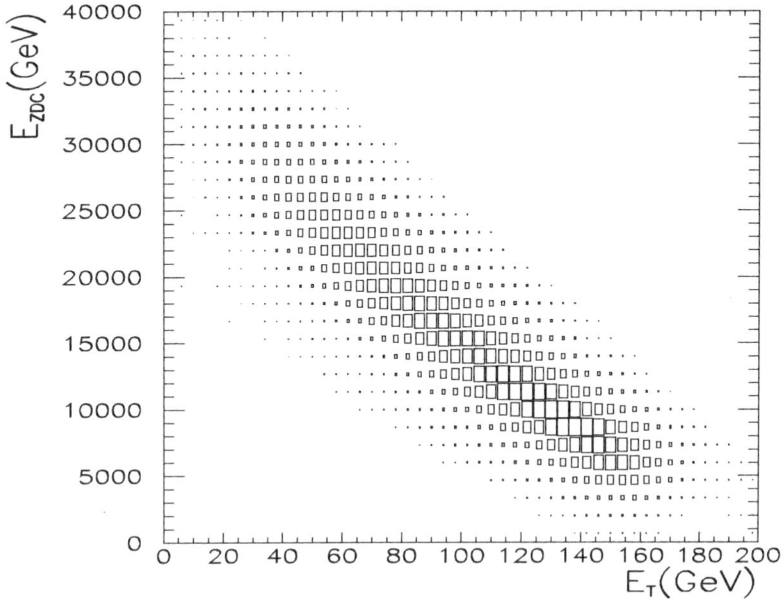

Figure 3.8: The correlation between neutral transverse energy (x-axis) and the forward energy in the zero degree calorimeter (y-axis). Note that the total $E_{\rm ZDC}$ is given, not the value per nucleon. The data were taken by the NA50 collaboration [49]. Reprinted from Ref. [49] with permission from Elsevier.

particles. Thus, the neutrino carries 'missing' energy since its energy is undetected. In heavy-ion collisions, on the other hand, the final state cannot be quantified precisely enough to define missing energy.

The produced E_T distribution can be calculated using models of the probability to produce some transverse energy at a given impact parameter or for some number of nucleon participants (also sometimes called the number of wounded nucleons). Each nucleon-nucleon interaction results in some final state particle production with accompanying E_T. The total E_T is obtained by summing the E_T from all nucleon-nucleon collisions. The maximum possible E_T at impact parameter b is the difference between the nucleus-nucleus center-of-mass energy and the mass of the system,

$$E_T^{\max} = \sqrt{S_{AA}} - (A_{\rm part} + B_{\rm part})m_N \tag{3.91}$$

$$= \left([A^2_{\rm part} + B^2_{\rm part} + 2(A_{\rm part}E_{\rm lab})(B_{\rm part}/m_N)]^{1/2} - (A_{\rm part} + B_{\rm part})\right)m_N \ . \tag{3.92}$$

We discuss two models of the E_T distribution here.

In the first, we assume that $p_w(E_T^{(i)})$ is the probability of producing total $E_T^{(i)}$ with the $i^{\rm th}$ participant. Then the probability distribution for producing total E_T from $N_{\rm part}$ at impact parameter b is the convolution of $p_w(E_T^{(i)})$ over $1 < i < N_{\rm part}$ [50, 51],

$$P(E_T; N_{\rm part}) = \int dE_T^{(1)} \ldots dE_T^{(N_{\rm part})} p_w(E_T^{(1)}) \ldots p_w(E_T^{(N_{\rm part})}) \times \delta\left(E_T - E_T^{(1)} - \ldots - E_T^{(N_{\rm part})}\right) \ . \tag{3.93}$$

The final E_T distribution sums over the number of participants [50],

$$P_{AB}(E_T) = \sum_{N_{\rm part}=1}^{N_{\rm part}^{\max}} P_{AB}(E_T; N_{\rm part}) P_{AB}(N_{\rm part}) \ , \tag{3.94}$$

where $P_{AB}(N_{\rm part})$ is the integral of Eq. (3.56) over d^2b and $N_{\rm part}(b)$ is calculated using Eq. (3.59). If

$$p_w\left(E_T^{(i)}\right) = \frac{\xi^d}{\Gamma(d)} \left(E_T^{(i)}\right)^{d-1} \exp(-\xi E_T^{(i)}) \tag{3.95}$$

where $\Gamma(d) = (d-1)!$ and ξ and d are parameters to be fit to data, then

$$P_{AB}(E_T; N_{\rm part}) = \frac{(\xi E_T)^{N_{\rm part} d}}{\Gamma(N_{\rm part} d)} \frac{\exp(-\xi E_T)}{E_T} \ . \tag{3.96}$$

EXAMPLE: Prove Eq. (3.96) by iteration for i =1, 2 and 3.

If $N_{\text{part}} = 1$,

$$\begin{aligned} P_{AB}(E_T;1) &= \frac{\xi^d}{\Gamma(d)} \int dE_T^{(1)} \left(E_T^{(1)}\right)^{d-1} \exp(-\xi E_T^{(1)})\delta\left(E_T - E_T^{(1)}\right) \\ &= \frac{\xi^d}{\Gamma(d)} E_T^{d-1} \exp(-\xi E_T) \;. \end{aligned} \tag{3.97}$$

For larger values of N_{part}, we take advantage of the definition

$$\int dy\, y^{a-1}(1-y)^{b-1} = \frac{\Gamma(a)\Gamma(b)}{\Gamma(a+b)} \;. \tag{3.98}$$

Thus for $N_{\text{part}} = 2$,

$$\begin{aligned} P_{AB}(E_T;2) &= \left(\frac{\xi^d}{\Gamma(d)}\right)^2 \int dE_T^{(1)}\, dE_T^{(2)} (E_T^{(1)} E_T^{(2)})^{d-1} \\ &\quad \times \exp[-\xi(E_T^{(1)} + E_T^{(2)})]\delta\left(E_T - E_T^{(1)} - E_T^{(2)}\right) \end{aligned}$$

After doing the integral over $E_T^{(2)}$ with the delta function, we let $y = E_T^{(1)}/E_T$ so that

$$\begin{aligned} P_{AB}(E_T;2) &= \left(\frac{\xi^d}{\Gamma(d)}\right)^2 E_T^{2d-1} \exp(-\xi E_T) \int dy y^{d-1}(1-y)^{d-1} \\ &= \frac{(\xi E_T)^{2d}}{\Gamma(2d)} \frac{1}{E_T} \exp(-\xi E_T) \;. \end{aligned} \tag{3.99}$$

The last equality was obtained using Eq. (3.98). Finally, for $N_{\text{part}} = 3$,

$$\begin{aligned} P_{AB}(E_T;3) &= \left(\frac{\xi^d}{\Gamma(d)}\right)^3 \int dE_T^{(1)}\, dE_T^{(2)}\, dE_T^{(3)} \left(E_T^{(1)} E_T^{(2)} E_T^{(3)}\right)^{d-1} \\ &\quad \times \exp[-\xi(E_T^{(1)} + E_T^{(2)} + E_T^{(3)})]\delta\left(E_T - E_T^{(1)} - E_T^{(2)} - E_T^{(3)}\right) \\ &= \left(\frac{\xi^d}{\Gamma(d)}\right)^3 \exp(-\xi E_T) \int dE_T^{(1)}\, dE_T^{(2)} \\ &\quad \times \left(E_T^{(1)}\right)^{d-1} \left(E_T^{(2)}\right)^{d-1} \left(E_T - E_T^{(1)} - E_T^{(2)}\right)^{d-1} \;. \end{aligned}$$

To do the integral over $E_T^{(2)}$, let $y_1 = E_T^{(2)}/(E_T - E_T^{(1)})$ so that, using Eq. (3.98), we have

$$P_{AB}(E_T;3) = \frac{\xi^{3d} E_T^{3d-1}}{\Gamma(d)\Gamma(2d)} \exp(-\xi E_T) \int dy_1 y_1^{d-1}(1-y_1)^{2d-1}$$

Using Eq. (3.98) a second time, we find

$$P_{AB}(E_T;3) = \frac{(\xi E_T)^{3d}}{\Gamma(3d)} \frac{1}{E_T} \exp(-\xi E_T) \,, \tag{3.100}$$

following Eq. (3.96).

This type of transverse energy calculation does not include correlations between the nucleons in the nucleus. The nucleons are, however, highly correlated since they are under the influence of both the Pauli exclusion principle (no two nucleons can be in the same spin and isospin state at the same spatial location) and short range repulsion due to the strong force. Correlations reduce the fluctuations relative to the Poisson statistics used above [52]. In principle, rescattering effects such as energy and momentum degradation and the difference between multiple collisions of bound and free nucleons should also be taken into account. Rescattering of the produced particles should only be possible after the secondary hadrons have a chance to form, called the formation time, about 1 fm in the rest frame of the hadron.

A Gaussian distribution better incorporates the correlations. Baym *et al.* [53] used such a distribution to fit the CERN SPS fixed-target data, asserting that

$$P_{AB}(E_T;N_{\rm coll}) \approx \exp\left[-\frac{(E_T - \overline{E}_T(b))^2}{(2\overline{E}_T^2(b)\Omega(b))}\right] \,. \tag{3.101}$$

For central collisions, $\overline{E}_T(0) = \epsilon_0 N_{\rm coll}(0)$ and $\Omega(0) = \omega/N_{\rm coll}(0)$. The parameters ϵ_0 and ω are the mean E_T produced per collision and the variance respectively.

Early experiments at the CERN SPS compared the dependence of E_T and $E_{\rm ZDC}$ to $N_{\rm coll}(b)$ and $N_{\rm part}(b)$ and found that the data were more compatible with E_T and $E_{\rm ZDC}$ being proportional to the number of participants [54]. The Gaussian parameterization of E_T can still be written as in Eq. (3.101) but now with

$$\overline{E}_T(b) = \epsilon_0 N_{\rm part}(b) \tag{3.102}$$

$$\Omega(b) = \omega/N_{\rm part}(b) \tag{3.103}$$

where ϵ_0 is instead the energy per participant. The parameters ϵ_0 and Ω are determined for each experiment. While they should, in principle, be the same for any AB collision system since $N_{\rm part}(b)$ is independent of detector specifics, there may be some variation. Differences can arise due to detector acceptances, inefficiencies, calibrations and other details.

Using this method, the minimum bias cross section as a function of E_T is

$$\begin{aligned} \frac{d\sigma}{dE_T} &= \int d^2b P_{AB}(E_T;b) \\ &= \int \frac{d^2b}{\sqrt{2\pi \overline{E}_T^2(b)\Omega(b)}} \exp\left[-\frac{(E_T-\overline{E}_T(b))^2}{2\overline{E}_T^2(b)\Omega(b)}\right] . \end{aligned} \quad (3.104)$$

The NA50 collaboration specialized in making measurements of lepton pairs, such as those produced in the Drell-Yan process. Drell-Yan production is a 'hard' process since the Drell-Yan cross section can be calculated in perturbative QCD, as discussed in Chapter 2. Production of hard processes is expected to grow as the number of collisions so that the Drell-Yan cross section in a nucleus-nucleus collision as a function of E_T is

$$\frac{d\sigma}{dE_T} = \int d^2b \int d^2s T_A(s) T_B(|\vec{b}-\vec{s}|) P_{AB}(E_T;b) \ . \quad (3.105)$$

The Drell-Yan E_T distribution is shown in Fig. 3.9. Since it is weighted by $T_{AB}(b)$, the distribution turns over at low E_T (large impact parameter). The dashed lines show

$$\frac{d\sigma}{d^2b dE_T} = \int d^2s T_A(s) T_B(|\vec{b}-\vec{s}|) P_{AB}(E_T;b) \quad (3.106)$$

for impact parameter intervals of 1.25 fm. Most of the contribution to the E_T distribution is at low E_T while central collisions are relatively rare. Since the highest E_T events are beyond the maximum $\overline{E}_T$, defined by the beginning of the drop off or the 'tail' of the distribution, these events are due to energy fluctuations. The Gaussians are widest

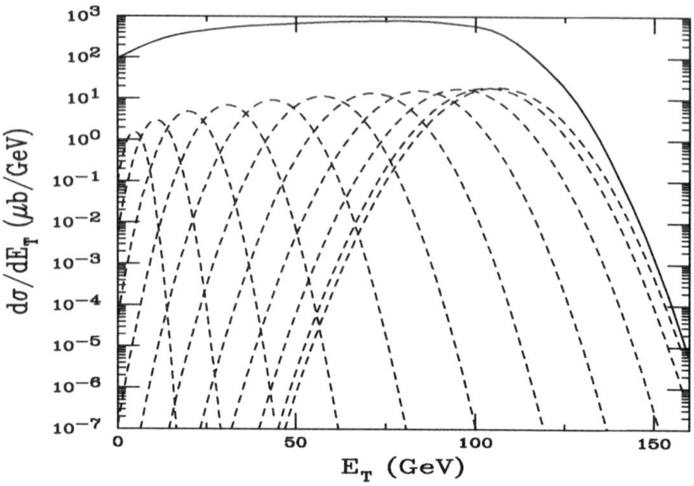

Figure 3.9: The Drell-Yan cross section in the mass range $2.9 < M < 4.5$ GeV as a function of E_T using Eq. (3.105). The parameters are $\epsilon_0 = 0.27$ GeV and $\omega = 3.2$. Some of the individual Gaussians contributing to the total, calculated using $d\sigma/d^2b$, are shown in the dashed curves.

and closest together in the tail of the distribution where fluctuations dominate. They become narrower at low E_T.

At collider energies, where hard processes become more important, transverse energy and multiplicity are no longer expected to be solely proportional to $N_{\rm part}$ but to have some component proportional to $N_{\rm coll}$ as well. The hard contribution to particle production at the CERN SPS is negligible due to the low energy. On the other hand, while the first data from RHIC could be described by including only $N_{\rm part}$, it could equally well be described by up to a 37% contribution to particle production by hard processes, proportional to $N_{\rm coll}$ [55]. At still higher energies, such as at the LHC, particle production should be dominated by the hard component.

3.4.3 Relating centrality to impact parameter

Measurements such as the E_T, $E_{\rm ZDC}$ and multiplicity distributions are often divided into bins of centrality, such as the central 10% of all collisions. It is then necessary to relate the centrality bin to an impact parameter range using a model. Once that is done, for processes which can be calculated using perturbative QCD, the hard probes, the impact parameter can be used to predict the production rate of that probe for a given b. The most typical model used to relate the centrality to impact parameter is the Glauber model we have discussed in this chapter. Indeed, in the recent RHIC experiments, rather than present the centrality dependence of an observable using E_T or multiplicity, the results are often presented directly as a function of $N_{\rm part}$.

Minimum bias (inclusive) data probe the total cross section, including both hard and 'soft' (nonperturbative) contributions. Thus this cross section is related to the geometric cross section,

$$\sigma_{\rm geo}(b_c) = 2\pi \int_0^{b_c} db\, b\, [1 - \exp(-\sigma_{\rm inel} T_{\rm AB}(b))] \tag{3.107}$$

where b_c is a 'central' impact parameter. The inelastic cross section, $\sigma_{\rm inel}$, is ≈ 32 mb at SPS energies and grows with energy, as was shown in Chapter 2. It is expected to be ~ 60 mb at LHC energies. The fraction of the geometric cross section contained in the interval $b < b_c$ is [56]

$$f_{\rm geo} = \frac{\sigma_{\rm geo}(b_c)}{\sigma_{\rm geo}} . \tag{3.108}$$

In central collisions, where $T_{\rm AB}$ is large, the impact parameter dependence is simple, $\sigma_{\rm geo}(b_c) \propto b_c^2$. However, in peripheral collisions where the nuclear overlap becomes small, $\sigma_{\rm geo}(b_c)$ deviates from the trivial b_c^2 scaling. Deviations from this scaling do not occur until $b_c \approx 2R_A$ in symmetric systems. Figure 3.10 shows the numerical result, Eq. (3.107), relative to the total geometric cross section, $b_c \to \infty$, for the symmetric systems O+O, Cu+Cu and Au+Au. The dotted curve shows the result for d+Au interactions as a comparison. The growth of the d+Au geometric cross section with b_c is intermediate to the O+O and Cu+Cu results. While $\sigma_{\rm inel} = 32$ mb is used in Eq. (3.107), using a higher cross section such as 60 mb, appropriate for the LHC, makes only a negligible difference in the most peripheral collisions.

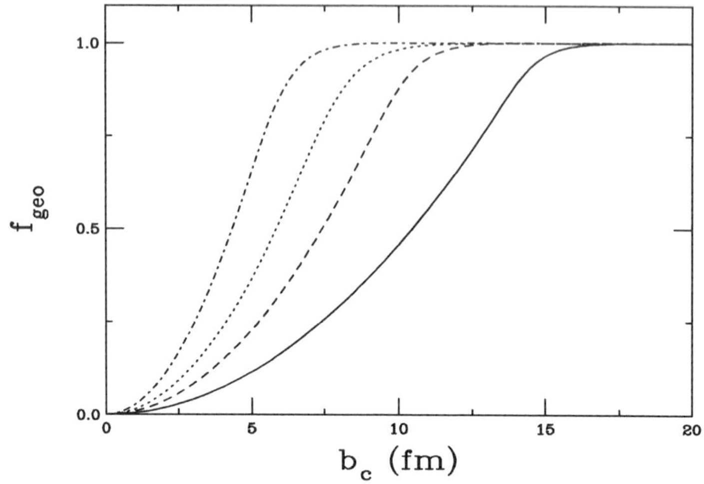

Figure 3.10: The fraction of the total geometrical cross section as a function of impact parameter cut b_c for several AA systems. The results are for d+Au (dotted), O+O (dot-dashed), Cu+Cu (dashed), and Au+Au (solid) interactions.

On the other hand, the hard cross section is expected to grow with the number of collisions so that in the same central interval, $b < b_c$, the fraction of the hard cross section encompassed is

$$f_{AB} = \frac{2\pi}{AB} \int_0^{b_c} db\, b\, T_{AB}(b) \ . \tag{3.109}$$

Figure 3.11 shows the increase of f_{AB} with b_c for the same systems as in Fig. 3.10. Note that $f_{\rm AA} \approx 1$ when $b_c \approx 2R_A$. If the bin is the 10% most central events, $f_{AB} = 0.1$, then $b_c = 2.05$ fm in Au+Au collisions and $b_c = 1.05$ fm in O+O collisions. If, instead, a tighter cut is chosen, *e.g.* $f_{AB} = 0.01$, then $b_c = 0.52$ fm in Au+Au and $b_c = 0.33$ fm in O+O collisions. Interestingly enough, here the d+Au result is very similar to the Cu+Cu result, showing somewhat slower growth of the hard cross sections in d+Au collisions relative to the growth of the geometric cross section. The hard fraction, f_{AB}, grows much faster than the fraction of the geometric cross section, $f_{\rm geo}$. Indeed at $b_c \approx 2R_A$ where $f_{AB} \sim 1$, $f_{\rm geo} \approx 0.75$.

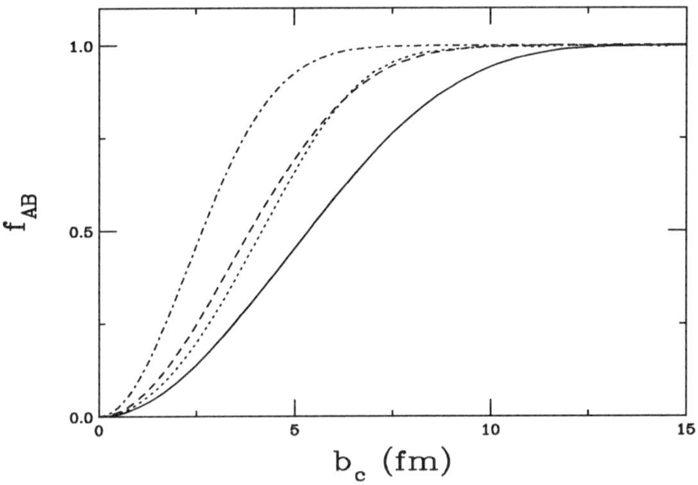

Figure 3.11: The central fraction of the hard cross section as a function of impact parameter cut b_c for several symmetric systems. The results are for d+Au (dotted), O+O (dot-dashed), Cu+Cu (dashed), and Au+Au (solid) interactions.

Figure 3.12 shows f_{AB} relative to $f_{\rm geo}$ for the same systems as in Figs. 3.10 and 3.11. The hard fraction grows more slowly relative to the geometric fraction in smaller systems such as O+O while the growth of the hard fraction is slowest of all in the d+Au system.

Table 3.2 gives the total geometric cross section and the value of $T_{AB}(0)$ for several systems. Note that while the geometric cross section is on the order of barns for all systems, the nuclear overlap function has units of inverse millibarns since 1 fm^2 = 10 mb.

The number of hard probes (such as Drell-Yan dileptons or $c\overline{c}$ pairs) produced in a central collision can then be calculated. First, b_c is obtained from the geometric cross section and then, using this b_c, f_{AB} can be found. Assuming no nuclear effects on the hard cross section, the average rate of hard production at $b = 0$ in Au+Au collisions is

$$\overline{N}_{AB}^{\rm hard}(0) = \sigma_{pp}^{\rm hard} T_{AB}(0) \tag{3.110}$$

where $\sigma_{pp}^{\rm hard}$ is the total production cross section for a particular hard process in pp interactions. The rate in the impact parameter interval

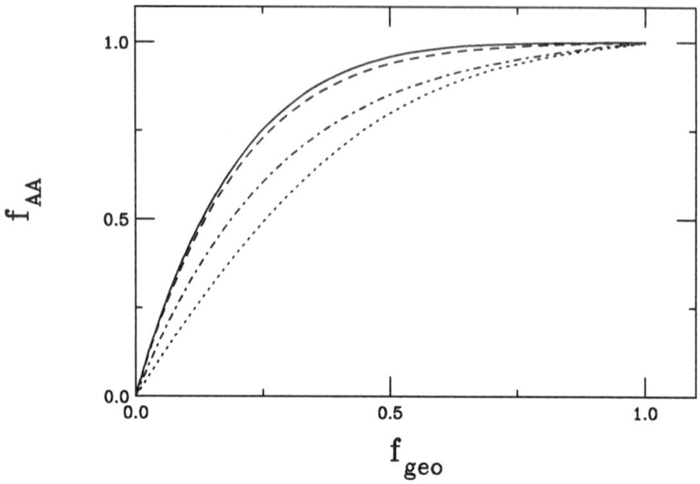

Figure 3.12: The fraction of the hard cross section as a function of the total geometrical cross section. The results are for d+Au (dotted), O+O (dot-dashed), Cu+Cu (dashed), and Au+Au (solid) interactions.

Table 3.2: Values of the geometric cross section and nuclear overlap integral at $b = 0$ for several colliding systems.

$A + B$	$\sigma_{\rm geo}$ (b)	$T_{AB}(0)$ (mb^{-1})
O+O	1.18	0.79
Ca+Ca	2.45	2.92
Cu+Cu	3.46	5.77
Au+Au	6.88	29.32
Pb+Pb	7.39	30.42

$0 < b < b_c$ is the ratio of the hard to geometric cross sections integrated over b,

$$\overline{N}_{AB}^{\rm hard}(b_c) = \frac{\int_0^{b_c} d\sigma_{AB}^{\rm hard}}{\sigma_{\rm geo}(b_c)} = \frac{\sigma_{pp}^{\rm hard}}{\sigma_{\rm geo}} \frac{ABf_{AB}}{f_{\rm geo}} \; . \tag{3.111}$$

To be specific, the average number of $c\overline{c}$ pairs produced at $b = 0$ in an Au+Au collision is

$$\overline{N}^{c\bar{c}}_{AB}(0) = \sigma^{c\bar{c}}_{pp} T_{AB}(0) \ . \tag{3.112}$$

At $\sqrt{S_{NN}} = 200$ GeV, $\sigma^{c\bar{c}}_{pp} = 0.35$ mb [57] and $T_{AB}(0) = 29.3$/mb, resulting in ≈ 10 $c\bar{c}$ pairs per Au+Au collision at $b = 0$. With $f_{\rm geo} = 0.1$, the average number of $c\bar{c}$ pairs produced over $0 < b < b_c$ is

$$\overline{N}^{c\bar{c}}_{AB}(b_c) = \frac{\sigma^{c\bar{c}}_{pp}}{\sigma_{\rm geo}} \frac{ABf_{AB}}{0.1} \ . \tag{3.113}$$

Since the central 10% of the geometric cross section corresponds to 40% of the hard cross section, there are ≈ 8 $c\bar{c}$ pairs in the 10% most central events.

Exercises

1. Estimate the number of participants at $b = 0$ for Si+Au, O+U, S+U and Pb+Pb collisions assuming that the nucleus has a sharp surface. For each case, give the number of participants in A, in B and the sum.

2. Estimate the nuclear overlap function at $b = 0$ for the same systems as above.

3. Using Eq. (3.76), evaluate $D^{(n)}(x_n)$ for $n = 4$ and 5.

4. If $a = 0$ in Eq. (3.89), calculate $\langle E_{\rm ZDC} \rangle$ at $b = 0$ and $b \gg R_A$ for the same systems as above assuming fixed-target kinematics and $p_{\rm lab} = 200$ GeV. (Assume that the first named nucleus is the projectile.)

5. Calculate the maximum possible E_T at $b = 0$ for the same systems as above.

6. Calculate $P_{AB}(E_T; N_{\rm part})$ for $N_{\rm part} = 4$ and 5.

7. If the $b\overline{b}$ production cross section is 0.19 mb in a pp collision at 5.5 TeV, when $f_{\rm geo} = 0.125$, find f_{AB} from Fig. 3.12 for Au+Au collisions to calculate the number of $b\overline{b}$ pairs in a central event.

Chapter 4

Thermodynamics

4.1 Introduction

Based on lattice QCD studies, as will be discussed in Chapter 6, there is a change in the state of matter (a phase transition) from a hadronic system to a nearly free gas of quarks and gluons, the quark-gluon plasma. Thus in this chapter, we consider the properties of an ideal gas of quarks and gluons in thermal equilibrium and the phase transition between the quark-gluon plasma and the hadronic world. This is, of course, an oversimplification, based on one of the properties of quantum chromodynamics, asymptotic freedom, which has implications for how a system of strongly interacting partons would behave at high temperatures and densities. Of course, we do not necessarily expect that an ideal gas is the best way to describe the system since thermodynamics cannot describe the time evolution of the system, but this static example can be very instructive.

The phase transitions in nuclear matter can, for illustration, be compared to those in water. The phase diagram of water is shown in Fig. 4.1. The lines on the diagram are where the phases coexist with each other. In our normal, everyday experience, at one atmosphere of pressure the solid, liquid and gas phases can coexist. Examples of such coexistence in water are ice in a glass of water (solid/liquid coexistence) and steam over a pot of boiling water (liquid/gas coexistence). At the

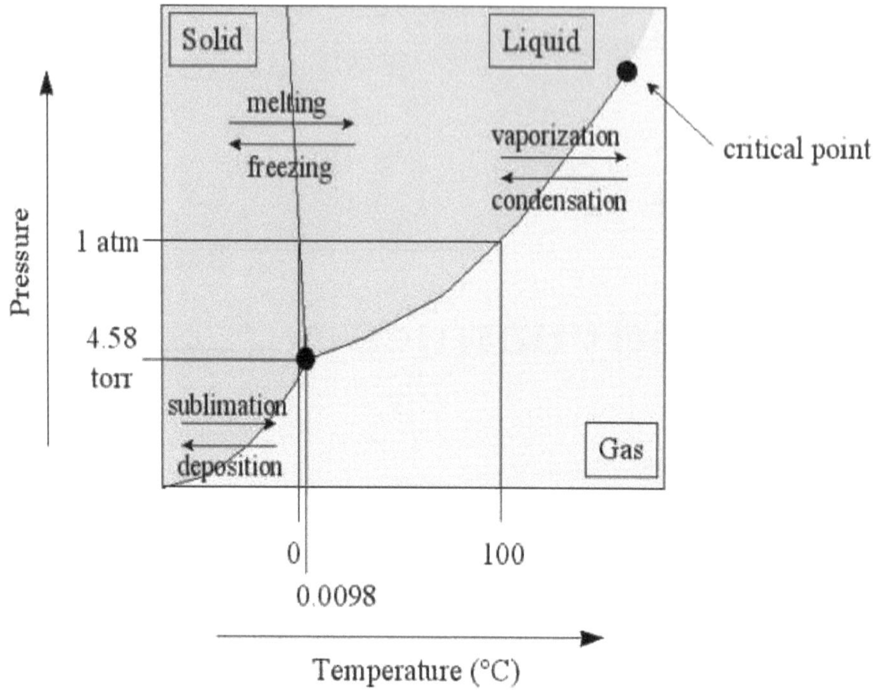

Figure 4.1: A simplified phase diagram of water [58]. Pressure in atmospheres is shown on the y-axis while temperature in Celsius is shown on the x-axis.

coexistence point, the temperature does not change but the volume of the system can increase. For example, the steam rising from boiling water escapes into a much larger volume since the gas phase fills the available space. Water can keep boiling away with steam rising off it until all the water in the pot is gone. Likewise, the temperature of the glass of ice water does not change until the ice is melted, at which point the temperature begins to increase. These changes in water are the manifestations of a first-order phase transition. In an ideal gas, $PV = nRT$ so that the coexistence of two phases at constant temperature also occurs at constant pressure. The volume is, however, expanding, as shown in Fig. 4.2.

In a first-order phase transition, while the two phases coexist, the

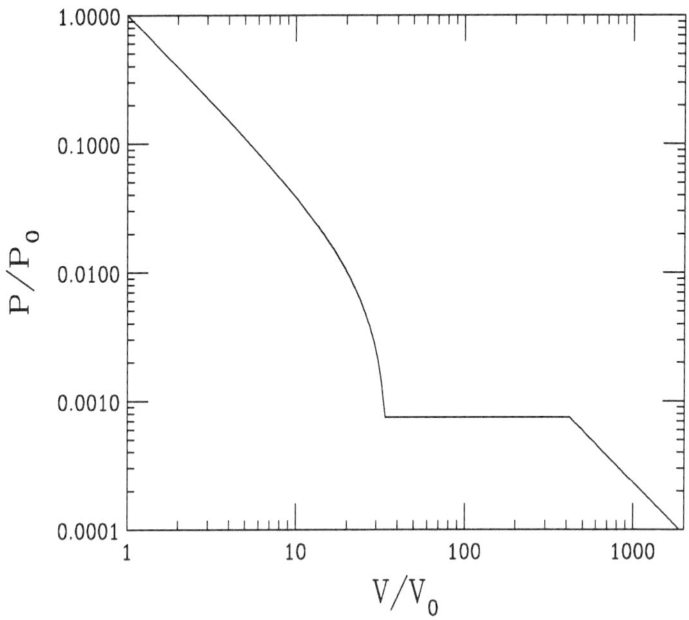

Figure 4.2: Pressure as a function of volume in a first-order phase transition.

pressure, temperature and chemical potential are equal at the phase boundary. There is, however, a discontinuity in the energy density of the two phases before and after the transition. This discontinuity, the latent heat of the transition, is characteristic of first-order phase transitions. Since the energy release due to the latent heat is not instantaneous, the phases can coexist over a fairly long time scale such as that needed to convert an entire pot of water to steam by boiling or for ice cubes to melt in a glass of cold water. These types of transitions are not easily studied because the dynamics are typically rather violent. Not all phase transitions have a discontinuity in the energy density across the transition and are thus not first order. In other types of transitions, some derivative of the free energy is continuous at the phase boundary and all higher derivatives are zero. The order of the lowest zero derivative is the order of the transition. The next higher order transition is second order, *i.e.* the second derivative of the free energy is zero. The usual nomenclature is for all transitions higher than first order to be termed continuous transitions. All transitions higher than first order involve no latent heat. We will discuss phase transitions in more detail later in the chapter.

In nuclear matter, several phases exist as well. At zero temperature we have normal nuclear matter with an energy density of ≈ 0.16 GeV/fm^3 for a nucleon at rest.

EXAMPLE: Show this.

The energy density, ϵ, is the energy per volume. The rest mass of a nucleon is 0.931 GeV, about 1 GeV. The density of a spherical nucleus, is $\rho_0 \approx 0.16$ fm^{-3}, as we also saw in Chapter 3. Thus the energy density in a cold nucleus is $\approx m_N/V$ or about 0.16 GeV/fm^3.

A schematic nuclear matter phase diagram for two massless light quarks and a heavy strange quark is shown in Fig. 4.3. The quark matter phase can be reached either by compressing nuclear matter to high densities while at rather low temperatures (along the baryon density axis) or by heating the vacuum (along the temperature axis). The interesting phases at high baryon density and low temperature were described much later than those at high temperature and low baryon density. See Ref. [59] for a review. The high density phases are expected to play a key role in the interior of neutron stars, including keeping the interior temperature elevated for a longer time than expected [60]. See the book by Weber [61] for an extensive review of neutron stars and the astrophysical effects of quark matter. On the other hand, the high temperature, heated-vacuum quark-matter phase was first produced in nature at the Big Bang. The line of first-order transitions ends somewhere near the high temperature axis, indicated on the diagram by the label 'tricritical point' but whether the experimentally accessible region is on one side or the other of this endpoint remains controversial. The location of the endpoint and whether the region to the left of the endpoint has a second-order transition or a smooth crossover depends on the masses of the up, down and strange quarks.

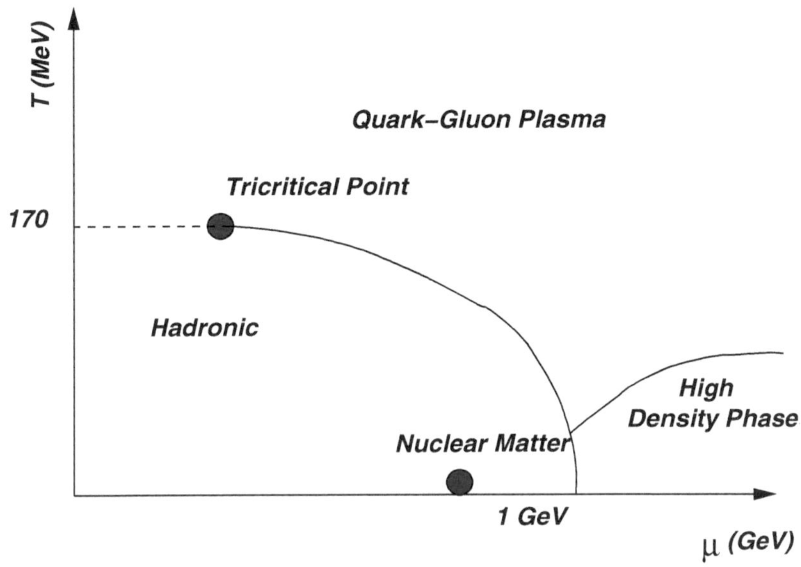

Figure 4.3: Schematic phase diagram for massless up and down quarks, after Ref. [59]. The solid curve moving toward the high temperature axis is a line of first-order transitions that ends at the tricritical point, beyond which the transition is second order. The critical temperature determined from finite temperature QCD studies is indicated on the y-axis. Normal nuclear matter is indicated by the point along the x-axis.

Phase transitions usually involve a change in the symmetry of the system. To take water as an example again, the liquid phase has no order and flows to fill the shape of the container. The solid phase of water, ice, tends to exhibit a symmetry. Snowflakes in particular are ice crystals of great symmetry. The condensation of water into snow in the atmosphere makes the solid phase ordered.

Phase transitions in nuclear matter also involve symmetries. In the early universe, the quarks were massless and there were an equal number of quarks and antiquarks. In the universe today, after the transition to normal nuclear matter, there is a small difference in the up and down quark masses – as well as an even larger difference between the light quark masses and those of the other quarks, even between the light quarks and the strange quark. This mass difference is caused by chiral symmetry breaking. (Chiral symmetry is the 'handedness' symmetry. Your body exhibits chiral symmetry since it is left/right symmetric. You can break this symmetry by wearing a watch or ring on one hand and not the other.) At high temperatures chiral symmetry is expected to be restored, as we will discuss in a later chapter. Today's universe is baryon dominated with quarks bound into normal matter while antibaryons do not exist in nature and must be produced in the laboratory. We expect that in heavy-ion collisions, there should be an equal number of quarks and antiquarks produced in the central rapidity region, as already discussed in Chapter 2.

The experimental trend from the AGS to the SPS to RHIC and the LHC has been toward the high temperature, low density axis. Thus we will emphasize this area of the phase diagram. The new lower energy facility, FAIR, at the Gesellschaft für Schwerionenforschung (GSI) will, however, be better able to explore the high density region [62].

4.2 Review of thermodynamics

The first law of thermodynamics tells us that heat is a form of energy. The second law can be stated in a number of equivalent ways but it basically says that the entropy of a closed system tends to remain constant or increase when a constraint internal to that system is removed. The entropy of a system is related to the number of states in the system. A closed system holds constant all external parameters that may influence the system such as energy, volume and particle number. Thus, if one of these is no longer held fixed but increases, the number of possible states and thus the entropy will increase as well. A closed system is equally likely (probable) to be in any state accessible to it. States that are not accessible have zero probability.

If a system has Ω accessible quantum states and $\mathcal{S}$ is a state label, the probability for the system to be in state $\mathcal{S}$ is

$$P(\mathcal{S}) = \frac{1}{\Omega} \tag{4.1}$$

while $P(\mathcal{S}) = 0$ if the state is not accessible. The sum of the probabilities must add to unity. If the system is not closed but has some functional dependence on a quantity X such as energy, E, volume, V, or particle number, N, the average value of such a quantity is

$$\langle X \rangle = \sum_{\mathcal{S}} X(\mathcal{S}) P(\mathcal{S}) = \sum_{\mathcal{S}} X(\mathcal{S}) \frac{1}{\Omega} , \tag{4.2}$$

called the ensemble average. An ensemble of systems is constructed so that each system in the ensemble is in one of the accessible quantum states. This ensemble thus represents the real system and would consist of Ω systems in our example.

Now if two systems are brought into thermal contact, energy can be transferred between the systems, forming a larger closed system, $\mathcal{S} = \mathcal{S}_1 + \mathcal{S}_2$, with constant energy $E = E_1 + E_2$ and particle number $N = N_1 + N_2$. The constant energy and particle number can be shared any number of ways between the two systems. However, the most probable way to share the energy is such that the number of accessible states is maximized in the combined system. If system $\mathcal{S}_1$ has energy E_1, then system $\mathcal{S}_2$ has energy $E - E_1$. The number of states, Ω, of the combined system is

$$\Omega(N, E) = \sum_{E_1} \Omega_1(N_1, E_1) \Omega_2(N_2, E - E_1) . \tag{4.3}$$

A configuration is the set of all states with E_1 and E_2. If this product is maximized for $E_1 = E_1'$, then this value of $\Omega_1 \Omega_2$ is the most probable configuration, $\Omega_1(N_1, E_1')\Omega_2(N_2, E - E_1')$. If the system is large, the most probable configuration describes many properties of the system. When two large systems in thermal contact are accurately described by the properties of the most probable configuration, the systems are in thermal equilibrium. To calculate the extremum (most probable configuration), the differential of $\Omega(N, E)$ must be zero for incremental differences in the energy,

$$d\Omega = \left(\frac{\partial \Omega_1}{\partial E_1}\right)_{N_1} \Omega_2 dE_1 + \Omega_1 \left(\frac{\partial \Omega_2}{\partial E_2}\right)_{N_2} dE_2 = 0 \tag{4.4}$$

while $dE_1 + dE_2 = 0$. Dividing by $\Omega_1 \Omega_2$ and using the fact that the energy, E, is constant, thermal equilibrium implies

$$\frac{1}{\Omega_1}\left(\frac{\partial \Omega_1}{\partial E_1}\right)_{N_1} = \frac{1}{\Omega_2}\left(\frac{\partial \Omega_2}{\partial E_2}\right)_{N_2} , \tag{4.5}$$

written equivalently as

$$\left(\frac{\partial \ln \Omega_1}{\partial E_1}\right)_{N_1} = \left(\frac{\partial \ln \Omega_2}{\partial E_2}\right)_{N_2} . \tag{4.6}$$

The entropy of a system is defined as

$$S(N, E) = \ln \Omega(N, E) \tag{4.7}$$

so that the condition for thermal equilibrium is finally

$$\left(\frac{\partial S_1}{\partial E_1}\right)_{N_1} = \left(\frac{\partial S_2}{\partial E_2}\right)_{N_2} . \tag{4.8}$$

While N is held fixed here, the entropy can also depend on the volume of the system, as we discuss later. The last step is to define the temperature. In thermal equilibrium, the temperatures of two systems in thermal contact are equal. Thus,

$$\left(\frac{\partial S}{\partial E}\right)_{N} = \frac{1}{T} , \tag{4.9}$$

where T is the fundamental temperature with dimensions of energy since the entropy is dimensionless. Equation (4.9) also holds at fixed volume.

The entropy increases with the number of accessible states. The total entropy always increases when two systems are brought into thermal contact. This is known as the law of increase of entropy — the entropy either remains constant or increases when a constraint internal to the system is removed. When two systems are brought into thermal contact, the new constraint is that $E_1 + E_2$ is constant, not E_1 and E_2 separately. If the systems are separated again before equilibrium is reached, the system will be in an intermediate configuration with an intermediate entropy. Thus the entropy is time dependent, as we will investigate further when we discuss hydrodynamics. Ways to increase the entropy relevant to heavy- ion collisions include adding particles, adding energy and increasing the volume.

If we have a system with energy E' in contact with a large reservoir of energy, $E - E'$ (so that the whole system has energy E), the number of states in that system is $\Omega(E - E')$. The system could equally likely have energy E'' with $\Omega(E - E'')$ states of energy E''. The ratio of probabilities then is the ratio of the number of states so that

$$\frac{P(E')}{P(E'')} = \frac{\Omega(E - E')}{\Omega(E - E'')} . \tag{4.10}$$

Since Eq. (4.7) tells us that $S = \ln \Omega$, we have $\Omega = \exp(S)$ and

$$\frac{P(E')}{P(E'')} = \exp[S(E - E') - S(E - E'')] . \tag{4.11}$$

From here we can find that

$$\frac{P(E')}{P(E'')} = \frac{\exp(-E'/T)}{\exp(-E''/T)} , \tag{4.12}$$

where the term $\exp(-E'/T)$ is the Boltzmann factor.

EXAMPLE: Show that Eq. (4.12) follows from Eq. (4.11).
Since $E \gg E', E''$, we can make a Taylor expansion around $S(E)$ so that

$$S(E - E') = S(E) - E'\left(\frac{\partial S}{\partial E}\right)_{V,N} + \cdots \approx S(E) - E'/T \tag{4.13}$$

where we have used Eq. (4.9). Upon substitution into Eq. (4.11) and subtracting, the $S(E)$ terms cancel, leaving us with $\exp[S(E-E')-S(E-E'')] \sim \exp[-(E' - E'')/T]$. Equation (4.12) then follows.

We now define the pressure in terms of a reversible process. In a reversible process, the system remains close to its equilibrium entropy value, $S(V)$, so that if the change applied to the system is reversed, it returns to its initial condition. The entropy of such a system is thus held fixed and the change is called isentropic. An example is an infinitesimal compression applied to all faces of a cube which changes its volume from V to $V - \Delta V$. The change in energy is

$$E(V - \Delta V) - E(V) = \Delta E = -\left(\frac{\partial E}{\partial V}\right)\Delta V . \tag{4.14}$$

If $V = xyz$, then $\Delta V = (\Delta x)yz + x(\Delta y)z + xy(\Delta z)$. Since, in a cube, $L = x = y = z$, $\Delta V = 3A\Delta L$ where $A = L^2$, the area of a face of the cube. The change in the energy due to an infinitesimal compression, ΔE, is defined as $P\Delta V$ where P is the pressure. Thus

$$P = -\left(\frac{\partial E}{\partial V}\right)_S \tag{4.15}$$

where the entropy is held fixed in a reversible process.

If the number of particles is fixed, $S(E, V)$ only and the differential of S is

$$dS(E,V) = \left(\frac{\partial S}{\partial E}\right)_V dE + \left(\frac{\partial S}{\partial V}\right)_E dV \ . \tag{4.16}$$

At fixed entropy, $dS = 0$. We can then write $dE \equiv (\delta E)_S$, $dV \equiv (\delta V)_S$ and $(\delta E)_S/(\delta V)_S \equiv (\partial E/\partial V)_S$. Thus it follows that $(\partial S/\partial V)_E = -(\partial S/\partial E)_V(\partial E/\partial V)_S = -(1/T)(-P)$ or

$$P = T\left(\frac{\partial S}{\partial V}\right)_E \ . \tag{4.17}$$

Then Eq. (4.16) becomes

$$dE = TdS - PdV \ . \tag{4.18}$$

The change in the energy in Eq. (4.18) is a combination of TdS, the heat added to the system, and $-PdV$, the work done on the system. One can associate the work done on the system with the 'useful' energy extractable from the system. Thus we can relate $-PdV$ with the 'free' energy, F, so that

$$dF \equiv dE - TdS \ . \tag{4.19}$$

Equation (4.9) tells us that, at fixed volume and temperature,

$$dE = TdS \ , \tag{4.20}$$

making

$$dF = 0 \ . \tag{4.21}$$

Thus the free energy is a minimum for a system in thermal contact with a heat bath if the volume is held fixed. The free energy is defined as

$$F = E - TS \ . \tag{4.22}$$

Finally, if the particle number is not held fixed, we can introduce the chemical potential, μ, which governs particle flow between systems in diffusive contact, just as temperature governs energy flow between systems in thermal contact. When two systems can exchange particles, they are said to be in diffusive contact and will come to diffusive equilibrium with no net particle flow between them.

In diffusive equilibrium of two systems with fixed total particle number, $N = N_1 + N_2$, the total free energy, $F = F_1 + F_2$, is a minimum at fixed temperature,

$$dF = \left(\frac{\partial F_1}{\partial N_1}\right)_T dN_1 + \left(\frac{\partial F_1}{\partial N_2}\right)_T dN_2 = 0 \ . \tag{4.23}$$

Since $dN = dN_1 + dN_2 = 0$, we substitute $-dN_2$ for dN_1 to obtain the equilibrium condition,

$$\left(\frac{\partial F_1}{\partial N_1}\right)_T = \left(\frac{\partial F_1}{\partial N_2}\right)_T \ . \tag{4.24}$$

The chemical potential is thus defined as

$$\mu \equiv \left(\frac{\partial F}{\partial N}\right)_{T,V} \ . \tag{4.25}$$

Then, in diffusive equilibrium, $\mu_1 = \mu_2$.

The entropy is thus fully a function of E, N and V, $S(E, N, V)$. The differential of the entropy is then

$$dS = \left(\frac{\partial S}{\partial E}\right)_{V,N} dE + \left(\frac{\partial S}{\partial V}\right)_{E,N} dV + \left(\frac{\partial S}{\partial N}\right)_{E,V} dN \ . \tag{4.26}$$

Assuming the volume and temperature remain fixed, $dV = 0$ and $dT = 0$, the middle term is dropped and $dS \equiv (\delta S)_T$, $dE \equiv (\delta E)_T$ and $dN = (\delta N)_T$. Thus

$$(\delta S)_T = \left(\frac{\partial S}{\partial E}\right)_N (\delta E)_T + \left(\frac{\partial S}{\partial N}\right)_E (\delta N)_T \ . \tag{4.27}$$

We divide both sides by $(\delta N)_T$ and identify $(\delta S)_T/(\delta N)_T \equiv (\partial S/\partial N)_T$ with $(\delta E)_T/(\delta N)_T \equiv (\partial E/\partial N)_T$ and let $(\partial S/\partial E)_N = 1/T$. Finally, we multiply all terms by the temperature and find

$$T\left(\frac{\partial S}{\partial N}\right)_T = \left(\frac{\partial E}{\partial N}\right)_T + T\left(\frac{\partial S}{\partial N}\right)_E . \tag{4.28}$$

According to the definition of μ in Eq. (4.25), we also have

$$\mu = -T\left(\frac{\partial S}{\partial N}\right)_E . \tag{4.29}$$

We now return to Eq. (4.26) and rewrite it using the definitions of T, P and μ in Eqs. (4.9), (4.17) and (4.29) respectively as

$$dS = \frac{1}{T}dE + \frac{P}{T}dV - \frac{\mu}{T}dN . \tag{4.30}$$

Multiplying through by T and moving dE to the left-hand side of Eq. (4.30), we obtain

$$dE = TdS - PdV + \mu dN , \tag{4.31}$$

the thermodynamic identity.

In heavy-ion collisions, it is often preferable to discuss the thermodynamic quantities per volume. Thus we define $\epsilon \equiv E/V$, $s \equiv S/V$ and $n \equiv N/V$ as the energy density, entropy density and number density respectively. Then $\epsilon = sT - P + \mu n$. If there is no constraint on N, $\mu = 0$ and $\epsilon + P = sT$.

Now that we have introduced the chemical potential, we can generalize the Boltzmann factor in Eq. (4.12) to include changes in particle number. In this case, in addition to energy, $E - E'$, the reservoir has $N - N'$ particles so that the number of states in the system is $\Omega(N - N', E - E')$ and the ratio of probabilities is

$$\begin{aligned} \frac{P(N', E')}{P(N'', E'')} &= \frac{\Omega(N - N', E - E')}{\Omega(N - N'', E - E'')} \\ &= \frac{\exp[S(N - N', E - E')]}{\exp[S(N - N'', E - E'')]} \\ &= \exp[S(N - N', E - E') - S(N - N'', E - E'')] . \end{aligned} \tag{4.32}$$

We Taylor expand S in terms of both variables now, giving

$$\begin{aligned} S(N-N',E-E') &= S(N,E) - N'\left(\frac{\partial S}{\partial N}\right)_E - E'\left(\frac{\partial S}{\partial E}\right)_N + \cdots \\ &\sim S(N,E) + \frac{N'\mu}{T} - \frac{E'}{T} . \end{aligned} \tag{4.33}$$

The difference in entropies becomes $\exp[(N'-N'')\mu/T - (E'-E'')/T]$. Thus the ratio of probabilities is finally

$$\frac{P(N',E')}{P(N'',E'')} = \frac{\exp[-(E'-N'\mu)/T]}{\exp[-(E''-N''\mu)/T]} . \tag{4.34}$$

The exponential term in Eq. (4.34) is known as the Gibbs factor for J. W. Gibbs who called it the grand canonical distribution. Systems in thermal and diffusive contact with a heat reservoir and can thus exchange energy and particles with the reservoir are then part of a grand canonical ensemble. The Boltzmann factor in Eq. (4.12) is for systems in thermal contact. Such systems are part of a canonical ensemble. To convert the ratio of probabilities to absolute probabilities, it is necessary to sum the Gibbs factors over all possible states with energy E and number of particles N,

$$Z_{\rm gen} = \sum_{N=0}^{\infty} \sum_{E(N)} \exp[-(E(N) - N\mu)/T] , \tag{4.35}$$

the grand partition function. Then

$$P(N',E') = \frac{1}{Z_{\rm gen}} \exp[-(E'-N'\mu)/T] \tag{4.36}$$

and the sum of the probabilities over all states and particle numbers is, by definition, unity.

The Gibbs factor is the classical limit of both the Fermi-Dirac and Bose-Einstein distributions. For spin 1/2 fermions, which obey the Pauli exclusion principle – an orbital can be occupied by either one or zero fermions of the same species – the Fermi-Dirac distribution is

$$f_{\rm FD}(E,\mu) = \frac{1}{\exp[(E-\mu)/T] + 1} \tag{4.37}$$

where μ is positive for particles and negative for antiparticles. Bosons, with integer spin, on the other hand can fill an orbital with any number of particles so that the Bose-Einstein distribution is

$$f_{\rm BE}(E,\mu) = \frac{1}{\exp[(E-\mu)/T]-1} \, . \tag{4.38}$$

EXAMPLE: Derive the distribution functions for both fermions and bosons.

The Fermi-Dirac and Bose-Einstein distributions can be derived from their Gibbs sums (the grand partition function). The distribution function is defined as the average number of particles that can occupy the system.

We start with fermions. Since only one particle can occupy a state, starting at a state with $N = 0$ and thus $E = 0$, we have

$$Z_{\rm FD} = 1 + \exp[-(E-\mu)/T] \, . \tag{4.39}$$

The average number of particles in the system, $\langle N \rangle$, is defined as

$$\begin{aligned} \langle N \rangle_{\rm FD} &= \frac{\sum_{N=0}^{1} \sum_{E(N)} N \exp[-(E-N\mu)/T]}{Z_{\rm FD}} \\ &= \frac{\exp[-(E-\mu)/T]}{1+\exp[-(E-\mu)/T]} \, . \end{aligned} \tag{4.40}$$

Multiplying the numerator and denominator by $\exp[(E-\mu)/T]$, we find

$$f_{\rm FD}(E,\mu) \equiv \langle N \rangle_{\rm FD} = \frac{1}{\exp[(E-\mu/T)]+1} \, . \tag{4.41}$$

On the other hand, any integer number of bosons can occupy a state. If the state contains N particles, each of energy E, then

$$Z_{\rm BE} = \sum_{N=0}^{\infty} \exp[-N(E-\mu)/T] = \sum_{N=0}^{\infty} (\exp[-(E-\mu)/T])^N \, . \tag{4.42}$$

Since $\sum_{N=0}^{\infty} x^N = 1/(1-x)$, the partition function is

$$Z_{\rm BE} = \frac{1}{1-\exp[-(E-\mu)/T]} \, . \tag{4.43}$$

To calculate the average number of particles in the state, we use the definition, $\langle N \rangle = T\partial \ln Z/\partial\mu$, derived below. Carrying out the derivative, we obtain

$$f_{\rm BE}(E,\mu) = \langle N \rangle_{\rm BE} = \frac{1}{\exp[(E-\mu)/T]-1} \ . \tag{4.44}$$

We note that in both cases, we can equally well write the distributions for antiparticles if we change the sign of the chemical potential.

We can write a generalized version of the partition function for fermions and bosons using Eqs. (4.39) and (4.43),

$$Z_{\rm gen} = (1 \mp \exp[-(E-\mu)/T])^{\mp 1} \tag{4.45}$$

where the $-$ is for bosons and the $+$ for fermions. Then

$$\ln Z_{\rm gen} = \mp \ln(1 \mp \exp[-(E-\mu)/T]) \ , \tag{4.46}$$

the form often used to calculate the thermodynamics quantities of pressure, energy density and number density. Taking the derivative of Eq. (4.46) with respect to the chemical potential, μ, gives

$$\frac{\partial \ln Z_{\rm gen}}{\partial \mu} = \mp \frac{1}{T} \frac{\exp[-(E-\mu)/T]}{1 \mp \exp[-(E-\mu)/T]} \ . \tag{4.47}$$

We bring the exponent in the numerator into the denominator and factor out -1 from the denominator for bosons so that the $\mp$ out in front of the right-hand side of Eq. (4.47) is changed to $+$ for bosons and fermions. Then multiplying both sides by T gives

$$\frac{T\partial \ln Z_{\rm gen}}{\partial \mu} = \frac{1}{\exp[(E-\mu)/T] \mp 1} = \langle N \rangle_{\rm gen} \ , \tag{4.48}$$

as we have already seen in Eqs. (4.41) and (4.44).

In the limit where $\exp[(E-\mu)/T] \gg 1$, the Gibbs factor becomes the distribution function. This is the classical limit. The classical distribution function, also known as the Maxwell-Boltzmann distribution, is then

$$f_{\rm MB}(E,\mu) \approx \exp[-(E-\mu)/T] = \lambda \exp(-E/T) \tag{4.49}$$

where $\lambda = \exp(\mu/T)$.

The three distribution functions are shown in Fig. 4.4 as a function of $(E-\mu)/T$. The Fermi-Dirac distribution has been extended to negative values to better show its asymptotic behavior. The Bose-Einstein and Maxwell-Boltzmann distributions are only shown for positive values since the Bose-Einstein distribution function blows up as $(E-\mu)/T \to 0$. The Maxwell-Boltzmann case has been included to show the asymptotic behavior at large $(E-\mu)/T$. The three curves come together at $(E-\mu)/T \approx 2$.

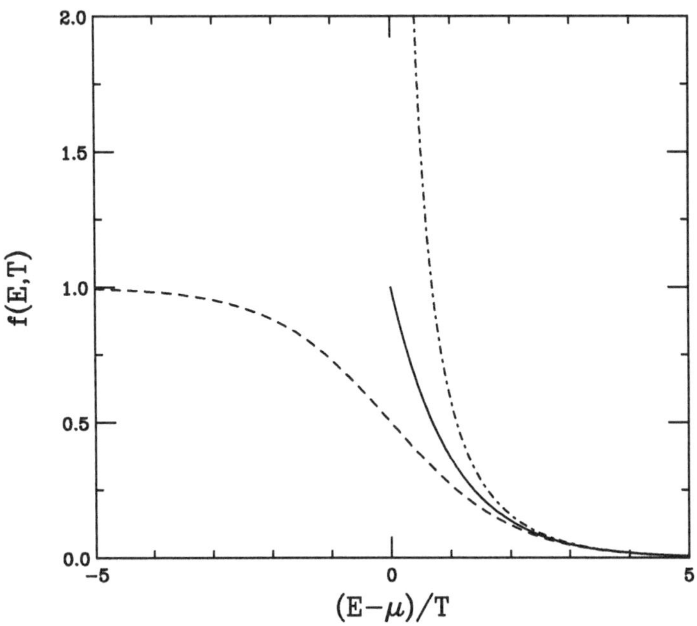

Figure 4.4: The Fermi-Dirac (dashed), Bose-Einstein (dot-dashed) and Maxwell-Boltzmann (solid) distributions are shown as functions of $(E-\mu)/T$.

The chemical potential of the Maxwell-Boltzmann distribution is found by assuming that the thermal average of the total number of particles is equal to the number of particles actually present in the system, $N = \langle N \rangle$, and

$$\langle N \rangle = \sum_s f_{\rm MB}(E_s) = \lambda \sum_s \exp(-E_s/T) = \lambda Z_1 \tag{4.50}$$

where Z_1 is the partition function for a single free particle of mass M in volume V. The energy of a single particle in a cubic box with side L ($V = L^3$) is

$$E_n = \frac{\hbar^2}{2M}\left(\frac{\pi}{L}\right)^2 (n_x^2 + n_y^2 + n_z^2)\ . \tag{4.51}$$

To sum over all states, s, of E involves the sum over the positive integers n_x, n_y and n_z. We replace the sums by integrals to obtain

$$\begin{aligned} Z_1 &= \int_0^\infty dn_x \int_0^\infty dn_y \int_0^\infty dn_z \exp\left[-\frac{\hbar^2\pi^2}{2ML^2T}(n_x^2 + n_y^2 + n_z^2)\right] \\ &= V\left(\frac{MT}{2\pi\hbar^2}\right)^{3/2} \end{aligned} \tag{4.52}$$

EXAMPLE: Show Eq. (4.52).
Let $t^2 = \hbar^2\pi^2/(2ML^2T)$ so that the argument of the exponential is $t^2(n_x^2 + n_y^2 + n_z^2)$. We can rewrite the sum of the integers n in the exponent as a product of exponentials so that

$$Z_1 = \left(\int_0^\infty dn \exp(-t^2n^2)\right)^3 .$$

Changing variables to $z = tn$ with $dn = dz/t$ and recalling that the integral over a Gaussian distribution is $\int_0^\infty dz \exp(-z^2) = \sqrt{\pi}/2$, we obtain

$$Z_1 = \frac{1}{t^3}\left(\int_0^\infty dz \exp(-z^2)\right)^3 = \left(\frac{\sqrt{\pi}}{2t}\right)^3 = V\left(\frac{MT}{2\pi\hbar^2}\right)^{3/2}$$

where we have replaced L^3 by V.

If we define the factor multiplying V in Eq. (4.52), $(MT/2\pi\hbar^2)^{3/2}$, as the quantum concentration, n_Q, since it has units of inverse volume (or density), then

$$Z_1 = n_Q V \tag{4.53}$$

where, in this case, the box contains only a single particle. Substituting the value of Z_1 into Eq. (4.50),

$$\langle N \rangle = N = \lambda n_Q V$$

where $\lambda = N/(n_Q V) = n/n_Q$ with $n = N/V$. Equating the value of λ to our previous definition with the chemical potential, $\lambda = \exp(\mu/T)$, we solve for the chemical potential to find

$$\mu = T \ln(n/n_Q) \ . \tag{4.54}$$

From the relationship between the free energy and the chemical potential in Eq. (4.25), $\mu = (\partial F/\partial N)_{T,V}$, we can determine the free energy,

$$F = \int_0^N dN \mu(N, T, V) = NT[\ln(n/n_Q) - 1] \ . \tag{4.55}$$

EXAMPLE: Use Eq. (4.55) to determine F above.
Using the definition of μ in Eq. (4.54), we can write the integral as

$$F = T \int_0^N dN[\ln N - \ln(V n_Q)] \ .$$

Recall that $\int dN \ln N = N \ln N - N$. With this result for the first part of the integral and noting that the second part is independent of N, we find

$$F = T[N \ln N - N - N \ln(V n_Q)] = NT[\ln(n/n_Q) - 1] \ .$$

The pressure is the derivative of the free energy with respect to the volume, $P = -(\partial F/\partial V)_{T,N} = NT/V$. Thus we obtain

$$PV = NT \ , \tag{4.56}$$

the ideal gas equation of state.

EXAMPLE: Determine the energy of an ideal gas.
The energy is related to the free energy by $E = F + TS = F - T(\partial F/\partial T)_{V,N}$. The derivative of the free energy with respect to temperature is

$$\frac{\partial F}{\partial T} = N[\ln N - \ln(V n_Q) - 1] - NT \frac{V}{n_Q V} \frac{\partial n_Q}{\partial T} \ . \tag{4.57}$$

Recall that $n_Q \propto T^{3/2}$ so that $\partial n_Q/\partial T = (3/2)n_Q/T$. The last term of $\partial F/\partial T$ is then $3N/2$. Multiplying the derivative by T, the first term in $T\partial F/\partial T$ cancels F itself, leaving

$$E = \frac{3}{2}NT \; . \tag{4.58}$$

The grand partition function derived in Eq. (4.35) can be used to determine the energy and number densities. In general, the partition function can be written in terms of quantum mechanical operators, the Hamiltonian, $\hat{H}$, or energy operator and the number, $\hat{N}$, operator where $\hat{H}|a\rangle = E_a|a\rangle$ and $\hat{N}|a\rangle = N_a|a\rangle$. The partition function is then the sum of the expectation values of the thermodynamical potential,

$$Z = \sum_a \langle a| \exp\left[-\frac{1}{T}(\hat{H} - \mu\hat{N})\right] |a\rangle \tag{4.59}$$

$$\equiv \mathrm{Tr}\left[\exp\left\{-\frac{1}{T}(\hat{H} - \mu\hat{N})\right\}\right] \tag{4.60}$$

where Tr in the second expression denotes the trace. (The sum of the diagonal components of a matrix is its trace.) The expression for Z is general and can be extended to many particles of any type.

Assuming the system can be described by a grand canonical ensemble, in thermal and diffusive equilibrium, the energy and particle number fluctuate around their average values, $\langle E\rangle$ and $\langle N\rangle$. The energy density and number density can then be obtained from these averages,

$$n = \frac{\langle N\rangle}{V} = \frac{1}{V}\frac{\mathrm{Tr}\left[\hat{N}\exp -(\hat{H} - \mu\hat{N})/T\right]}{\mathrm{Tr}\left[\exp -(\hat{H} - \mu\hat{N})/T\right]} \tag{4.61}$$

$$= \frac{T}{V}\frac{\partial \ln Z}{\partial \mu} \; , \tag{4.62}$$

$$\epsilon = \frac{\langle E\rangle}{V} = \frac{1}{V}\frac{\mathrm{Tr}\left[\hat{H}\exp -(\hat{H} - \mu\hat{N})/T\right]}{\mathrm{Tr}\left[\exp -(\hat{H} - \mu\hat{N})/T\right]} \tag{4.63}$$

$$= \frac{T^2}{V}\frac{\partial \ln Z}{\partial T} + \mu n \; . \tag{4.64}$$

It can be shown that the free energy at $\mu = 0$, $F = E - TS$, is related to the partition function by

$$F = -T \ln Z \tag{4.65}$$

since

$$dS(E,V) = \left(\frac{\partial S}{\partial E}\right)_V dE + \left(\frac{\partial S}{\partial V}\right)_E dV = \frac{dE}{T} + \frac{PdV}{T} \; . \tag{4.66}$$

Multiplying both sides by T, we have $TdS = dE + PdV$. At fixed volume, this reduces to $dE = TdS$, Eq. (4.20). The differential of the free energy is

$$dF = dE - TdS - SdT \; . \tag{4.67}$$

Making the substitution from Eq. (4.66) above, we obtain

$$dF = -PdV - SdT \; . \tag{4.68}$$

Thus we can define the entropy and the pressure as differentials of the free energy with respect to temperature and volume, respectively,

$$S = -\left(\frac{\partial F}{\partial T}\right)_V \tag{4.69}$$

$$P = -\left(\frac{\partial F}{\partial V}\right)_T \; . \tag{4.70}$$

Using Eq. (4.69) for the entropy, we can rewrite the free energy as

$$F = E + T\left(\frac{\partial F}{\partial T}\right)_V \; , \tag{4.71}$$

equivalent to the second-order differential equation,

$$E = -T^2 \frac{\partial (F/T)}{\partial T} \; . \tag{4.72}$$

EXAMPLE: Show the equivalence of Eqs. (4.71) and (4.72).
The derivative of F/T with respect to T is

$$\frac{\partial (F/T)}{\partial T} = \frac{1}{T}\frac{\partial F}{\partial T} - \frac{F}{T^2} \; .$$

Multiplying both sides by $-T^2$ gives

$$-T^2\frac{\partial(F/T)}{\partial T} = -T\frac{\partial F}{\partial T} + F\ ,$$

the same as Eq. (4.71) when E is replaced by Eq. (4.72).

Comparing this result with Eq. (4.64) when $\mu = 0$ gives

$$\frac{F}{T} = -\ln Z \tag{4.73}$$

or $F = -T\ln Z$, as in Eq. (4.65). From this it follows that

$$P = -\left(\frac{\partial(-T\ln Z)}{\partial V}\right)_T = \frac{T\ln Z}{V} = -\frac{F}{V}\ , \tag{4.74}$$

$$S = \frac{\partial(T\ln Z)}{\partial T}\ . \tag{4.75}$$

We note one final useful piece of information before deriving the thermodynamic identity, $E = TS - PV + \mu N$, from the Gibbs free energy below. When $\mu = 0$, $dE = TdS + SdT - PdV - VdP$ for $\mu = 0$. Equation (4.20) tells us that $dE = TdS$ at fixed volume ($dV = 0$) so that $VdP = SdT$. (We can alternatively arrive at this last relation if we replace dE above by $TdS - PdV$, as in Eq. (4.18).) Dividing VdP by dE, we obtain

$$\frac{VdP}{dE} = \frac{dP}{d\epsilon} = \frac{SdT}{TdS} = \frac{d\ln T}{d\ln S} \equiv c_s^2 \tag{4.76}$$

where c_s^2 is the square of the speed of sound in the medium. As we will see later in this chapter, in an ideal gas of massless particles, $c_s^2 = 1/3$.

The free energy associated with the Gibbs factor and the grand partition function when $\mu \neq 0$ is

$$G \equiv E - TS + PV\ , \tag{4.77}$$

also known as the thermodynamic potential or the Gibbs free energy. The differential of G is

$$dG = dE - TdS - SdT + PdV + VdP\ . \tag{4.78}$$

If the system is in thermal and mechanical contact, $dT = 0$ and $dP = 0$. Then using the thermodynamic identity, Eq. (4.31),

$$dG = \mu dN \ . \tag{4.79}$$

If no particles can be exchanged, $dN = 0$ and $dG = 0$. This is the equilibrium condition for systems in thermal and mechanical contact. When $dG = 0$ as well, the systems are also in diffusive contact. The extremum of G is a minimum since G decreases with increasing entropy, $-TS$. Substituting the thermodynamic identity into Eq. (4.78), we have

$$\begin{aligned} dG &= \mu dN - SdT + VdP \\ &= \left(\frac{\partial G}{\partial N}\right)_{T,V} dN + \left(\frac{\partial G}{\partial T}\right)_{N,V} dT + \left(\frac{\partial G}{\partial P}\right)_{N,T} dP \ . \end{aligned} \tag{4.80}$$

From Eq. (4.80) we identify

$$\mu = \left(\frac{\partial G}{\partial N}\right)_{T,V} , \qquad -S = \left(\frac{\partial G}{\partial T}\right)_{N,V} , \qquad V = \left(\frac{\partial G}{\partial P}\right)_{N,T} . \tag{4.81}$$

Since $dP = dT = 0$, the temperature and pressure are extensive (unchanging) quantities in G. Because, however, the energy, entropy and volume increase linearly with the number of particles, the Gibbs free energy is also directly proportional to N and

$$G(N, P, T) = N\mu(P, T) \ . \tag{4.82}$$

Equating the Gibbs free energy in Eq. (4.82) with that in Eq. (4.77), the energy in the system, E, is

$$E = TS + \mu N - PV \ . \tag{4.83}$$

or

$$\epsilon = Ts + \mu n - P \tag{4.84}$$

after dividing by the volume.

EXAMPLE: Show that Eqs. (4.61), (4.64), (4.74) and (4.75) satisfy Eq. (4.84).

We rewrite the components of the thermodynamic identity in terms of their definitions with respect to $\ln Z$ so that

$$\frac{T^2}{V}\frac{\partial \ln Z}{\partial T} + \mu n = \frac{T}{V}\frac{\partial (T \ln Z)}{\partial T} + \mu n - \frac{T \ln Z}{V} \, .$$

The μn terms cancel. Then expanding the entropy term, we have

$$\frac{T^2}{V}\frac{\partial \ln Z}{\partial T} = \frac{T^2}{V}\frac{\partial \ln Z}{\partial T} + \frac{T \ln Z}{V} - \frac{T \ln Z}{V} \, , \tag{4.85}$$

and Eq. (4.84) is satisfied.

Before going on to derive specific results for the energy density, pressure, entropy and other thermodynamic quantities in systems specific to nuclear collisions, we will discuss phase transitions in a rather general way.

4.3 Phase transitions

At the beginning of this chapter we showed an isotherm for the quark-gluon plasma as an example in Fig. 4.2, the pressure as a function of volume at constant temperature. As we discussed before, water and water vapor may coexist in part of an isotherm only if the temperature is less than or equal to some critical temperature, T_c. Above T_c only one phase exists. We call the phase above T_c the fluid phase. (Such nomenclature is consistent with the idea that the matter produced in collisions at RHIC is a 'perfect fluid' [64].) Phase coexistence can only occur if the two phases which are in thermal, diffusive and mechanical contact are in equilibrium, $T_1 = T_2$, $\mu_1 = \mu_2$ and $P_1 = P_2$. In our ideal case, phase 1 would be the quark-gluon plasma and phase 2 would be a hadron gas.

The phases do not coexist over the entire pressure and temperature plane. If, *e.g.* $\mu_{\rm QGP} < \mu_{\rm HG}$, the plasma phase alone is stable while if $\mu_{\rm QGP} > \mu_{\rm HG}$, the hadron phase is stable. Having derived the thermodynamic quantities in terms of the grand partition function in the previous section, we can derive the coexistence curve for two phases in equilibrium, the pressure as a function of temperature.

For small deviations in pressure and temperature around their equilibrium values, the coexistence condition for the chemical potential may be written as

$$\mu_1(P+dP,T+dT) = \mu_2(P+dP,T+dT) \ . \tag{4.86}$$

Expanding both sides, we have

$$\begin{aligned} \mu_1(P,T) + \left(\frac{\partial \mu_1}{\partial P}\right)_T dP + \left(\frac{\partial \mu_1}{\partial T}\right)_P dT = \\ \mu_2(P,T) + \left(\frac{\partial \mu_2}{\partial P}\right)_T dP + \left(\frac{\partial \mu_2}{\partial T}\right)_P dT \ . \end{aligned} \tag{4.87}$$

Since the coexistence condition tells us that $\mu_1(P,T) = \mu_2(P,T)$, these terms cancel in Eq. (4.87). We can rearrange the remaining terms to obtain the derivative dP/dT,

$$\frac{dP}{dT} = \frac{(\partial \mu_2/\partial T)_P - (\partial \mu_1/\partial T)_P}{(\partial \mu_1/\partial P)_T - (\partial \mu_2/\partial P)_T} \ . \tag{4.88}$$

Given that $G(N,P,T) = N\mu(P,T)$, using the definitions in Eq. (4.81) at constant N, we have

$$\left(\frac{\partial \mu}{\partial P}\right)_T = \frac{1}{N}\left(\frac{\partial G}{\partial P}\right)_T = \frac{V}{N} \ , \tag{4.89}$$

$$\left(\frac{\partial \mu}{\partial T}\right)_V = \frac{1}{N}\left(\frac{\partial G}{\partial T}\right)_V = -\frac{S}{N} \ . \tag{4.90}$$

Note that while the total number of particles in the system, $N = N_1 + N_2$, is a constant, the number of particles in the two phases changes as the volume changes. Thus the differential in Eq. (4.88) becomes

$$\frac{dP}{dT} = \frac{S_1/N_1 - S_2/N_2}{V_1/N_1 - V_2/N_2} \ . \tag{4.91}$$

The numerator is the increase of entropy caused by moving one particles from phase 1 to phase 2 while the denominator is the increase in the volume cause by transferring a particle from phase 1 to phase 2. Note that even though $dP/dT \neq 0$, the two phases continue to coexist. We can relate the numerator to the amount of heat needed to transfer one particle from phase 1 to phase 2 by multiplying it by the temperature. This quantity,

$$L = T\left(\frac{S_1}{N_1} - \frac{S_2}{N_2}\right) , \tag{4.92}$$

is the latent heat of the system. The latent heat is a measure of the entropy difference between two systems. First-order phase transitions are characterized by latent heat.

Phase transitions were described systematically by Landau using mean-field theory. We know that the free energy at constant temperature and volume is minimized in equilibrium. However, there are a number of variables that it could be minimized with respect to. Instead of considering a variety of possible variables, Landau assumed that the system could be characterized by a single variable, ξ, called the order parameter. In thermal equilibrium, $\xi = \xi_0(T)$, the value of ξ that minimizes the free energy. Then, limiting ourselves to $\mu = 0$, the Landau free energy, F_L, is

$$F_L(\xi, T) \equiv E(\xi, T) - TS(\xi, T) . \tag{4.93}$$

When $\xi = \xi_0$, in equilibrium, F_L is a minimum, equal to the free energy $F(T)$, or

$$F(T) = F_L(\xi_0, T) \leq F_L(\xi, T) . \tag{4.94}$$

At constant temperature, the Landau free energy may have more than one minimum in ξ. The lowest minimum is the equilibrium value.

Here we restrict ourselves to the case where F_L is an even function of ξ and can be expanded in powers of ξ^{2n}. The temperature dependence is in the expansion coefficients so that

$$\begin{aligned} F_L(\xi, T) &= g_0(T) + \frac{1}{2}g_2(T)\xi^2 + \frac{1}{4}g_4(T)\xi^4 \\ &\quad + \frac{1}{6}g_6(T)\xi^6 + \cdots . \end{aligned} \tag{4.95}$$

The simplest phase transition occurs when, neglecting higher terms, the second coefficient, g_2, changes sign at $T = T_0$. If g_2 is linear in T, it can be defined schematically as $A(T - T_0)$ so that the sign change at $T = T_0$ is obvious. We assume g_4 is constant and all higher terms are negligible. Such simplistic assumptions can only hold over a limited temperature range. Then

$$F_L(\xi, T) = g_0(T) + \frac{1}{2}A(T - T_0)\xi^2 + \frac{1}{4}g_4\xi^4 \,. \tag{4.96}$$

The minimum of the Landau free energy with respect to ξ is

$$\left(\frac{\partial F_L}{\partial \xi}\right)_T = A(T - T_0)\xi + g_4\xi^3 = 0 \tag{4.97}$$

The condition in Eq. (4.97) is satisfied if either $\xi = 0$ or

$$\xi^2_{\min} = -\frac{A}{g_4}(T - T_0) \,. \tag{4.98}$$

If A and g_4 are both positive, $\xi = 0$ is the minimum for $T > T_0$ where $F(T) = g_0(T)$. The second root corresponds to a minimum in the Landau free energy for $T < T_0$. Note that $\xi^2 \to 0$ as $T \to T_0$ from below. Substituting Eq. (4.98) for ξ^2 in Eq. (4.96), we have

$$\begin{aligned} F_L(\xi^2_{\min}, T) = F(T) &= g_0(T) - \frac{A^2}{2g_4}(T - T_0)^2 + \frac{g_4}{4}\frac{A^2}{g_4^2}(T_0 - T)^2 \\ &= g_0(T) - \frac{A}{4g_4}(T_0 - T) \,. \end{aligned} \tag{4.99}$$

The derivative of the free energy with respect to T is

$$\frac{\partial F}{\partial T} = \frac{\partial g_0(T)}{\partial T} + \frac{A^2}{2g_4}(T_0 - T) \tag{4.100}$$

and, at $T = T_0$, $\partial F/\partial T \neq 0$ but is continuous. Thus, this transition is known as second order, or continuous, and there is no latent heat associated with the transition. This is the case, for example, for ferromagnets and superconductors.

A schematic diagram of $F_L(\xi^2, T)$ as a function of ξ^2 for three different temperatures: $T > T_0$, $T = T_0$ and $T < T_0$ is shown in Fig. 4.5.

The values of ξ^2 that minimize F_L are indicated by the arrows. Note how, as $T \to T_0$ from below, the value of ξ_0^2 moves to the left to coincide with $\xi^2 = 0$ at $T = T_0$. Thus the equilibrium value of ξ^2 increases the more the temperature decreases below T_0.

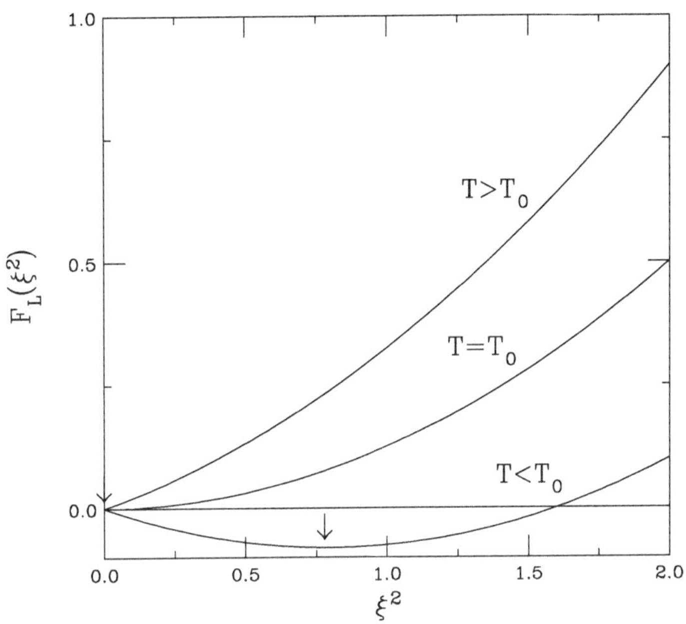

Figure 4.5: The Landau free energy, F_L, as a function of ξ^2 for different values of T in the case of a second-order phase transition. The three cases shown are $T > T_0$, $T = T_0$ and $T < T_0$. For $T < T_0$, there is a second, nonzero minimum value of $\xi^2 = \xi_0^2$ for which F_L is minimized. Both $\xi^2 = \xi_0^2$ and $\xi^2 = 0$ are indicated by the vertical arrows. See also Ref. [63].

The water to ice transition at constant (*e.g.* atmospheric) pressure mentioned at the beginning of the chapter, is first order, involving latent heat. In this case, the Landau theory can describe a first-order transition when g_4 is negative, written here as $-|g_4|$, and g_6 is positive. The Landau free energy is then

$$F_L(\xi, T) = g_0(T) + \frac{A}{2}(T - T_0)\xi^2 - \frac{1}{4}|g_4(T)|\xi^4 + \frac{1}{6}g_6\xi^6 \ . \quad (4.101)$$

Taking the derivative of $F_L(\xi, T)$ with respect to ξ^2 to find the minima, we obtain

$$\frac{\partial F_L}{\partial \xi} = A(T - T_0)\xi - |g_4(T)|\xi^3 + g_6\xi^5 = 0 \ . \tag{4.102}$$

Again, one root is found at $\xi = 0$ while another exists at

$$A(T - T_0) - |g_4(T)|\xi^2 + g_6\xi^4 = 0$$

which has a root at

$$\xi^2 = \frac{1}{2A(T - T_0)}\left[|g_4(T)| \pm \sqrt{|g_4(T)|^2 - 4Ag_6(T - T_0)}\right] . \tag{4.103}$$

In this case, the free energies will be equal to the values for $\xi = 0$ and $\xi \neq 0$ for some value of $T \neq T_0$. This value of T is the critical temperature, T_c. Since $T_c \neq T_0$, ξ^2 does not approach zero continuously at T_c. In this case, a first-order transition, there is latent heat.

Figure 4.6 shows F_L as a function of ξ^2 for various values of the temperature. For $T > T_c$, the critical temperature, the only minimum in F_L occurs at $\xi^2 = 0$. At $T = T_c$, on the other hand, there are two equal minima,

$$F_L(\xi^2 = 0, T = T_c) = F_L(\xi^2 = \xi_0^2, T = T_c) \ . \tag{4.104}$$

Below T_c there is only one minimum, at finite ξ_0^2. As the temperature drops below T_c, the value of ξ_0^2 increases. Thus below T_c there is a discontinuous jump in the order parameter, resulting in a discontinuity in the minimum F_L. This discontinuity is a sign of a first-order transition with latent heat.

We now discuss one more model of phase transitions specifically for the liquid-gas transition. This model is particularly useful for applications to heavy-ion collisions. We begin by modifying the ideal gas equation of state in Eq. (4.56), $PV = NT$, into the van der Waals equation of state which accounts for the finite volume of particles in a gas and the interactions between them. Taking these aspects into account, the free energy of an ideal gas, Eq. (4.55), becomes

$$F_{\rm vdW} = -NT[\ln(n_Q V_{\rm vdW}/N) + 1] + \Delta F_{\rm int} \tag{4.105}$$

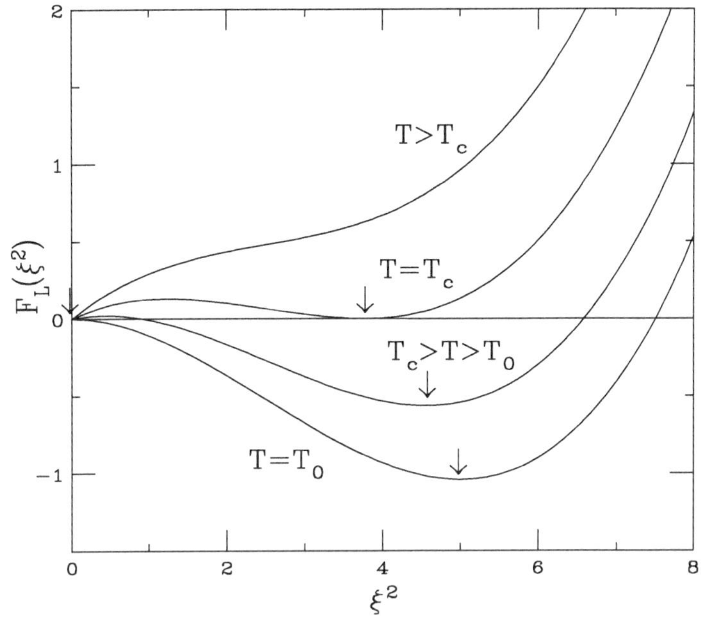

Figure 4.6: The Landau free energy, F_L, as a function of ξ^2 for different values of T in the case of a first-order transition. The four cases shown are $T > T_c$, $T = T_c$, $T_c > T > T_0$ and $T = T_0$. For $T < T_0$, there is a second, nonzero minimum value of ξ^2 for which F_L is minimized. This double minimum as well as the values of ξ^2 for $T < T_c$ are indicated by the vertical arrows. See also Ref. [63].

where we have factored out -1 from Eq. (4.55). Here $V_{\rm vdW}$ is the available volume and $\Delta F_{\rm int}$ is the free energy due to internal interactions in the gas.

First, the volume available is not V, the total volume of the system, but the volume that is not taken up by the particles in the gas. If the gas consists of N particles, each with volume b, then a volume Nb is not available as free volume. The volume Nb is known as 'excluded volume'. Thus the van der Waals volume is

$$V_{\rm vdW} = V - Nb \,. \tag{4.106}$$

We have called quarks 'point particles' which would have no size and would then have no excluded volume. However, normal hadrons, atoms and molecules do have finite sizes and would be subject to volume exclusion.

If the interactions between particles are unknown but not perturbatively describable, an approximation is often used. In the case of the van der Waals gas, we assume that the particles are uniformly distributed throughout the volume so that the point-to-point density, $n = N/V$, can be replaced by the average, or mean, value. Thus, such methods are referred to as mean-field methods. Correlations between particles and fluctuations in the density and the interaction potential are assumed to be negligible. The interaction potential energy for two particles a distance r apart is denoted $\phi(r)$. The average value of the interaction of a single particle at $r = 0$ with the $N - 1$ remaining particles of an N particle system is

$$\Delta\phi_{\rm int} = \int_b^\infty dV n\phi_{\rm int}(r) \ .$$

Note that the volume b of the particle is excluded from the range of integration over r. By the mean-field approximation, n is uniform over V and can be removed from the integral. We then have

$$\Delta\phi_{\rm int} = n\int_b^\infty dV\phi_{\rm int}(r) \equiv -2na \tag{4.107}$$

where, by definition, $\int_b^\infty dV\phi_{\rm int}(r) = -2a$. The factor of two is included by convention. The change in the free energy is defined as

$$\Delta F_{\rm int} = \frac{1}{2}N\Delta\phi_{\rm int} = -Nna = -\frac{N^2 a}{V} \ . \tag{4.108}$$

The factor of $1/2$ allows the interaction between two particles to be counted only once in the total energy. Since the interactions around each particle are counted, not including the factor of $1/2$ would be equivalent to doubling the energy of the total 'bond' between two particles.

The van der Waals free energy is obtained by substituting Eqs. (4.106) and (4.108) into Eq. (4.105). Then

$$F_{\rm vdW} = -NT[\ln(n_Q(V - Nb)/N) + 1] - \frac{N^2 a}{V^2} \ . \tag{4.109}$$

The pressure of the van der Waals gas is

$$P_{\rm vdW} = -\left(\frac{\partial F}{\partial V}\right)_{T,N} = \frac{NT}{V-Nb} - \frac{N^2 a}{V^2} \ . \tag{4.110}$$

Note that the interaction contribution to the pressure, N^2a/V^2, may be thought of as the internal pressure due to interactions between particles. Rearranging terms, we find the van der Waals equation of state,

$$\left(P_{\rm vdW} + \frac{N^2 a}{V^2}\right)(V - Nb) = NT \ . \tag{4.111}$$

The critical point for a van der Waals gas at constant temperature occurs at a horizontal point of inflection where, by definition,

$$\left(\frac{\partial P_{\rm vdW}}{\partial V}\right)_T = 0 \ , \qquad \left(\frac{\partial^2 P_{\rm vdW}}{\partial V^2}\right)_T = 0 \ . \tag{4.112}$$

At this point in the PV plane, $T = T_c$, there is no separation between the gas and liquid phases. Also for temperatures greater than the critical temperature, T_c, no phase separation exists.

EXAMPLE: Show that the critical pressure, volume and temperature for a van der Waals gas are

$$P_c = \frac{a}{27b^2} \ , \tag{4.113}$$

$$V_c = 3Nb \ , \tag{4.114}$$

$$T_c = \frac{8a}{27b} \ . \tag{4.115}$$

The critical point is determined by the definition of the horizontal inflection point in Eq. (4.112). Taking the first and second derivatives of P_{vdW}, we obtain

$$\left(\frac{\partial P_{\rm vdW}}{\partial V}\right)_T = -\frac{NT}{(V-Nb)^2} + \frac{2N^2 a}{V^3} = 0 \ , \tag{4.116}$$

$$\left(\frac{\partial^2 P_{\rm vdW}}{\partial V^2}\right)_T = \frac{2NT}{(V-Nb)^3} - \frac{6N^2 a}{V^4} = 0 \ . \tag{4.117}$$

Rearranging terms and dividing Eq. (4.116) by Eq. (4.117), we first obtain the critical volume,

$$\frac{1}{2}(V - Nb) = \frac{1}{3}V \longrightarrow V_c = 3Nb\ . \tag{4.118}$$

Then, putting T alone on the left-hand side of Eq. (4.116), we can solve for T_c using the value of V_c in Eq. (4.118),

$$T_c = \frac{1}{N}(V_c - Nb)^2 \frac{2N^2 a}{V_c^3} = \frac{8}{27}\frac{a}{b}\ . \tag{4.119}$$

Finally, using V_c and T_c,

$$P_c = \frac{NT_c}{V_c - Nb} - \frac{N^2 a}{V_c^2} = \frac{a}{27b^2}\ . \tag{4.120}$$

Substituting the critical values, Eqs. (4.118)-(4.120), into the van der Waals equation of state and defining $\hat{P} = P_{\mathrm{vdW}}/P_c$, $\hat{V} = V/V_c$ and $\hat{T} = T/T_c$, the equation of state becomes

$$\left(\hat{P} + \frac{3}{\hat{V}}\right)\left(\hat{V} - \frac{1}{3}\right) = \frac{8}{3}\hat{T}\ . \tag{4.121}$$

Then $\hat{P}$ is written as

$$\hat{P} = \frac{8\hat{T}}{3(\hat{V} - 1/3)} - \frac{3}{\hat{V}^2}\ . \tag{4.122}$$

The result shows that, if a gas obeys a van der Waals equation of state, it looks like any other gas with a similar equation of state. This is known as the law of corresponding states. The ratio P/P_c as a function of V/V_c is shown in Fig. 4.7 for several different values of T/T_c. The dotted curve is $T/T_c = 0.7$, the dashed, $T/T_c = 0.85$, the solid, $T/T_c = 1$ and the dot-dashed $T/T_c = 1.15$. The disappearance of the obvious minimum for $T/T_c < 1$ is clear, as is the horizontal inflection point at $T/T_c = 1$. Only the values of a and b differ between gases. These values can be obtained by fits to the observed P_c and T_c for a particular gas.

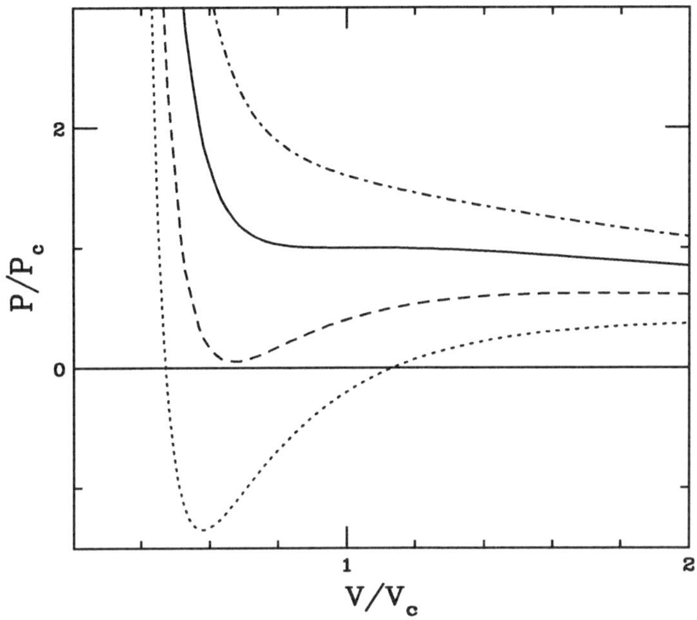

Figure 4.7: The ratio of the van der Waals pressure to the critical pressure as a function of the ratio V/V_c for curves of constant temperature for $T/T_c = 0.7$ (dotted), 0.85 (dashed), 1 (solid) and 1.15 (dot-dashed). See also Ref. [63].

Recall that we had three criteria for a phase transition: constant pressure, temperature and chemical potential across a phase boundary. We have so far only shown two of these three criteria for the van der Waals case. The Gibbs free energy of the van der Waals gas is

$$\begin{aligned} G(T,V,N) &= F_{\mathrm{vdW}} + P_{\mathrm{vdW}} V \\ &= -NT[\ln(n_Q(V-Nb)/N)+1] \\ &\quad -\frac{N^2 a}{V} + \frac{NTV}{V-Nb} - \frac{N^2 a V}{V^2} \\ &= -NT[\ln(n_Q(V-Nb)/N)+1] \\ &\quad -\frac{2N^2 a}{V} + \frac{NTV}{V-Nb} \,. \end{aligned} \tag{4.123}$$

Unfortunately, Eq. (4.123) is given as a function of T, N and V rather than as a function of T, N and P, a more natural form for obtaining the chemical potential through $G(T, P, N)/N$.

Figure 4.8 shows, in the PV plane, the regions of the van der Waals equation of state. For $V < V_1$, only the liquid phase exists while for $V > V_2$, only the gas phase exists. For $V_1 < V < V_2$, the phases coexist. The value of V_1 can be determined from the condition that $\mu_l = \mu_g$ along the line between V_1 and V_2. This is true if the shaded area below the line is equal to the shaded area above the line. Here the proportion of the liquid phase to the gas phase must be such that the sum of their volumes is equal to the available volume.

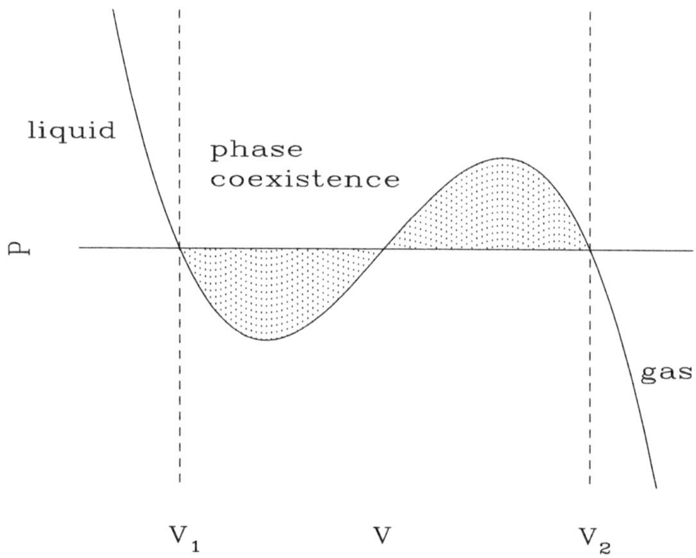

Figure 4.8: The van der Waals pressure as a function of volume for a constant temperature $T < T_c$. When $V < V_1$, only the liquid phase is present while for $V > V_2$, only the gas phase is present. For intermediate volumes, the two phases coexist. See also Ref. [63].

4.4 Phase transitions in nuclear physics

In this section, we evaluate the partition function in Eq. (4.60) for three cases: two extreme assumptions as well as a more realistic one. In the first case, we neglect the particle masses and assume the temperature is high, as in the central rapidity region of a high energy heavy-ion collision. The second case is for $T \to 0$ when massive nucleons are the only relevant degrees of freedom, as might be the case at the core of a neutron star. Finally, we take a more realistic example of finite temperature and finite particle mass.

The generalized $\ln Z$ for particles and antiparticles, either bosons or fermions, is [65]

$$\ln Z(T,\mu,V) = \frac{gV}{2\pi^2 T}\int_0^\infty \frac{dk\,k^4}{3E} \times \left[\frac{1}{\exp[(E-\mu)/T]\mp 1} + \frac{1}{\exp[(E+\mu)/T]\mp 1}\right] \tag{4.124}$$

where $E = (k^2+m^2)^{1/2}$ and g is the degeneracy factor. We take the positive sign for fermions and the negative sign for bosons.

EXAMPLE: Derive Eq. (4.124) from Eq. (4.46).

This equation arises from Eq. (4.46) after the antiparticle contribution (the second term in Eq. (4.124)), is included. For a particle or antiparticle in a three-dimensional box with momentum vector (k_x, k_y, k_z) in volume $L_x L_y L_z$ with *e.g.* $k_x L_x = 2\pi\hbar K_x$ where K_x is the mode number, the total partition function is equal to the sum over the modes in all dimensions. This is equivalent to the integral

$$\begin{aligned}\ln Z(T,\mu) &= g\int d^3r\frac{d^3k}{(2\pi)^3}[\ln Z(E,T,\mu)+\ln Z(E,T,-\mu)] \\ &= gV\int\frac{d^3k}{2\pi^2}[\ln Z(E;T,\mu)+\ln Z(E;T,-\mu)]\,. \end{aligned}\tag{4.125}$$

Since the partition function depends only on the magnitude of k and not its angular distribution, we have $d^3k = 4\pi k^2 dk$. Because $\ln Z$ is a function of $E = \sqrt{k^2+m^2}$ rather than k, we can further write $d^3k = 4\pi k E dE$ since $kdk = EdE$. Then we have

$$\ln Z(T,\mu) = \frac{gV}{2\pi^2}\int dE\,k\,E\,[\ln Z(E;T,\mu)+\ln Z(E;T,-\mu)]\,. \tag{4.126}$$

Integration by parts gives, *e.g.* for the particle case,

$$\int_0^\infty dE\, E\sqrt{E^2-m^2}\,\ln Z(E;T,\mu) =$$
$$\frac{k^3}{3}\ln Z(E,T,\mu)\bigg|_0^\infty + \int \frac{dE(E^2-m^2)^{3/2}}{3T}\frac{\mp\exp[-(E-\mu)/T]}{\exp[-(E-\mu)/T]\mp 1}\ .$$

The first term is zero. The second becomes

$$\int \frac{dEk^3}{3T}\frac{\mp 1}{1\mp\exp[(E-\mu)/T]} = \int \frac{dkk^4}{3TE}\frac{1}{\exp[(E-\mu)/T]\mp 1}$$

where we have changed the integration variable back to k. Finally,

$$\ln Z(T,\mu) = \frac{gV}{2\pi^2 T}\int \frac{dkk^4}{3E}\left[\frac{1}{\exp[(E-\mu)/T]\mp 1}\right] \qquad (4.127)$$

for particles only. The integral is now written in terms of the distribution functions, Eqs. (4.41) and (4.44). Including the antiparticle contribution, we obtain Eq. (4.124).

The most useful form of the free energy for calculating the thermodynamic quantities, ϵ, P, n and s in Eqs. (4.64), (4.74), (4.62) and (4.75) is with the partition function calculated as in Eq. (4.125) so that

$$\begin{aligned} F &= -T\ln Z(T,\mu) \\ &= -\frac{gVT}{2\pi^2}\int d^3k[\ln Z(E;T,\mu) + \ln Z(E;T,-\mu)]\ . \end{aligned} \qquad (4.128)$$

It is more convenient to write the free energy in terms of the distribution function,

$$f(E;T,\mu) = (\exp[(E-\mu)/T]\mp 1)^{-1} \qquad (4.129)$$

and

$$f(E;T,-\mu) = (\exp[(E+\mu)/T]\mp 1)^{-1} \qquad (4.130)$$

for antiparticles. We note that

$$1\pm f(E;T,\mu) = f'(E;T,\mu) = \frac{1}{1\mp\exp[-(E-\mu)/T]} \qquad (4.131)$$

where the + is for bosons and the − for fermions. Then the free energy in Eq. (4.128) can also be written as

$$F = -\frac{gVT}{2\pi^2}\int d^3k[\ln f'(E;T,\mu) + \ln f'(E;T,-\mu)] \ . \tag{4.132}$$

The pressure is the most straightforward quantity to obtain since $P = -\partial F/\partial V$. Using the form for $\ln Z(E;T,\mu)$ in Eq. (4.127), we have

$$P = \frac{gV}{2\pi^2}\int_0^\infty \frac{dk\,k^4}{3E}[f(E;T,\mu) + f(E;T,-\mu)] \tag{4.133}$$

We now turn to the number density. We note that, while $\mu = 0$ in some cases, the number density in those cases does not likewise vanish. Indeed, it can be quite large. The number density may be calculated by first taking the derivative of the free energy with respect to μ and then, if needed, letting $\mu \to 0$. Starting from Eq. (4.125), we have, for particles only,

$$\begin{aligned} n &= -\frac{\partial F}{\partial \mu} = \mp\frac{gT}{2\pi^2}\int dE\,k\,E\,\frac{\partial \ln f'(E;T,\mu)}{\partial \mu} \\ &= \mp\frac{gT}{2\pi^2}\int \frac{dE\,k\,E}{T}\frac{\exp[-(E-\mu)/T]}{\exp[-(E-\mu)/T]\mp 1} \\ &= \frac{g}{2\pi^2}\int \frac{dE\,k\,E}{\exp[(E-\mu)/T]\mp 1} \\ &= \frac{g}{2\pi^2}\int dk\,k^2 f(E;T,\mu) \ . \end{aligned} \tag{4.134}$$

To calculate the total number of particles and antiparticles, we add the contributions from Eq. (4.134) to its antiparticle counterpart with $f(E;T,-\mu)$. We note, however, that since the antiparticle chemical potential is of opposite sign, the net particle number would be

$$n_{\rm net} = \frac{g}{2\pi^2}\int dk\,k^2\,[f(E;T,\mu) - f(E;T,-\mu)] \ . \tag{4.135}$$

One typically expects a net baryon number density in the system.

We can now calculate the energy density since it depends on n,

$$\begin{aligned}
\epsilon &= -gT^2\frac{\partial(F/T)}{\partial T} + \mu n \\
&= \frac{gT^2}{2\pi^2}\int dE\, k\, E\frac{\partial}{\partial T}[\ln f'(E;T,\mu) + \ln f'(E;T,-\mu)] \\
&\quad + \frac{g}{2\pi^2}\int dE\, k\, E\,[\mu f(E;T,\mu) - \mu f(E;T,-\mu)] \\
&= \frac{g}{2\pi^2}\int dE\, k\, E\,[(E-\mu)f(E;T,\mu) + (E+\mu)f(E;T,-\mu) \\
&+ \quad \mu f(E;T,\mu) - \mu f(E;T,-\mu)]\ . \qquad (4.136)
\end{aligned}$$

The terms proportional to μ in Eq. (4.136) cancel each other, leaving

$$\epsilon = \frac{g}{2\pi^2}\int dk\, k^2\, E\,[f(E;T,\mu) + f(E;T,-\mu)]\ . \qquad (4.137)$$

Note that for massless particles, $E = k$ in Eqs. (4.133) and (4.137), so that $\epsilon = 3P$.

The entropy density is somewhat more complicated. To simplify the calculation, we will only show the result for particles. The antiparticle contribution can be derived identically. We note that the relationship between f and f' given in Eq. (4.131) also implies

$$f(E;T,\mu) = \exp[-(E-\mu)/T]f'(E;T,\mu)\ . \qquad (4.138)$$

From Eq. (4.131) we also note that $f'(E;T,\mu) \mp f(E;T,\mu) = 1$. Then

$$\begin{aligned}
s &= \frac{\partial(T\ln Z)}{\partial T} = \frac{g}{2\pi^2}\int dk\, k^2\frac{\partial(T\ln f'(E;T,\mu))}{\partial T} \\
&= \frac{g}{2\pi^2}\int dk\, k^2\left(\ln f'(E;T,\mu) - f(E;T,\mu)\left(\frac{E-\mu}{T}\right)\right)
\end{aligned}$$

where the second term in Eq. (4.139) is the derivative of $f'(E;T,\mu)$ with respect to T. Multiplying the first term by unity in the form $f'(E;T,\mu) \mp f(E;T,\mu)$, we then have

$$\begin{aligned}
s &= \frac{g}{2\pi^2}\int dk\, k^2\Bigg(f'(E;T,\mu)\ln f'(E;T,\mu) \\
&\quad \mp f(E;T,\mu)\left(\ln f'(E;T,\mu) - \frac{E-\mu}{T}\right)\Bigg)\ . \qquad (4.139)
\end{aligned}$$

Equation (4.138) tells us that

$$\ln f(E;T,\mu) = \ln f'(E;T,\mu) - \left(\frac{E-\mu}{T}\right) . \tag{4.140}$$

After inserting Eq. (4.140) into Eq. (4.139), we obtain the final result for the particle entropy density,

$$\begin{aligned} s &= \frac{g}{2\pi^2}\int dkk^2[f'(E;T,\mu)\ln f'(E;T,\mu) \\ &\quad \mp f(E;T,\mu)\ln f(E;T,\mu)] . \end{aligned} \tag{4.141}$$

The expression for the antiparticle entropy density is the same except for the sign change of μ.

4.4.1 Case I: massless partons, massless pions

In this case, the integral in Eq. (4.124) can be calculated exactly for fermions and for bosons with zero chemical potential, $\mu = 0$. To do the calculations, we bring the temperature factor in the denominator of the right-hand side of Eq. (4.60) and calculate $T\ln Z$. In the massless case, we have

$$\begin{aligned} T\ln Z(T,\mu,V) &= \frac{gV}{6\pi^2}\int_0^\infty dk\, k^3 \\ &\times \left[\frac{1}{\exp[(k-\mu)/T]\mp 1} + \frac{1}{\exp[(k+\mu)/T]\mp 1}\right] . \end{aligned} \tag{4.142}$$

We begin with bosons. We can take $\mu = 0$ for bosons since there is no limit on the boson number. The result for massless bosons is

$$T\ln Z|_{\text{bosons}} = \frac{g_b V}{90}\pi^2 T^4 \tag{4.143}$$

EXAMPLE: Show this.
When $\mu = 0$, we begin with

$$T\ln Z = \frac{g_b V}{6\pi^2}\int_0^\infty \frac{dk\, k^3}{\exp(k/T)-1} . \tag{4.144}$$

We first make a change of variable in the integral from k to $y = k/T$ so that the integral becomes

$$T \ln Z = \frac{g_b V}{6\pi^2} T^4 \int_0^\infty \frac{dy\, y^3}{\exp(y) - 1} \,. \tag{4.145}$$

To simplify the integration, we rewrite the Bose-Einstein distribution as

$$\begin{aligned} \frac{1}{\exp(y)-1} &= \frac{\exp(-y)}{1-\exp(-y)} \\ &= \exp(-y)[1 + \exp(-y) + \exp(-2y) + \cdots] \,. \end{aligned} \tag{4.146}$$

The second equality in Eq. (4.146) is obtained by Taylor expansion of the denomination. We can then replace the denominator in Eq. (4.145) by the sum of resulting series of exponentials,

$$\frac{1}{\exp(y)-1} = \exp(-y) \sum_{n=0}^{\infty} \exp(-ny) \,. \tag{4.147}$$

The integral then becomes

$$T \ln Z = \frac{g_b V}{6\pi^2} T^4 \sum_{n=0}^{\infty} \int_0^\infty dy\, y^3 \exp(-(n+1)y) \,. \tag{4.148}$$

Successively integrating by parts four times gives

$$\begin{aligned} T \ln Z &= \frac{g_b V}{6\pi^2} T^4 \sum_{n=0}^{\infty} \frac{3}{n+1} \int_0^\infty dy y^2 \exp[-(n+1)y] \\ &= \frac{g_b V}{6\pi^2} T^4 \sum_{n=0}^{\infty} \frac{3 \cdot 2}{(n+1)^2} \int_0^\infty dy y \exp[-(n+1)y] \\ &= \frac{g_b V}{6\pi^2} T^4 \sum_{n=0}^{\infty} \frac{3 \cdot 2 \cdot 1}{(n+1)^3} \int_0^\infty dy \exp[-(n+1)y] \\ &= \frac{g_b V}{6\pi^2} T^4 \Gamma(4) \sum_{n=0}^{\infty} \frac{1}{(n+1)^4} \,. \end{aligned} \tag{4.149}$$

Since $n! = \Gamma(n+1)$, we have replaced $3 \cdot 2 \cdot 1$ by $\Gamma(4) = 3!$. The infinite sum over n of $1/(n+1)^4$ can be defined in terms of the Riemann zeta function,

$$\zeta(x) = \sum_{m=1}^{\infty} \frac{1}{m^x} \to \sum_{n=0}^{\infty} \frac{1}{(n+1)^x} \,. \tag{4.150}$$

The relevant values of the Riemann zeta function for our purposes are

$$\zeta(2) = \frac{\pi^2}{6}\,, \tag{4.151}$$

$$\zeta(3) = 1.20205....\ , \tag{4.152}$$

$$\zeta(4) = \frac{\pi^4}{90}\,. \tag{4.153}$$

We will find that $\zeta(2)$ appears in our calculation of the fermion contribution to the equation of state while $\zeta(3)$ will be a factor in the number density. The value of $\zeta(4)$ is inserted into Eq. (4.149) to obtain the result of Eq. (4.143).

Note that our discussion has been completely general and can apply to any massless bosons. We could refer to either a system of gluons or massless pions since both are bosons. The only difference is in the number of degrees of freedom, g_b. In the pion case, there exists 3 pion degrees of freedom, π^+, π^- and π^0 so that $g_\pi = 3$. On the other hand, the gluon has 8 colors and two spin degrees of freedom since there are only transverse gluons and not longitudinal ones. Thus $g_g = 2\times 8 = 16$.

We now turn to the case of massless fermions. Here we can only count the quark and antiquark contributions since there is no thermal scenario where we can consider the lightest baryon (fermion), the nucleon, as massless. We also cannot assume that the quark chemical potential, μ_q, will be equal to zero, $\mu_q = 0$, as in the boson case. We do, however, make the assumption that the quark and antiquark chemical potentials are equal to each other since $q\overline{q}$ pairs will be produced by interactions in the ideal gas we treat here. The integral we need to solve to obtain the equation of state for fermions is

$$T \ln Z = \frac{g_f V}{6\pi^2} \int_0^\infty dk\, k^3 \times \left[\frac{1}{\exp[(k-\mu_q)/T]+1} + \frac{1}{\exp[(k+\mu_q)/T]+1} \right]\,. \tag{4.154}$$

After integration, we will find

$$T \ln Z = g_f V \left\{ \frac{7\pi^2}{360} T^4 + \frac{\mu_q^2}{12} T^2 + \frac{\mu_q^4}{24\pi^2} \right\}\,. \tag{4.155}$$

EXAMPLE: Show this.

It is best to take one term at a time to begin with. We start with the quarks. First we change integration variables to $y = (k - \mu_q)/T$ so that the lower limit of the integral becomes $-\mu_q/T$ and $k = yT + \mu_q$. The integral over y from $-\mu_q/T$ to ∞ can be broken into two separate integrals, one from $-\mu_q/T$ to 0 and the other from 0 to ∞, as before. Thus the integral over the quark part is

$$\int_0^\infty \frac{dk\, k^3}{\exp[(k-\mu_q)/T]+1} = T\left\{\int_{-\mu_q/T}^{0} \frac{dy(\mu_q+yT)^3}{\exp(y)+1} + \int_0^\infty \frac{dy(\mu_q+yT)^3}{\exp(y)+1}\right\} \tag{4.156}$$

The antiquarks may be treated similarly. In this case, we change integration variables to $y = (k + \mu_q)/T$ so that the lower limit of the integral becomes μ_q/T and $k = yT - \mu_q$. Now when we separate the integral into two parts, we have a difference instead of a sum since we subtract an integral over 0 to μ_q/T from an integral over 0 to ∞ so that

$$\int_0^\infty \frac{dk\, k^3}{\exp[(k+\mu_q)/T]+1} = T\left\{\int_0^\infty \frac{(yT-\mu_q)^3\, dy}{\exp(y)+1} - \int_0^{\mu_q/T} \frac{(yT-\mu_q)^3\, dy}{\exp(y)+1}\right\}. \tag{4.157}$$

To make progress, we would like to make the antiquark integrals look more like the quark integrals. To do this, note that, by adding and subtracting $\exp(y)$ in the numerator of $1/(\exp(y) + 1)$, after some manipulation this fraction becomes $1-1/(\exp(-y)+1)$. Then the right-hand side of Eq. (4.157) is

$$T\left\{\int_0^\infty \frac{dy\,(yT-\mu_q)^3}{\exp(y)+1} - \left(\int_0^{\mu_q/T} dy\,(yT-\mu_q)^3 - \int_0^{\mu_q/T} \frac{dy\,(yT-\mu_q)^3}{\exp(-y)+1}\right)\right\}. \tag{4.158}$$

We now make the substitution $y \to -y$ so that $dy \to -dy$ and $(yT - \mu_q)^3 \to (-yT - \mu_q)^3 = -(yT + \mu_q)^3$. The upper limits of the finite-range integrals then becomes $-\mu_q/T$. The last step is to switch the limits and change the sign of the integrals so that we finally have

$$T\Big\{-\int_{-\mu_q/T}^{0}\frac{dy\,(yT+\mu_q)^3}{\exp(y)+1}+ \int_0^\infty\frac{dy\,(yT-\mu_q)^3}{\exp(y)+1}+\int_{-\mu_q/T}^{0}dy\,(yT+\mu_q)^3\Big\}\,. \tag{4.159}$$

We have arranged the terms in Eq. (4.159) so that the first two terms coincide with similar terms in Eq. (4.156). The first term in Eq. (4.159) then cancels the first term in Eq. (4.156) when we sum the quark and antiquark contributions. We are left with the combined result

$$T\ln Z=\frac{g_fV}{6\pi^2}\Big\{\int_{-\mu_q/T}^{0}dy\,(\mu_q+yT)^3+ \int_0^\infty dy\,\frac{[(yT+\mu_q)^3+(yT-\mu_q)^3]}{\exp(y)+1}\Big\}\,. \tag{4.160}$$

The first integral is straightforward, giving

$$\int_{-\mu_q/T}^{0}dy\,(\mu_q+yT)^3=2\frac{\mu_q^4}{4T}\,. \tag{4.161}$$

The sum of the two terms in the second integral leaves

$$2T\int_0^\infty dy\,\frac{T^3y^3+3\mu_q^2Ty}{\exp(y)+1}\,. \tag{4.162}$$

We treat the $1/(\exp(y)+1)$ term similarly to the expansion in Eq. (4.147) except that now we have an expansion with an alternating sign with every other term is multiplied by -1, giving instead

$$\frac{1}{\exp(y)+1}=\sum_{n=0}^{\infty}(-1)^n\exp[-(n+1)y]\,. \tag{4.163}$$

After this expansion, we are left with

$$2T\sum_{n=0}^{\infty}(-1)^n\int_0^\infty dy\,(T^3y^3+3\mu_q^2Ty)\exp[-(n+1)y]\,. \tag{4.164}$$

The first integral in Eq. (4.164), proportional to y^3, is the same as that in Eq. (4.148) for the boson case except for the $(-1)^n$ factor and will be proportional to $\zeta(4)$, as before. The second, proportional to y, involves only two integrations by parts and will be proportional to $\zeta(2)$. The factor of $(-1)^n$ will make the integrals only proportional to the zeta function, not equal to it. We use the identity

$$\sum_{n=0}^{\infty}(-1)^n \frac{1}{(n+1)^x} = 1 - \frac{1}{2^x} + \frac{1}{3^x} - \frac{1}{4^x} + \cdots = \left(1 - \frac{2}{2^x}\right)\zeta(x) \quad (4.165)$$

to evaluate the sums. For $x = 4$, the result is $(1 - 1/8)\zeta(4) = 7\zeta(4)/8$ while for $x = 3$, $(1 - 1/4)\zeta(3) = 3\zeta(3)/4$ and for $x = 2$, $(1 - 1/2)\zeta(2) = \zeta(2)/2$. Then Eq. (4.164) becomes

$$\begin{aligned}
2T^4\Gamma(4)\sum_{n=0}^{\infty}(-1)^n \frac{1}{(n+1)^4} + 6\mu_q^2 T^2\Gamma(2)\sum_{n=0}^{\infty}(-1)^n \frac{1}{(n+1)^2} &= \\
2 \cdot 6T^4\left(1 - \frac{2}{2^4}\right)\zeta(4) + 6\mu_q^2 T^2\left(1 - \frac{2}{2^2}\right)\zeta(2) & \\
= \frac{7}{8}\frac{12}{90}\pi^4 T^4 + \frac{\pi^2}{2}\mu_q^2 T^2 \, . & \quad (4.166)
\end{aligned}$$

Putting Eqs. (4.161) and (4.166) together, we obtain Eq. (4.155).

The number of quark degrees of freedom, g_f, depends on the number of 'active' quarks in the system. For example, in the initial state, if the temperature is relatively low, less than the mass of the strange quark, $T < m_s$, then only the up and down flavors are active. However, if $T > m_s$, the strange quark could also be counted as massless. Thus g_f depends on n_f, the number of active quark flavors. In addition, there are two spin states for fermions and three colors. We do not count quarks and antiquarks as separate degrees of freedom here because the partition function, Eq. (4.155), already includes both contributions. For two active quark flavors, $g_q = 2(\text{flavors}) \times 2(\text{spins}) \times 3(\text{colors}) = 12$.

Summing the quark, antiquark and gluon contributions to the quark-gluon gas, we obtain the total partition function for a system of massless fermions and bosons,

$$\begin{aligned}
T \ln Z|_{\rm qgp} &= \frac{g_g V}{90\pi^2}T^4 + \frac{g_f V}{12}\left(\frac{7\pi^2}{30}T^4 + \mu_q^2 T^2 + \frac{1}{2\pi^2}\mu_q^4\right) \\
&= V\left(\frac{37\pi^2}{90}T^4 + \mu_q^2 T^2 + \frac{\mu_q^4}{2\pi^2}\right) . \quad (4.167)
\end{aligned}$$

On the hadron side, we account only for massless pions with

$$T \ln Z|_{\pi} = \frac{\pi^2 V}{30} T^4 \; . \tag{4.168}$$

From the partition function, we can calculate the thermodynamic properties of the quark-gluon system. We begin with the net quark number density, calculated as in Eq. (4.62),

$$n_q = \frac{T}{V} \frac{\partial \ln Z|_{\rm qgp}}{\partial \mu_q} = 2\mu_q (T^2 + \frac{\mu_q^2}{\pi^2}) \; . \tag{4.169}$$

There are three quarks per baryon so that $n_q = 3n_{\rm bar}$.

To calculate the total number density in an ideal gas of massless particles, we use Eq. (4.134). For massless gluons with $\mu = 0$, we have

$$n_g = \frac{g_g}{2\pi^2} \int \frac{dk\, k^2}{\exp(k/T) - 1} \; . \tag{4.170}$$

Changing variables and Taylor expanding the denominator, we find

$$n_g = \frac{g_g T^3}{2\pi^2} \sum_{n=0}^{\infty} \int dy\, y^2 \exp[-(n+1)y] \; . \tag{4.171}$$

After integration, we are left with

$$n_g = \frac{g_g T^3}{2\pi^2} \Gamma(3) \sum_{n=0}^{\infty} \frac{1}{(n+1)^3} = \frac{g_g T^3}{2\pi^2} \Gamma(3) \zeta(3) \; . \tag{4.172}$$

Finally, inserting $\Gamma(3) = 2$ and $g_g = 16$,

$$n_g = \frac{16\zeta(3)}{\pi^2} T^3 \; . \tag{4.173}$$

The quark and antiquark contributions are obtained from

$$\begin{aligned} n_q &= \frac{g_f T}{2\pi^2} \Bigg\{ \int_{-\mu_q/T}^{0} dy\, (\mu_q + yT)^2 \\ &+ \int_0^{\infty} dy\, \frac{[(yT + \mu_q)^2 + (yT - \mu_q)^2]}{\exp(y) + 1} \Bigg\} \; , \end{aligned} \tag{4.174}$$

following the same steps beginning with Eq. (4.134) to obtain the analogy of Eq. (4.160). The first term in Eq. (4.174) yields

$$\int_{-\mu_q/T}^{0} dy\,(\mu_q + yT)^2 = \frac{\mu_q^3}{3T}\,.$$

The second term,

$$2\int_0^\infty dy\,\frac{y^2T^2+\mu_q^2}{\exp(y)+1}\,,$$

becomes, similar to Eq. (4.166) but with one lower power of y,

$$2\Gamma(3)T^2\sum_{n=0}^{\infty}\frac{(-1)^n}{(n+1)^3} + 2\mu_q^2\sum_{n=0}^{\infty}\frac{(-1)^n}{n+1}\,.$$

The sum in the first term gives $(1-1/4)\zeta(3)$ while the second gives zero since $1-2/2=0$ from Eq. (4.165). Thus the total number of quarks and antiquarks is

$$n_q = \frac{g_f}{\pi^2}\left(\frac{3\zeta(3)}{2}T^3 + \frac{\mu_q^3}{6}\right)\,. \tag{4.175}$$

For a 2-quark flavor plasma, $g_f = 12$, as before, giving

$$n_q = \frac{18\zeta(3)}{\pi^2}T^3 + 2\mu_q^3\,. \tag{4.176}$$

The total parton number density (quarks, antiquarks and gluons) in the quark-gluon plasma is

$$n_{\rm qgp} = \frac{34\zeta(3)}{\pi^2}T^3 + 2\mu_q^3\,. \tag{4.177}$$

The energy density is calculated according to Eq. (4.64), giving

$$\begin{aligned}\epsilon_{\rm qgp} &= \frac{T^2}{V}\frac{\partial}{\partial T}\ln Z|_{\rm qgp} + \mu_q n_q \\ &= T^2\left(\frac{37\pi^2}{30}T^2 + \mu_q^2 - \frac{\mu_q^4}{2\pi^2T^2}\right) + 2\left(\mu_q^2T^2 + \frac{\mu_q^4}{\pi^2}\right) \\ &= \frac{37\pi^2}{30}T^4 + 3\mu_q^2T^2 + \frac{3}{2}\frac{\mu_q^4}{\pi^2}\,. \end{aligned} \tag{4.178}$$

We calculate the pressure from Eq. (4.74), $P_{\rm qgp} = T \ln Z|_{\rm qgp}/V$, to find

$$P_{\rm qgp} = \frac{37\pi^2}{90} T^4 + \mu_q^2 T^2 + \frac{\mu_q^4}{2\pi^2} \ . \tag{4.179}$$

Note that this is just $\epsilon_{\rm qgp}/3$, as might be expected from the definition of the speed of sound in medium, $dP/d\epsilon = c_s^2 = 1/3$, where the last equality is the result for an ideal gas.

Finally, the entropy density is

$$s_{\rm qgp} = \frac{\partial(T \ln Z|_{\rm qgp})}{\partial T} = \frac{74\pi^2}{45} T^3 + 2\mu_q T \ . \tag{4.180}$$

The ratio of the terms proportional to T^3 in the entropy and number densities is

$$\frac{s_{\rm qgp}}{n_{\rm qgp}} = \frac{74\pi^4}{45 \cdot 34\zeta(3)} = 3.92 \ . \tag{4.181}$$

Since other values for the boson and fermion degrees of freedom would give different ratios, we can write, for a massless gas with $\mu = 0$,

$$s = \xi n \ . \tag{4.182}$$

In this case, $\xi = 3.92$.

On the hadron side, the energy density, pressure, number density and entropy density are

$$\epsilon_\pi = \frac{\pi^2}{10} T^4 \ , \tag{4.183}$$

$$P_\pi = \frac{\pi^2}{30} T^4 \ , \tag{4.184}$$

$$n_\pi = \frac{3\zeta(3)}{\pi^2} T^3 \ , \tag{4.185}$$

$$s_\pi = \frac{2\pi^2}{15} T^3 \ . \tag{4.186}$$

Note that in the case of the pion gas, the ratio of the entropy to number densities gives $\xi = 3.6$, only $\sim 9\%$ smaller than the quark-gluon plasma ratio even though the number of degrees of freedom in a plasma is larger.

Comparing the pressure of the quark-gluon gas, Eq. (4.179), and that of the massless pion gas, Eq. (4.184), it is clear that the criteria for a phase transition, equal pressure across the transition region, will not be satisfied by these contributions alone. Another term is needed to equalize the pressure and make the transition possible. This additional pressure can be obtained by assuming that the quarks, whether confined to a single hadron or to the larger quark-gluon plasma, are contained within a field of constant, positive potential energy per unit volume, referred to as B for 'bag'. This potential energy was introduced in the MIT model of hadron structure, referred to as the 'MIT bag model' [66].

The bag model includes both asymptotic freedom and confinement, the two main properties of QCD. Inside the bag, as long as the quark constituents of the hadron do not come too near the walls of the bag, they can be treated as free fields, the property of 'asymptotic freedom'. However, when the quarks approach the bag surface, they are reflected back into the bag interior and so are confined to the bag interior, hence 'confinement'.

The main idea is that, wherever quark and gluon fields exist, such as inside a hadron, there is also a positive energy density, B, that sets the confinement scale. The world of the bag model has two phases: the vacuum, which expels quarks and gluons, and the hadrons, where the quarks and gluons are quasi-free in the interior. The term quasi-free is used because the quarks and gluons are ultimately confined to the interior of the bag. Dynamically, B is a pressure and may be thought of as the latent heat released when a hadron returns to the vacuum. Thus field theory inside a bag is that of a theory restricted inside a finite spatial domain by a pressure.

Any field theory has equations of motion related to its Lagrangian where, classically, $L = T - V$ with T the kinetic and V the potential energy of the theory. In the bag, the equations of motion are modified by the presence of the pressure relative to the free field but must still obey energy and momentum conservation. The equations of motion are obtained from the energy-momentum tensor, $T^{\mu\nu}$. In the free field theory, $T^{\mu\nu} = T_0^{\mu\nu}$. The bag pressure introduces a term $-g^{\mu\nu}B$, giving us overall

$$T_{\text{bag}}^{\mu\nu} = (T_0^{\mu\nu} - g^{\mu\nu}B)\theta_V \tag{4.187}$$

where the theta function, θ_V, is zero outside the bag but equal to unity inside. There are boundary conditions that must be satisfied both in the interior of the bag (inside the volume V) and on the surface of the bag (surface area δV). Inside the bag, the equations of motion are the same as for the free field because the bag pressure B is constant. The quarks in the bag are fermions and obey the Dirac equation,

$$i\partial\!\!\!/\psi(x) = 0 \ , \tag{4.188}$$

when the quarks are massless. Energy and momentum conservation specifies that

$$\partial_\mu T_{\text{bag}}^{\mu\nu}(x) = 0 \ . \tag{4.189}$$

The derivative of $T_{\text{bag}}^{\mu\nu}$ includes both the derivative of the dynamical contributions and the derivative of the theta function at the surface,

$$\partial_\mu T_{\text{bag}}^{\mu\nu} = \partial_\mu(T_0^{\mu\nu} - g^{\mu\nu}B)\theta_V + (T_0^{\mu\nu} - g^{\mu\nu}B)\partial_\mu\theta_V \ . \tag{4.190}$$

EXAMPLE: Show that the first term in Eq. (4.190) vanishes.
The Lagrangian for a massless, noninteracting Dirac field is

$$L_0 = \frac{i}{2}[\overline{\psi}\gamma^\lambda\partial_\lambda\psi - (\partial_\lambda\overline{\psi})\gamma^\lambda\psi] = \frac{i}{2}[\overline{\psi}\partial\!\!\!/\psi - \partial\!\!\!/\overline{\psi}\psi] \tag{4.191}$$

where ψ and $\overline{\psi}$ are the fermion and anti-fermion fields, ∂_λ is the covariant derivative and γ^λ are the Dirac γ matrices. (For more detailed discussions of Dirac fields, see Ref. [4].) The energy-momentum tensor for the Dirac field is

$$\begin{aligned} T_0^{\mu\nu} &= \frac{\partial L_0}{\partial(\partial^\mu\psi)}\partial^\nu\psi + \frac{\partial L_0}{\partial(\partial^\mu\overline{\psi})}\partial^\nu\overline{\psi} - g^{\mu\nu}L_0 \\ &= \frac{i}{2}\overline{\psi}\gamma^\mu\partial^\nu\psi - \frac{i}{2}(\partial^\nu\overline{\psi})\gamma^\mu\psi - g^{\mu\nu}L_0 \ . \end{aligned} \tag{4.192}$$

The L_0 contribution to $T_0^{\mu\nu}$ is equal to zero by virtue of the Dirac equation, Eq. (4.188), since $\partial\!\!\!/\psi = 0$ and $\partial\!\!\!/\overline{\psi} = 0$. The derivative of the remaining part of $T_0^{\mu\nu}$ is

$$\begin{aligned}\partial_\mu T_0^{\mu\nu} &= \frac{i}{2}(\partial_\mu\overline{\psi}\gamma^\mu\partial^\nu\psi + \overline{\psi}\partial_\mu\gamma^\mu\partial^\nu\psi - \partial_\mu\partial^\nu\overline{\psi}\gamma^\mu\psi - \partial^\nu\overline{\psi}\partial_\mu\gamma^\mu\psi) \\ &= \frac{i}{2}(\partial\!\!\!/\overline{\psi}\partial^\nu\psi + \overline{\psi}\partial\!\!\!/\partial^\nu\psi - \partial\!\!\!/\overline{\psi} - \partial^\nu\overline{\psi}\partial\!\!\!/\psi)\ . \end{aligned} \tag{4.193}$$

The first and last terms of Eq. (4.193) are zero by the Dirac equation. The other two terms are also zero using the Dirac equation after the substitution $\partial\!\!\!/\partial^\nu = \partial^\nu\partial\!\!\!/$ is made. Finally, since B is a constant, $\partial_\mu B = 0$.

The derivative of θ_V at the surface is equal to $\mathbf{n}_\mu\delta_V$ where $\mathbf{n}_\mu$ is the interior covariant unit vector normal to the surface of V and δ_V is zero away from the surface. Finally, to satisfy energy-momentum conservation, we need

$$\mathbf{n}_\mu(T_0^{\mu\nu} - g^{\mu\nu}B) = 0 \tag{4.194}$$

on the surface of the bag. Using the expression for $T_0^{\mu\nu}$ in Eq. (4.192), the boundary condition on the surface becomes

$$\frac{i}{2}(\overline{\psi}\mathbf{n}\!\!\!/\partial^\nu\psi - \partial^\nu\overline{\psi}\mathbf{n}\!\!\!/\psi) - \mathbf{n}^\nu B = 0\ . \tag{4.195}$$

The boundary condition given above is not unique because, for any given combination of gamma matrices, Γ, we can let

$$i\mathbf{n}\!\!\!/\psi = \Gamma\psi \tag{4.196}$$

on the surface. It is also true that [67]

$$\Gamma\mathbf{n}\!\!\!/\Gamma = \mathbf{n}^2\mathbf{n}\!\!\!/ = \mathbf{n}\!\!\!/\ . \tag{4.197}$$

The local rest frame defines $\mathbf{n}^\mu = (0, \vec{n})$ with $\mathbf{n}^2 = 1$. The most general Γ that satisfies Eqs. (4.196) and (4.197) is $\Gamma = \exp(i\alpha\gamma_5/2)$ where γ_5 is a product of gamma matrices that anticommutes with other gamma matrices, $\{\gamma_5, \gamma^\mu\} = 0$.

EXAMPLE: Given the anticommuting property of γ_5, show that Eq. (4.197) is satisfied by $\Gamma = \exp(i\alpha\gamma_5/2)$.

We expand Γ, keeping only the first two terms in the expansion since the rest will not contribute to the product in Eq. (4.197) so that $\Gamma = 1 + i\alpha\gamma_5/2 + \cdots$. We also write $\not{\mathbf{n}}$ as $\mathbf{n}_\mu\gamma^\mu$ on the left-hand side of Eq. (4.197). Then

$$\begin{aligned}
\Gamma\not{\mathbf{n}}\Gamma &= \left(1 + \frac{i\alpha}{2}\gamma_5 + \cdots\right)\mathbf{n}_\mu\gamma^\mu\left(1 + \frac{i\alpha}{2}\gamma_5 + \cdots\right) \\
&= \mathbf{n}^\mu\gamma^\mu + \frac{i\alpha}{2}\mathbf{n}_\mu(\gamma_5\gamma^\mu + \gamma^\mu\gamma_5) \\
&= \not{\mathbf{n}} + \frac{i\alpha}{2}\mathbf{n}_\mu\{\gamma_5, \gamma^\mu\} \\
&= \not{\mathbf{n}} \; .
\end{aligned}$$

as in Eq. (4.197).

The bag should also satisfy current conservation, $\partial_\mu \mathbf{j}^\mu = 0$ where, for fermion fields,

$$\mathbf{j}^\mu = \overline{\psi}\gamma^\mu\psi \; , \tag{4.198}$$

in general and

$$\mathbf{j}^\mu_{\text{bag}} = \overline{\psi}\gamma^\mu\psi\theta_V \tag{4.199}$$

on the bag surface. Thus current conservation implies

$$\begin{aligned}
\partial_\mu \mathbf{j}^\mu_{\text{bag}} &= [\partial_\mu(\overline{\psi}\gamma^\mu\psi)]\theta_V + \overline{\psi}\gamma^\mu\psi\partial_\mu\theta_V \\
&= \overline{\psi}\not{\mathbf{n}}\psi = 0
\end{aligned} \tag{4.200}$$

on the surface of the bag. The derivative of Eq. (4.198) is zero,

$$\begin{aligned}
\partial_\mu \mathbf{j}^\mu &= (\partial_\mu\overline{\psi})\gamma^\mu\psi + \overline{\psi}\gamma^\mu\partial_\mu\psi \\
&= (\not{\partial}\overline{\psi})\psi + \overline{\psi}(\not{\partial}\psi) = 0
\end{aligned} \tag{4.201}$$

by the Dirac equation for massless particles.

We multiply the left-hand sides of Eq. (4.196) by $\overline{\psi}$ so that

$$i\overline{\psi}\not{\mathbf{n}}\psi = \overline{\psi}\Gamma\psi \; . \tag{4.202}$$

From the arguments of current conservation we have $\overline{\psi}\not{\mathbf{n}}\psi = 0$ in Eq. (4.200) which tells us that, on the surface of the bag,

$$\overline{\psi}\Gamma\psi = \overline{\psi}\exp(i\alpha\gamma_5/2)\psi = 0 \; . \tag{4.203}$$

The original assumption, Eq. (4.196), gives us

$$i\not{n}\psi = \exp(i\alpha\gamma_5/2)\psi \tag{4.204}$$

while Eq. (4.195) gives us another boundary condition. The first term in Eq. (4.195) becomes zero using the Dirac equation after rewriting $\not{n}\partial^\nu\psi$ as $\mathbf{n}^\nu\not{\partial}\psi = 0$. Multiplying both sides of Eq. (4.195) by $n_n u$ and using Eq. (4.204), we find

$$-\frac{1}{2}(\mathbf{n}\cdot\partial\overline{\psi})\exp(i\alpha\gamma_5/2)\psi = B\ . \tag{4.205}$$

There are many solutions to Eqs. (4.204) and (4.205) depending on the value of α. The set of solutions is over-complete if α is arbitrary but, by choosing a value of α, we violate chiral symmetry which says that if α is a solution to the equations, α' is an equally good solution. For the basic bag model, chiral symmetry is violated and $\alpha \equiv 0$ so that

$$i\not{n}\psi = \psi\ , \tag{4.206}$$

$$\frac{1}{2}(\mathbf{n}\cdot\partial\overline{\psi})\psi = B\ . \tag{4.207}$$

We will discuss more about chiral symmetry in a later chapter.

Now that we have the boundary conditions for the behavior of the quark fields inside and at the surface of the bag, we discuss the basic model of hadron structure in the bag. The energy of the quarks inside the bag includes several terms that are proportional to $1/R$ where R is the bag radius. These terms include the kinetic energy, the zero-point energy, spin-orbit energy and a chromoelectric energy, all of which we represent by $E_{\rm kin} = C_H/R$ where C_H is a constant containing the coefficients of all the $1/R$ terms. We write C_H since, although C_H is constant, it is a different constant for each hadron. There is another term that contributes to the energy, as we have already mentioned. The internal pressure of the quarks is balanced by an external pressure to keep the bag from collapsing,

$$E_{\rm vac} = BV = \frac{4\pi}{3}BR^3\ . \tag{4.208}$$

The total energy of the bag, the hadron mass in the MIT bag model, is the sum of all contributions, abbreviated here as

$$M_H = E_{\text{bag}} = E_{\text{vac}} + E_{\text{kin}} = \frac{4\pi}{3} B R^3 + \frac{C_H}{R} . \tag{4.209}$$

The bag radius is obtained by the condition

$$\frac{\partial E_{\text{bag}}}{\partial R} = 0 \tag{4.210}$$

which arises from Eq. (4.205), the bag boundary condition on the surface.

The MIT bag model, as we stated earlier, was proposed as a model of hadron structure. Thus the unknown zero point energy, along with the bag constant, B, were fixed by fitting the lower lying hadron masses and using these values to predict the mass spectra of other hadrons. The bag radius, determined from the condition in Eq. (4.210), is

$$R = \left(\frac{C_H}{4\pi B} \right)^{1/4} \tag{4.211}$$

so that $B^{1/4}$ has units of energy. This value of R can be then put back into Eq. (4.209) to fix the hadron mass.

We have spent some time on the discussion of the bag model because the quark-gluon plasma must satisfy asymptotic freedom and, ultimately, confinement. We do not observe free quarks coming from the plasma in a heavy-ion collision. Thus the plasma can be thought of as being inside a larger version of the bag and we add a term to the free energy proportional to the bag pressure,

$$T \ln Z|_{\text{vac}} = -BV . \tag{4.212}$$

The pressure from the bag is

$$P_{\text{vac}} = \frac{T}{V} \ln Z|_{\text{vac}} = -B . \tag{4.213}$$

Thus the pressure in the quark-gluon plasma is

$$\begin{aligned} P_{\text{QGP}} &= P_{\text{qgp}} + P_{\text{vac}} \\ &= \frac{37\pi^2}{90} + \mu_q^2 T^2 + \frac{\mu_q^4}{2\pi^2} - BV . \end{aligned} \tag{4.214}$$

The energy density due to the bag is then

$$\epsilon_{\text{vac}} = \frac{T^2}{V}\frac{\partial(-BV/T)}{\partial T} = B \tag{4.215}$$

so that the total energy density from the quark-gluon plasma is

$$\begin{aligned}\epsilon_{\text{QGP}} &= \epsilon_{\text{qgp}} + \epsilon_{\text{vac}} \\ &= \frac{37\pi^2}{30}T^4 + 3\mu_q^2T^2 + \frac{3}{2}\frac{\mu_q^4}{\pi^2} + B \; .\end{aligned} \tag{4.216}$$

The entropy density is unchanged by the presence of the bag since $\partial(T\ln Z|_{\text{vac}})/\partial T = 0$. Thus $s_{\text{QGP}} = s_{\text{qgp}}$.

With the additional term due to the bag pressure, it is possible to equalize the pressures at the critical temperature and make a first-order phase transition occur. At the critical point for a transition from a massless quark-gluon plasma to a gas of massless pions,

$$P_c = P_\pi = P_{\text{QGP}} \; , \tag{4.217}$$

$$T_c = T_\pi = T_{\text{QGP}} \; . \tag{4.218}$$

We have not included the chemical potential in this case because the pion chemical potential is equal to zero and, in that case, the quark-gluon plasma chemical potential would also have to be zero in the massless case.

We can calculate the critical temperature for a given B by setting $P_\pi = P_{\text{QGP}}$ with $\mu_q = 0$. In the case where the number of degrees of freedom on either side of the transition is unspecified,

$$\frac{g_\pi\pi^2}{90}T_c^4 = \frac{g_{qg}\pi^2}{90}T_c^4 - B \; , \tag{4.219}$$

leading to

$$T_c = \left(\frac{90B}{(g_{qg} - g_\pi)\pi^2}\right)^{1/4} \; . \tag{4.220}$$

In the case of two massless quark flavors and a massless pion gas, $g_{qg} = 37$ and $g_\pi = 3$, leading to

$$T_c = \left(\frac{90}{34\pi^2}\right)^{1/4} B^{1/4} = 104 \text{ MeV} \tag{4.221}$$

for $B^{1/4} = 0.145$ GeV. Note that this value of T_c is rather low, lower than the transition temperature of 170 MeV obtained by lattice QCD studies. Turning the argument around, for a fixed $T_c = 170$ MeV, we can obtain the value of $B^{1/4}$ needed,

$$\begin{aligned} B &= \frac{34\pi^2}{90} T_c^4 , \\ B^{1/4} &= 0.236 \,\text{GeV} . \end{aligned} \tag{4.222}$$

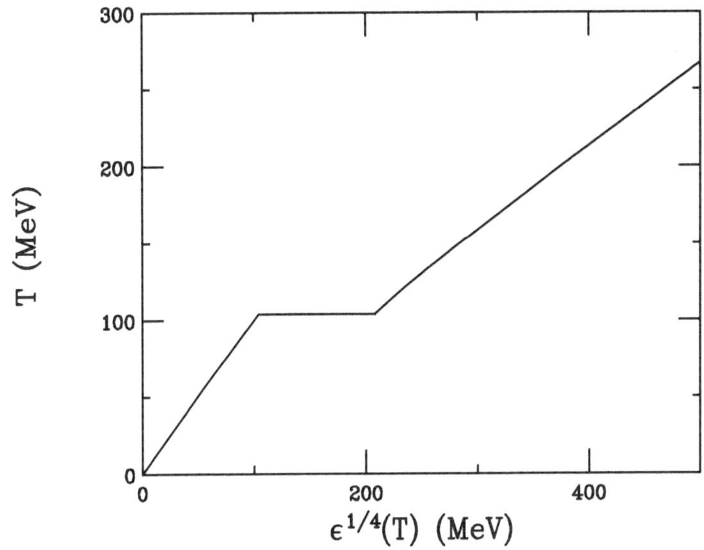

Figure 4.9: The temperature is shown versus energy density in a first-order phase transition with $T_c = 104$ MeV.

Figure 4.9 shows the one-fourth power of the energy density (x-axis) as a function of temperature (y-axis) for $T_c = 104$ MeV with $B^{1/4} = 0.145$ GeV. The transition region is the horizontal line at $T = T_c$. The calculation in the hadron phase is shown all the way down to $T = 0$ although, realistically, for $T < m_\pi$ the pion mass should be taken into account, as we discuss later. The MeV units on the x-axis are obtained by taking the one-fourth power of the energy density with ϵ in units of MeV^4. However, the energy density is also often expressed in units of MeV/fm^3 or GeV/fm^3, a more natural way of speaking about an energy per volume, obtained by dividing the result in units of MeV^4 by $(\hbar c)^3$.

Since this is a first-order transition, there is a finite latent heat in the system, defined in Eq. (4.92) as proportional to the increase in the entropy when one particle is transferred from one phase to the other. The energy density in a given phase is

$$\epsilon = Ts + \mu N - P \; . \tag{4.223}$$

In this case, the chemical potential is zero on both sides of the transition while the temperature and pressure remain fixed. Thus

$$\epsilon_{\rm QGP} - \epsilon_\pi = T(s_{\rm QGP} - s_\pi) \; , \tag{4.224}$$

so that the latent heat can also be expressed as the difference in the energy densities of the two phases:

$$L = \epsilon_{\rm QGP} - \epsilon_\pi = \frac{37\pi^2}{30} T_c^4 + B - \frac{3\pi^2}{30} T_c^4 \; . \tag{4.225}$$

for a two-flavor quark-gluon plasma. Inserting the result for T_c in Eq. (4.220), the latent heat of the system is

$$\begin{aligned} L &= \frac{34\pi^2}{30}\left(\frac{90}{34\pi^2}\right) B + B \\ &= 4B \; . \end{aligned} \tag{4.226}$$

The latent heat of the transition in the massless case is $L = 0.0124$ GeV^4 = 1.6 GeV/fm^3 for $B^{1/4} = 0.236$ GeV and $L = 0.00176$ GeV^4 = 0.23 GeV/fm^3 for $B^{1/4} = 0.145$ GeV.

4.4.2 Case II: zero temperature nucleon gas

The phase transition between a quark-gluon plasma and a nucleon gas at zero temperature is relevant for the interior of neutron stars where the density is high and the temperature low. While we would still have a massless quark-gluon gas on the plasma side, the pressure, energy density and number density would be reduced to

$$P_{\rm QGP} = \frac{\mu_q^4}{2\pi^2} - B\,, \tag{4.227}$$

$$\epsilon_{\rm QGP} = \frac{3\mu_q^4}{2\pi^2} + B\,, \tag{4.228}$$

$$n_{\rm QGP} = \frac{2\mu_q^3}{\pi^2}\,, \tag{4.229}$$

$$s_{\rm QGP} = 0\,. \tag{4.230}$$

The entropy density is zero when $T = 0$ since Eq. (4.180) has no temperature independent component.

On the hadron side at $T = 0$ there is a degenerate Fermi gas with nucleons only since there are no antibaryons in the system. Thus we keep only the particle term in Eq. (4.124). Since the nucleon mass, M, is much larger than the temperature, we cannot neglect the mass and keep $E = (k^2 + M^2)^{1/2}$. The expression we have to evaluate is

$$T \ln Z|_N = \frac{g_N V}{6\pi^2} \int_0^\infty \frac{dk k^4}{\sqrt{k^2+M^2}} \frac{1}{\exp[(\sqrt{k^2+M^2}-\mu)/T]+1}\,. \tag{4.231}$$

We will find

$$T \ln Z|_N = \frac{g_N V}{6\pi^2}\frac{M^4}{4}\left[r(r^2 - \frac{5}{2})\sqrt{r^2-1} + \frac{3}{2}\ln(r+\sqrt{r^2-1})\right] \tag{4.232}$$

where $r = \mu/M$.

EXAMPLE: Derive Eq. (4.232).
To do the integral, we first change variables from k to z where $z = \sqrt{k^2+M^2}$. Then $k = \sqrt{z^2-M^2}$ and $dk = zdz/(z^2-M^2)$ so that

$$T \ln Z|_N = \frac{g_N V}{6\pi^2} \int_0^\infty \frac{dz\,(z^2-M^2)^{3/2}}{\exp[(z-\mu)/T]+1}\,. \tag{4.233}$$

At low temperatures where $\mu \gg T$, we can apply the Sommerfeld expansion:

$$\begin{aligned} &\int_0^\infty \frac{dz\, f(z)}{1+\exp[(z-\mu)/T]} \\ &= \int_0^\mu f(z)\, dz + \frac{\pi^2 T^2}{6} \left.\frac{df}{dz}\right|_{z=\mu} + \frac{7\pi^4 T^4}{360} \left.\frac{d^3 f}{dz^3}\right|_{z=\mu} + \dots \end{aligned} \tag{4.234}$$

where $f(z) = (z^2 - M^2)^{3/2}$. When $T = 0$, only the first term remains. Note also that while the limits of integration on the right-hand side of Eq. (4.234) are $0 < z < \mu$, the lower bound on z must be M to obtain a result. Thus

$$T \ln Z|_N = \frac{g_N V}{6\pi^2} \int_M^\mu dz\, (z^2 - M^2)^{3/2} \,. \tag{4.235}$$

We can already see that, in this case, the nucleon chemical potential must be relatively large.

The indefinite integral of $(z^2 - M^2)^{3/2}$ can be found in any standard table of integrals. The result is

$$\begin{aligned} \int_M^\mu dz (z^2 - M^2)^{3/2} &= \frac{1}{8} \Big(-3M^2 z \sqrt{z^2 - M^2} + 2z(z^2 - M^2)^{3/2} \\ &\quad + 3M^4 \ln(z + \sqrt{z^2 - M^2}) \Big) \Big|_M^\mu \,. \end{aligned} \tag{4.236}$$

The first two terms on the right-hand side of Eq. (4.236) vanish for the lower limit, $z = M$, leaving

$$\begin{aligned} &\frac{1}{8} \Big(-3M^2 \mu \sqrt{\mu^2 - M^2} + 2\mu(\mu^2 - M^2)^{3/2} \\ &\quad + 3M^4 \ln \left(\frac{\mu + \sqrt{\mu^2 - M^2}}{M} \right) \Big) \,. \end{aligned} \tag{4.237}$$

There is a common factor of $(\mu^2 - M^2)^{1/2}$ in the first two terms. Factoring this out, these terms reduce to $\mu(2\mu^2 - 5M^2)(\mu^2 - M^2)^{1/2}$. After taking a common factor of $2M^4$ from Eq. (4.237), we are left with

$$\begin{aligned} &\frac{M^4}{4} \Big(\frac{\mu}{M} \left(\frac{\mu^2}{M^2} - \frac{5}{2} \right) \left(\frac{\mu^2}{M^2} - 1 \right)^{1/2} \\ &\quad + \frac{3}{2} \ln \left(\frac{\mu}{M} + \frac{\sqrt{\mu^2 - M^2}}{M} \right) \Big) \,. \end{aligned} \tag{4.238}$$

Replacing μ/M with r and substituting Eq. (4.238) for the integral in Eq. (4.233), we obtain Eq. (4.232).

The degeneracy factor, g_N, for the nucleon includes two degrees of freedom for spin and two for isospin (proton and neutron) so that $g_N = 2(\text{spin}) \times 2(\text{isospin}) = 4$. This factor of four cancels that in the denominator of Eq. (4.238).

We can now calculate the thermodynamic quantities, P_N, n_N and ϵ_N for a nucleon gas at $T = 0$. The pressure is simply $T \ln Z|_N/V$ or

$$P_N = \frac{M^4}{6\pi^2}[r(r^2-1)^{1/2}(r^2-\frac{5}{2}) + \frac{3}{2}\ln(r+\sqrt{r^2-1})] \ . \tag{4.239}$$

The number density is

$$n_N = \frac{T}{V}\frac{\partial \ln Z|_N}{\partial \mu} = \frac{2M^3}{3\pi^2}(r^2-1)^{3/2} \ . \tag{4.240}$$

EXAMPLE: Calculate n_N.
Since we have written $T\ln Z|_N$ in terms of $r = \mu/M$, we change the derivative with respect to μ to a derivate with respect to r,

$$n_N = \frac{T}{VM}\frac{\partial \ln Z|_N}{\partial r} \ . \tag{4.241}$$

The derivative of $T \ln Z|_N$ with respect to r gives

$$\begin{aligned} n_N &= \frac{M^3}{6\pi^2}\Bigg((r^2-\frac{5}{2})(r^2-1)^{1/2} + 2r^2(r^2-1)^{1/2} \\ &\quad + r^2(r^2-\frac{5}{2})(r^2-1)^{-1/2} + \frac{3}{2}\frac{1+r(r^2-1)^{-1/2}}{r+(r^2-1)^{1/2}}\Bigg) \ . \end{aligned} \tag{4.242}$$

Note that we have one less power of M in n_N due to the M in the denominator of the derivative with respect to r in Eq. (4.241). Removing a common factor of $(r^2-1)^{-1/2}$ from the expression, multiplying all terms and combining, we have

$$n_N = \frac{M^3}{6\pi^2}\frac{1}{\sqrt{r^2-1}}(4r^4 - 8r^2 + 4) \ . \tag{4.243}$$

The last factor above is just $4(r^2-1)^2$ which cancels the square root in the denominator, leaving us with

$$n_N = \frac{2M_3}{3\pi^2}(r^2-1)^{3/2} = \frac{2}{3\pi^2}(\mu^2 - M^2)^{3/2} \ . \tag{4.244}$$

The energy density is

$$\epsilon_N = \frac{T^2}{V}\frac{\partial \ln Z|_N}{\partial T} + \mu n_N \ . \tag{4.245}$$

However, since $T \ln Z|_N$ is not an explicit function of temperature, to obtain the energy density, we replace $\ln Z|_N$ in the derivative by VP_N/T. Then the first term in Eq. (4.245) is just $-P_N$ and the energy density becomes

$$\begin{aligned} \epsilon_N &= \mu n_N - P_N \\ &= \frac{2\mu}{3\pi^2}(\mu^2 - M^2)^{3/2} - P_N \ . \end{aligned} \tag{4.246}$$

We now turn to the transition between the quark-gluon plasma and the nucleon gas phases. The latent heat across the $T = 0$ transition is

$$L = \epsilon_{\text{QGP}} - \epsilon_N = \frac{3\mu_q^4}{2\pi^2} + B - \frac{2\mu}{3\pi^2}(\mu^2 - M^2)^{3/2} + P_N \ . \tag{4.247}$$

Since the pressure must be equal across the transition, we can replace P_N in Eq. (4.247) with P_{QGP} to obtain

$$\begin{aligned} L &= \frac{3\mu_q^4}{2\pi^2} + B - \frac{2\mu}{3\pi^2}(\mu^2 - M^2)^{3/2} + \frac{\mu_q^4}{2\pi^2} - B \\ &= \frac{2\mu_q^4}{\pi^2} - \frac{2\mu}{3\pi^2}(\mu^2 - M^2)^{3/2} \ . \end{aligned} \tag{4.248}$$

Note that the expression for the latent heat in Eq. (4.248) contains two chemical potentials, μ_q for the quarks and μ for the nucleons. However, the Gibbs criteria for the phase transition specifies that the chemical potentials must be equal across the boundary. We can determine the relationship between these two chemical potentials by recalling that, for a first-order transition, there will be a mixed phase, as in the case of water to ice, where some of the degrees of freedom will be in quarks and some in hadrons. We must conserve baryon number at all times. Thus, if the net number of quarks (quarks – antiquarks) is N_q and the net number of baryons (baryons – antibaryons) is N_b, since there are three quarks per baryon, baryon number conservation gives us

$$\frac{1}{3}N_q + N_b \equiv \text{constant} . \tag{4.249}$$

Differentiating the above with no net change in baryon number,

$$\frac{1}{3}dN_q + dN_b = 0 ,$$

or,

$$dN_q = -3dN_b . \tag{4.250}$$

Equation (4.250) tells us that increasing the net baryon number by one decreases the net quark number by 3. From Eq. (4.79), we know that the derivative of the Gibbs free energy with respect to particle number is the chemical potential. Thus we have

$$\frac{\partial G}{\partial N_q} = \mu_q , \qquad \frac{\partial G}{\partial N_b} = \mu . \tag{4.251}$$

Therefore, across a first-order transition with fixed temperature and pressure and with the Gibbs free energy a function of N_q and N_b, $G(N_q, N_b)$,

$$\frac{\partial G}{\partial N_q}dN_q + \frac{\partial G}{\partial N_b}dN_b = 0 . \tag{4.252}$$

Substituting Eq. (4.250) in for dN_q, we have

$$-3\frac{\partial G}{\partial N_q}dN_b + \frac{\partial G}{\partial N_b}dN_b = 0 . \tag{4.253}$$

or, using Eq. (4.251),

$$3\mu_q = \mu \equiv \mu_c \tag{4.254}$$

across the transition. Using this relation in Eq. (4.248), we have

$$
\begin{aligned}
L &= \frac{2}{\pi^2}\left(\frac{\mu_c}{3}\right)^4 - \frac{2\mu}{3\pi^2}(\mu_c^2 - M^2)^{3/2} \\
&= \frac{2\mu_c}{3\pi^2}\left(\left(\frac{\mu_c}{3}\right)^3 - (\mu_c^2 - M^2)^{3/2}\right) . \qquad (4.255)
\end{aligned}
$$

The latent heat must be positive, $L \geq 0$, for the transition to proceed. We already know that the lower limit for μ_c must be the nucleon mass, M. The criteria in Eq. (4.255) implies an upper limit on μ_c of $3M/2\sqrt{2}$.

EXAMPLE: Show that the upper limit of μ_c is $3M/2\sqrt{2}$.
From Eq. (4.255), we have

$$\left(\frac{\mu_c}{3}\right)^3 - (\mu_c^2 - M^2)^{3/2} \geq 0 . \qquad (4.256)$$

Taking both sides to the power 2/3, we have

$$\frac{\mu_c^2}{9} \geq \mu_c^2 - M^2 .$$

Rearranging terms and taking the square root of both sides, we have

$$\frac{3M}{2\sqrt{2}} \geq \mu_c . \qquad (4.257)$$

The range on μ_c,

$$\frac{3M}{2\sqrt{2}} \geq \mu_c \geq M , \qquad (4.258)$$

puts a rather narrow limit on the chemical potential across the transition. From this limit and the requirement that $P_{\rm QGP} = P_N$, we can determine a range of acceptable values for the bag constant, B. We write $P_{\rm QGP}$ in terms of $r_c = \mu_c/M$,

$$P_{\rm QGP} = \frac{M^4}{6\pi^2}\frac{r_c^4}{27} - B . \qquad (4.259)$$

The expression for the bag constant as a function of r_c is

$$B = \frac{M^4}{8\pi^2}\left(\frac{r_c^4}{27} - r_c(r_c^2 - \frac{5}{2})(r^2-1)^{1/2} - \frac{3}{2}\ln(r_c + (r_c^2-1)^{1/2})\right) . \tag{4.260}$$

Inserting the lower limit of Eq. (4.258), $\mu_c = M$ or $r_c = \mu_c/M = 1$, gives the lower limit on B,

$$B = \frac{1}{2\pi^2}\left(\frac{M}{4}\right)^4 . \tag{4.261}$$

The upper limit on μ_c, $3M/\sqrt{8}$ or $r_c = \sqrt{9/8}$, gives the upper limit on B,

$$B = \frac{M^4}{8\pi^2}\left[\frac{3}{4} - \ln 2\right] . \tag{4.262}$$

The possible range of B is thus

$$\frac{M^4}{8\pi^2}\left[\frac{3}{4} - \ln 2\right] \geq B \geq \frac{1}{2\pi^2}\left(\frac{M}{4}\right)^4 . \tag{4.263}$$

If we take $M = 0.938$ GeV, the proton mass, then

$$0.154 \geq B^{1/4} \geq 0.148 \text{ GeV} , \tag{4.264}$$

similar to that of the original MIT bag model, $B^{1/4} = 0.145$ GeV. Taking a higher nucleon mass, such as obtained by averaging the masses of the nucleon and the Delta resonance ($M_\Delta = 1.232$ GeV), $M = 1.11$ GeV, one obtains a higher value of B,

$$0.182 \geq B^{1/4} \geq 0.176 \text{ GeV} . \tag{4.265}$$

While this range of values is a bit larger, it is not as large as the value needed to give $T_c = 170$ MeV in Eq. (4.222), $B^{1/4} = 0.236$ GeV.

4.4.3 Case III: finite temperature, finite mass

We finally examine the case where both the temperature of the system and the mass of the hadronic constituents are finite. At this level, the hadron phase can be treated as an ideal gas of mesons, baryons and antibaryons. If the initial temperature of the system was only high enough for hadron gas formation, instead of a quark-gluon plasma, the degrees of freedom of the hadron gas are likely to be only pions and nucleons since $T_c \sim 1.3 m_\pi$ when $T_c = 170$ MeV and the nucleons are already present during the initial collisions. Other mesons and baryons, all with masses greater than T_c will be produced in abundances relative to their statistical weights. Their contributions will be negligible in this case. However, hadronization of a quark-gluon plasma, where we still assume massless partons, will produce a range of particles containing up, down and strange quarks. Thus we take the more general case here.

We begin with the meson contribution, with $\mu_{\rm mes} = 0$ for bosons,

$$T \ln Z|_{\rm mes} = \frac{g_{\rm mes} V}{6\pi^2} \int_0^\infty \frac{dk\, k^4}{\sqrt{k^2+m^2}} \frac{1}{\exp[(\sqrt{k^2+m^2})/T] - 1} \tag{4.266}$$

where m is the meson mass. After integration, we obtain

$$T \ln Z|_{\rm mes} = \frac{g_{\rm mes} V T^2 m^2}{6\pi^2} \sum_{n=1}^{\infty} \frac{1}{n^2} K_2 \left(\frac{nm}{T} \right) . \tag{4.267}$$

EXAMPLE: Derive Eq. (4.267).
We begin by making a change of variable in Eq. (4.266), defining $y = \sqrt{k^2+m^2}/T$ and $dy = k dk / y T^2$ with $k = \sqrt{y^2 T^2 - m^2} = T\sqrt{y^2 - (m/T)^2}$. We can also rewrite the Bose-Einstein distribution as in Eq. (4.147),

$$\frac{1}{\exp(y) - 1} = \sum_{n=0}^{\infty} \exp[-(n+1)y] . \tag{4.268}$$

Making these replacements in Eq. (4.266), we are left with

$$T \ln Z|_{\rm mes} = \frac{g_{\rm mes} V T^4}{6\pi^2} \sum_{n=0}^{\infty} \int_{m/T}^{\infty} dy \left(y^2 - \frac{m^2}{T^2} \right)^{3/2} \exp[-(n+1)y] \tag{4.269}$$

where the lower limit of the integral is now m/T.

After some rearrangement, we can make the integral in Eq. (4.269) look suggestively like the integral definition of integer-order modified Bessel functions of the second kind, $K_n(x)$:

$$K_l(x) = \frac{2^l l!}{(2l)!} x^{-l} \int_x^\infty dt (t^2 - x^2)^{l-1/2} \exp(-t) \ . \tag{4.270}$$

To use the definition of $K_l(x)$, we change variables again so that $t = (n+1)y$ and $dy = dy/(n+1)$. The 3/2 power in Eq. (4.269) gives $l = 2$. Replacing y by $t/(n+1)$ in the expression $(y^2 - m^2/T^2)^{3/2}$, we find $(t^2 - x^2)^{3/2}/(n+1)^3$ with $x = (n+1)m/T$. After all these substitutions, we have

$$T \ln Z|_{\text{mes}} = \frac{g_{\text{mes}} V T^4}{6\pi^2} \sum_{n=0}^{\infty} \frac{1}{(n+1)^4} \int_x^\infty dt (t^2 - x^2)^{3/2} \exp(-t) \tag{4.271}$$

where

$$\begin{aligned} \int_x^\infty dt (t^2 - x^2)^{3/2} \exp(-t) &= \frac{x^2 4!}{2^2 \cdot 2!} K_2(x) \\ &= \frac{3(n+1)^2 m^2}{T^2} K_2\left(\frac{(n+1)m}{T}\right) . \end{aligned} \tag{4.272}$$

Finally,

$$T \ln Z|_{\text{mes}} = 3 \frac{g_{\text{mes}} V T^2 m^2}{6\pi^2} \sum_{n=0}^{\infty} \frac{1}{(n+1)^2} K_2\left(\frac{(n+1)m}{T}\right) . \tag{4.273}$$

After shifting the first term of the sum from $n = 0$ to $n = 1$ with the corresponding shift from $(n+1)^2$ to n^2 in the sum, we obtain Eq. (4.267).

The pressure due to the mesons is

$$P_{\text{mes}} = \frac{T \ln Z|_{\text{mes}}}{V} = \frac{g_{\text{mes}} m^2 T^2}{2\pi^2} \sum_{n=1}^{\infty} \frac{1}{n^2} K_2\left(\frac{nm}{T}\right) . \tag{4.274}$$

There is no number density for the mesons since $\mu_{\text{mes}} = 0$. The energy density due to the mesons is

$$\epsilon_{\text{mes}} = 3P_{\text{mes}} + \frac{g_{\text{mes}} m^3 T}{2\pi^2} \sum_{n=1}^{\infty} \frac{1}{n} K_1\left(\frac{nm}{T}\right) . \tag{4.275}$$

EXAMPLE: Prove Eq. (4.275).
The energy density is

$$\epsilon_{\text{mes}} = \frac{T^2}{V}\frac{\partial \ln Z|_{\text{mes}}}{\partial T} = \frac{T^2}{V}\frac{\partial}{\partial T}\left[\frac{g_{\text{mes}} m^2 T}{2\pi^2}\sum_{n=1}^{\infty}\frac{1}{n^2}K_2\left(\frac{nm}{T}\right)\right] \tag{4.276}$$

since $\mu_{\text{mes}} = 0$. We have

$$\begin{aligned} \epsilon_{\text{mes}} &= 3\frac{g_{\text{mes}} m^2 T^2}{6\pi^2}\sum_{n=1}^{\infty}\frac{1}{n^2}K_2\left(\frac{nm}{T}\right) \\ &+ \frac{g_{\text{mes}} m^2 T^3}{2\pi^2}\sum_{n=1}^{\infty}\frac{1}{n^2}\frac{\partial K_2(nm/T)}{\partial T} . \end{aligned} \tag{4.277}$$

The first term is the same as the meson contribution to the pressure in Eq. (4.274). We concentrate now on the second term, the derivative of K_2. We change variables to $x = nm/T$ so that $1/\partial T = (x^2/nm)/\partial x$ and

$$\frac{\partial K_2(nm/T)}{\partial T} = -\frac{nm}{T^2}\frac{\partial K_2(x)}{\partial x} . \tag{4.278}$$

Using the identity for the derivative of a modified Bessel function,

$$K_l'(x) = -K_{l-1}(x) - \frac{l}{x}K_l(x) , \tag{4.279}$$

Eq. (4.278) becomes

$$\frac{\partial K_2(nm/T)}{\partial T} = \frac{nm}{T^2}\left(K_1\left(\frac{nm}{T}\right) + \frac{2T}{nm}K_2\left(\frac{nm}{T}\right)\right) . \tag{4.280}$$

Putting Eq. (4.280) into Eq. (4.277), we find

$$\begin{aligned} \epsilon_{\text{mes}} &= 3\frac{g_{\text{mes}} m^2 T^2}{2\pi^2}\sum_{n=1}^{\infty}\frac{1}{n^2}K_2\left(\frac{nm}{T}\right) \\ &+ \frac{g_{\text{mes}} m^3 T}{2\pi^2}\sum_{n=1}^{\infty}\frac{1}{n}K_1\left(\frac{nm}{T}\right) . \end{aligned} \tag{4.281}$$

This is then Eq. (4.275) since the first term in Eq. (4.281) is simply twice the meson pressure.

Following the same arguments as above, the meson contribution to the entropy density is

$$\begin{aligned} s_{\text{mes}} &= \frac{\partial}{\partial T}\left[\frac{g_{\text{mes}} m^2 T^2}{2\pi^2} \sum_{n=0}^{\infty} \frac{1}{n^2} K_2\left(\frac{nm}{T}\right)\right] \\ &= \frac{3}{T} P_{\text{meson}} + \frac{g_{\text{mes}} m^2}{2\pi^2} \sum_{n=1}^{\infty} \frac{1}{n} K_1\left(\frac{nm}{T}\right) . \end{aligned} \tag{4.282}$$

Note that these results are appropriate for any meson of mass m with g_{mes} degrees of freedom. It could be generalized to a sum over mesons i with mass m_i and g_i degrees of freedom.

We now turn to the baryon-antibaryon contribution to the hadron gas. In this case

$$\begin{aligned} T \ln Z|_{\text{bar}} &= \frac{g_{\text{bar}} V}{6\pi^2} \int_0^{\infty} \frac{dk\, k^4}{\sqrt{k^2+M^2}} \left[\frac{1}{\exp[(\sqrt{k^2+M^2}-\mu)/T]+1}\right. \\ &\left. + \frac{1}{\exp[(\sqrt{k^2+M^2}+\mu)/T]+1}\right] \end{aligned} \tag{4.283}$$

where M is the baryon mass. We can derive a result similar to Eq. (4.274) for the baryons except that now we will have a power of $(-1)^n$ and a finite number density since

$$\begin{aligned} &\frac{1}{\exp[(E \mp \mu)/T]+1} = \\ &\sum_{n=0}^{\infty} (-1)^n \exp\left[-\frac{(n+1)E}{T}\right] \exp\left[\pm\frac{(n+1)\mu}{T}\right] . \end{aligned} \tag{4.284}$$

After some manipulation, we arrive at

$$\begin{aligned} T \ln Z|_{\text{bar}} &= \frac{g_{\text{bar}} V M^2 T^2}{2\pi^2} \sum_{n=1}^{\infty} \frac{(-1)^{n-1}}{n^2} K_2\left(\frac{nM}{T}\right) \\ &\times [\exp(n\mu/T) + \exp(-n\mu/T)] \end{aligned} \tag{4.285}$$

with

$$\begin{aligned} P_{\text{bar}} &= \frac{g_{\text{bar}} M^2 T^2}{2\pi^2} \sum_{n=1}^{\infty} \frac{(-1)^{n-1}}{n^2} K_2\left(\frac{nM}{T}\right) \\ &\times [\exp(n\mu/T) + \exp(-n\mu/T)] . \end{aligned} \tag{4.286}$$

We could write the contribution from the baryon and antibaryon chemical potentials as a single factor,

$$\exp(n\mu/T) + \exp(-n\mu/T) = 2\cosh(n\mu/T) \ . \tag{4.287}$$

We choose to keep the two contributions separate as a reminder that there are two components to the total baryon-antibaryon contribution, making numerical evaluation more transparent as well.

The number density is simply obtained by taking the derivative of Eq. (4.285) with respect to μ,

$$\begin{aligned} n_{\text{bar}} &= \frac{g_{\text{bar}} M^2 T}{2\pi^2} \sum_{n=1}^{\infty} \frac{(-1)^{n-1}}{n} K_2\left(\frac{nM}{T}\right) \\ &\quad \times \left[\exp(n\mu/T) - \exp(-n\mu/T)\right] \ . \end{aligned} \tag{4.288}$$

The derivative of $\ln Z_{\text{bar}}$ in the energy density leads to three terms when the baryon chemical potential is included,

$$\begin{aligned} \epsilon_{\text{bar}} &= \frac{T^2}{V} \frac{\partial \ln Z_{\text{bar}}}{\partial T} + \mu n_{\text{bar}} \\ &= 3P_{\text{bar}} + \frac{g_{\text{bar}} M^3 T}{2\pi^2} \sum_{n=1}^{\infty} \frac{(-1)^{n-1}}{n} K_1\left(\frac{nM}{T}\right) \\ &\quad \times \left[\exp(n\mu/T) + \exp(-n\mu/T)\right] \\ &\quad - \frac{g_{\text{bar}} M^2 T \mu}{2\pi^2} \sum_{n=1}^{\infty} \frac{(-1)^{n-1}}{n} K_2\left(\frac{nM}{T}\right) \\ &\quad \times \left[\exp(n\mu/T) - \exp(-n\mu/T)\right] + \mu n_{\text{bar}} \ . \end{aligned}$$

The term proportional to M^2 arises from the derivative of $\exp(n\mu/T) + \exp(-n\mu/T)$ with respect to T and cancels μn_{bar}. Thus the final expression for the energy density is

$$\begin{aligned} \epsilon_{\text{bar}} &= 3P_{\text{bar}} + \frac{g_{\text{bar}} M^3 T}{2\pi^2} \sum_{n=1}^{\infty} \frac{(-1)^{n-1}}{n} K_1\left(\frac{nM}{T}\right) \\ &\quad \times \left[\exp(n\mu/T) - \exp(-n\mu/T)\right] \ . \end{aligned} \tag{4.289}$$

We can also generalize these results to any set of baryons and antibaryons. Finally, the baryon entropy density is

$$\begin{aligned} s_{\rm bar} &= \frac{3}{T}P_{\rm bar} + \frac{g_{\rm bar}M^3}{2\pi^2}\sum_{n=1}^{\infty}\frac{(-1)^{n-1}}{n}K_1\left(\frac{nM}{T}\right) \\ &\quad \times \left[\exp(n\mu/T) + \exp(-n\mu/T)\right] . \end{aligned} \tag{4.290}$$

For numerical purposes, it is perhaps more convenient to retain the integral form of Eqs. (4.285)-(4.289) [65]. This is most easily done by starting from Eq. (4.125). Let $u = k/\sqrt{k^2+M^2}$ so that $k = \sqrt{u^2M^2/(1-u^2)}$ and $kdk = M^2udu/(1-u^2)^2$. The arguments of the exponents in the Fermi-Dirac distributions are transformed as $(\sqrt{k^2+M^2}\mp\mu)/T = (M/\sqrt{1-u^2}\mp\mu)/T$. Then the baryon free energy is

$$\begin{aligned} F_{\rm bar} &= \frac{g_{\rm bar}VM^4}{6\pi^2}\int_0^1 \frac{du\, u^4}{(1-u^2)^3}\left[\frac{1}{\exp[(M/\sqrt{1-u^2}-\mu)/T]+1}\right. \\ &\quad \left. + \frac{1}{\exp[(M/\sqrt{1-u^2}+\mu)/T]+1}\right] \\ &= \frac{g_{\rm bar}VM^4}{6\pi^2}\int_0^1 \frac{du\, u^4}{(1-u^2)^3}[f_{\rm FD}(u;T,\mu) + f_{\rm FD}(u;T,-\mu)] \end{aligned} \tag{4.291}$$

where $f_{\rm FD}$ is the Fermi-Dirac distributions for baryons (positive μ) and antibaryons (negative μ). The baryon pressure is just Eq. (4.291) divided by the volume, V.

EXAMPLE: Derive the integral relations similar to Eq. (4.291) for the number density, energy density and the entropy density.
The number density is equal to

$$n_{\rm bar} = \frac{g_{\rm bar}T}{2\pi^2}\int dk\, k^2[f_{\rm FD}(E;T,\mu) - f_{\rm FD}(E;T,-\mu)] . \tag{4.292}$$

There is a difference of distributions instead of a sum since the antiparticle chemical potential is equal and opposite the particle chemical potential. Changing variables from k to u gives

$$n_{\rm bar} = \frac{g_{\rm bar}M^3}{2\pi^2}\int_0^1 \frac{du\, u^2}{(1-u^2)^{5/2}}[f_{\rm FD}(u;T,\mu) - f_{\rm FD}(u;T,-\mu)] . \tag{4.293}$$

The energy density will be

$$\begin{aligned}
\epsilon_{\rm bar} &= \frac{g_{\rm bar}}{2\pi^2}\int dk\, kE^2[f_{\rm FD}(E;T,\mu)+f_{\rm FD}(E;T,-\mu)] \\
&= \frac{g_{\rm bar}M^4}{2\pi^2}\int_0^1 \frac{u^2du}{(1-u^2)^3}[f_{\rm FD}(u;T,\mu)+f_{\rm FD}(u;T,-\mu)]\ . \qquad (4.294)
\end{aligned}$$

The entropy density for particles alone is

$$\begin{aligned}
s_{\rm bar} &= \frac{g_{\rm bar}}{2\pi^2}\int dk\, k^2[f'_{\rm FD}(E;T,\mu)\ln f'_{\rm FD}(E;T,\mu) \\
&\quad + f_{\rm FD}(E;T,\mu)\ln f_{\rm FD}(E;T,\mu)] \\
&= \frac{g_{\rm bar}M^3}{2\pi^2}\int_0^1 \frac{du\, u^2}{(1-u^2)^{5/2}}[f'_{\rm FD}(u;T,\mu)\ln f'_{\rm FD}(u;T,\mu) \\
&\quad + f_{\rm FD}(u;T,\mu)\ln f_{\rm FD}(u;T,\mu)]\ . \qquad (4.295)
\end{aligned}$$

To obtain the total entropy density, we need to include another integral with the antiparticles and chemical potential $-\mu$.

Both the meson and baryon contributions are included in the hadron gas. At the phase boundary, we have, at $T = T_c$ with again $\mu_c = 3\mu_q = \mu$,

$$P_{\rm QGP} = P_{\rm mes} + P_{\rm bar} \qquad (4.296)$$

with latent heat

$$L = \epsilon_{\rm QGP} - (\epsilon_{\rm mes} + \epsilon_{\rm bar})\ . \qquad (4.297)$$

Recall that both the pressure and energy density in the quark-gluon plasma include the bag constant, as in Eqs. (4.214) and (4.216) respectively.

Note that in all the cases studied, we have treated the quark-gluon plasma as an ideal gas of massless quarks and gluons while the hadron side has varied. In central heavy-ion collisions at high energies, the baryon chemical potential has been determined to be rather small and decreasing with increasing energy [68]. Thus the finite temperature case described here, Case III, with $\mu = 0$ on the hadron side is perhaps the most physical scenario for systems produced at heavy-ion colliders, especially at the LHC.

Ideal thermodynamics assumes a static system in thermal equilibrium. However, in a real heavy-ion collision the system must evolve with time. Thus a more realistic description awaits a description of the time dependence, introduced in the next chapter.

Exercises

1. Give the effective number of degrees of freedom in a quark-gluon plasma when a massless strange quark is included.

2. Calculate the critical temperature at the phase boundary between a quark-gluon plasma and a massless pion gas for $B^{1/4} = 200$ MeV and $\mu = 0$ (Case I).

3. Calculate $B^{1/4}$ in units of MeV for $T_c = 100$, 130 and 200 MeV.

4. Plot the temperature as a function of $\epsilon^{1/4}$ (in units of MeV or GeV) for $T_c = 100$, 130, 170 and 200 MeV, similar to Fig. 4.9. Assume that the hadron side is a massless pion gas with $\mu = 0$ (Case I). Calculate the latent heat of transition in each case in units of MeV^4 and $\mathrm{MeV/fm}^3$.

5. Fill in the details to derive Eqs. (4.286), (4.288), (4.289) and (4.290) for baryons in Case III, as done for mesons in Eqs. (4.274), (4.275) and (4.282).

6. Check that the thermodynamic identity, $\epsilon = Ts + \mu n - P$, is satisfied for the cases studied in this chapter: (a) the quark-gluon plasma with massless constituents; (b) massless pions (Case I); (c) massive baryons at $T = 0$ (Case II); and (d) massive mesons and baryons at finite temperature (Case III).

Chapter 5

Hydrodynamics

5.1 Introduction

In this chapter, we come to the evolution of the system with time. Thermodynamics studies static systems in equilibrium and, alone, can not tell us about how the system changes in space-time. The application of hydrodynamics requires a system in local equilibrium, that is, the system has to be describable as a complete entity like a liquid or gas rather than a collection of individual particles. These particles must be interacting with each other to reach equilibrium. The question is whether their interactions are frequent enough for equilibrium to be established.

One way to quantify the frequency of collisions is by comparing the mean-free path, λ, the average distance a particle travels between collisions, to the size, L, of the medium. The mean-free path is defined as

$$\lambda = \frac{1}{\rho\sigma} \tag{5.1}$$

where ρ is the density of the medium and σ is the interaction cross section. There are several different possible regimes. If $\lambda > L$, then no equilibration can occur because a particle leaves the medium before it has a chance to collide again. If $\lambda \sim L$, the system could perhaps be described by multiple collision models, like some of the event generators in use today. Finally, if $\lambda \ll L$, then it is not really feasible to speak of successive single collisions but to treat the system like a gas of particles. It is in this limit that the system can come to local thermal equilibrium and a hydrodynamical description is practical.

EXAMPLE: What is a typical mean-free path relative to the size of a large nucleus?
If we take the central nuclear density, $\rho_A \sim 0.16/\text{fm}^3$, and use the nucleon-nucleon inelastic cross section at $\sqrt{S_{NN}} = 200$ GeV, $\sigma_{\text{inel}} = 40$ mb $= 4$ fm^2, then

$$\lambda = \frac{1}{0.16 \times 4} \sim 1.6 \text{ fm} . \tag{5.2}$$

If $L \sim 2R_A$ where $R_A \sim 6 - 7$ fm, then a proton could collide about 8 times on its way through the nucleus at its center to $L > \lambda$. Experiments can test how well data agrees with a hydrodynamic expansion relative to multiple scattering models.

The hydrodynamic equations result from applying the constraints of energy and momentum conservation to a gas. To describe the gas itself, we need the thermodynamical concepts developed in the previous chapter. The initial conditions are crucial to the resulting description of the space-time evolution of the system. These initial conditions include the initial temperature and initial time after which the system can be treated hydrodynamically. We will discuss some estimates of the initial energy density of the system, related to the initial temperature. If the energy density is assumed to be the beam energy per volume, the energy density in a sphere of radius R is [65]

$$\epsilon = \frac{E}{(4/3)\pi R^3} . \tag{5.3}$$

The longitudinal dimension in a moving system may naively be thought to be contracted by the Lorentz boost so that the volume is instead $(4/3)\pi R^3/\gamma$ where $\gamma = E/m$. The energy density is then

$$\epsilon = \frac{E^2}{(4/3)\pi R^3 m} . \tag{5.4}$$

If we specify the conditions in the center-of-mass frame, since the value of ϵ in Eq. (5.4) is for a single collision partner, the total energy density includes contributions from both colliding objects so that the final energy density is

$$\epsilon = \frac{2E^2}{(4/3)\pi R^3 m} \; . \tag{5.5}$$

EXAMPLE: Calculate the energy density for pp and Pb+Pb collisions at $\sqrt{S_{NN}} = 20$ and 200 GeV using Eq. (5.5).
Since E is the energy of one beam in the center-of-mass frame, $E = \sqrt{S}/2$, or 10 and 100 GeV/nucleon at the lower and higher energies. In the case of the proton, $R = r_p \sim 0.8$ fm, while for the nucleus, $R = R_{\rm Pb} = r_0 A^{1/3} \sim 6.6$ fm. The *pp* energy densities are then

$$\epsilon_{pp} = \frac{2 \times (\sqrt{S}/2)^2}{(4/3)\pi r_p^3 m_p} \tag{5.6}$$

giving 102 and 10170 GeV/fm^3 for $\sqrt{S} = 20$ and 200 GeV respectively. In the case of AA collisions, we have

$$\epsilon_{AA} = \frac{2 \times (A\sqrt{S_{NN}}/2)^2}{(4/3)\pi R_A^3 A m_N} \; . \tag{5.7}$$

Note that now we have to include all appropriate factors of A, both for the total energy in the numerator and for the radius and mass in the denominator. These powers of A cancel each other, leaving us with

$$\epsilon_{AA} = \frac{2 \times (\sqrt{S_{NN}}/2)^2}{(4/3)\pi r_0^3 m_N} \; . \tag{5.8}$$

This last equality is very similar to the one in Eq. (5.6) except that we have r_0 in Eq. (5.8) and r_p in Eq. (5.6). (The difference between m_p and m_N is small.) Since $r_0 > r_p$, the energy density in an AA collision is 29.4 and 2940 GeV/fm^3 for $\sqrt{S_{NN}} = 20$ and 200 GeV respectively. The difference in the AA and *pp* energy densities is due only to the relative values of r_p and r_0.

The densities obtained from Eqs. (5.6)-(5.8) are huge, far beyond any reasonable assumptions about hadronic matter. If we had considered the energies available for AA collisions at the LHC, $\sqrt{S_{NN}} \geq 5.5$

TeV, where the Lorentz boost is even larger, the energy densities would be correspondingly higher. Even at the CERN SPS and BNL RHIC energies used above, the boost contributes a large factor. As a reality check on this estimate of the energy density, we can calculate the resulting spatial extent of the beams in the center-of-mass frame due to the boost. For a lead nucleus, $R_{\rm Pb}/\gamma \sim 0.6$ fm at $\sqrt{S_{NN}} = 20$ GeV and 0.06 fm at 200 GeV. The longitudinal extent of the proton would be reduced to $r_p/\gamma \sim 0.08$ fm and 0.008 fm respectively. Such numbers, with spatial dimensions much less than 1 fm, would violate the uncertainty principle, thus localizing the particle position and momentum simultaneously. (The momentum is localized since the particles are accelerated with a well-defined momentum.)

In fact, it is not really possible to pack all the partons, with energies less than the center-of-mass energy, in the proton into such a small distance. The bulk of the partons, the gluons, have only rather low momentum, as we saw in Chapter 2. For low momentum partons, the boost is rather irrelevant and these constituents of the hadron or nucleus are essentially at rest. The so-called 'Fermi momentum', $\Delta p \sim$ 200 MeV, then gives a spatial extent of ~ 1 fm by the uncertainty principle,

$$\Delta x \Delta p \sim \hbar c \; . \tag{5.9}$$

Without the boost then, the pp energy densities would be reduced to ~ 10 GeV/fm^3 at $\sqrt{S} = 20$ GeV and ~ 101 GeV/fm^3 at 200 GeV.

These densities are still rather high. A more empirically-based calculation of the energy density gives perhaps a more realistic result. The energy in the rapidity slice Δy is the product of the number of particles and their average energy,

$$E = \langle m_T \rangle \frac{dN}{dy} \Delta y \tag{5.10}$$

where $m_T = (p_T^2 + m^2)^{1/2}$ and dN/dy is the final particle multiplicity. The spatial volume element, as we will discuss in more detail later in the chapter, is

$$dV = \tau dy d^2 x \tag{5.11}$$

where τ is the proper time. If $dy = \Delta y$, integration over d^2x gives

$$V = \tau \Delta y A_{\rm eff} \tag{5.12}$$

where $A_{\rm eff} = \pi R_A^2$ and we let $\tau = \tau_0 \approx 1$ fm. Then the energy density at proper time τ_0, using Eqs. (5.10) and (5.12), is

$$\epsilon_0 = \frac{E}{V} = \frac{\langle m_T \rangle}{\tau_0 A_{\rm eff}} \frac{dN}{dy} \tag{5.13}$$

since rapidity bin size Δy cancels. For light hadrons, $\langle m_T \rangle \sim 0.5$ GeV. If $dN/dy \sim 4$ for a pp collision and $A_{\rm eff} \sim 2$ fm^2, $\epsilon_0 \sim 1$ GeV/fm^3. In a Pb+Pb collision at the same energy, $dN_{AA}/dy \sim A dN/dy$ and $A_{\rm eff} \sim \pi R_A^2 \sim 160$ fm^2, $\epsilon_0 \sim 2.6$ GeV/fm^3.

While this value of the energy density is still a factor of 10 or so larger than the energy density inside a nucleon, it is not so unreasonable. The formation time of ~ 1 fm is only a guess but can be justified by the uncertainty principle since $c\Delta t \sim \Delta x$ with $c \equiv 1$ and we have already calculated $\Delta x \sim 1$ fm. We will concentrate on a perfect fluid which has the same, constant temperature throughout the system. Thus there are no temperature gradients.

5.2 Energy-momentum tensor

To obtain the equations of motion of the systems, we begin with the energy-momentum tensor, $T^{\mu\nu}$. We start in the fluid rest frame with $T_R^{\mu\nu}$ where the subscript R denotes the rest frame.. We will then generalize to the moving frame by Lorentz transformation. The energy-momentum tensor is the four-momentum component in the μ direction per three-dimensional surface area perpendicular to the ν direction. Thus $T^{\mu\nu}$ has dimensions of energy per volume.

If we have differential four-momentum $\Delta \mathbf{p} = (\Delta E, \Delta p_x, \Delta p_y, \Delta p_z)$ with differential space-time four-vector $\Delta \mathbf{x} = (\Delta t, \Delta x, \Delta y, \Delta z)$, then, for $\mu = \nu = 0$,

$$T_R^{00} = \frac{\Delta E}{\Delta x \Delta y \Delta z} = \frac{\Delta E}{\Delta V} = \epsilon \;, \tag{5.14}$$

the energy density. We take $\mu = \nu = 1$ (the x coordinate) as an example of the spatial directions,

$$T_R^{00} = \frac{\Delta p_x}{\Delta t \Delta y \Delta z} , \tag{5.15}$$

where $\Delta p_x/\Delta t$ is the force in the x direction acting on a surface of area $\Delta y \Delta z$ perpendicular to the force. The force exerted per area is the pressure. For a perfect fluid, the pressure is the same in all directions or isotropic. Then

$$T_R^{ij} = P\delta^{ij} . \tag{5.16}$$

Thus, in the fluid rest frame,

$$T_R^{\mu\nu} = \begin{pmatrix} \epsilon & 0 & 0 & 0 \\ 0 & P & 0 & 0 \\ 0 & 0 & P & 0 \\ 0 & 0 & 0 & P \end{pmatrix} . \tag{5.17}$$

We now boost into the moving system. We Lorentz boost the velocities using the form,

$$\mathbf{u}^\mu = \Lambda^\mu_\nu \mathbf{u}^\nu_R \tag{5.18}$$

where $\mathbf{u}^\nu_R$ is the velocity unit vector in the rest frame of the system,

$$\mathbf{u}^\nu_R = (1, 0, 0, 0) , \tag{5.19}$$

and Λ^μ_ν is a Lorentz transformation. Thus the velocity in the boosted frame defines the $\nu = 0$ component of the transformation,

$$\mathbf{u}^\mu = \Lambda^\mu_0 \mathbf{u}^0_R = \Lambda^\mu_0 . \tag{5.20}$$

To find the components of the transformation in the spatial dimensions, we use

$$g^{\rho\sigma} = g^{\mu\nu} \Lambda^\rho_\mu \Lambda^\sigma_\nu . \tag{5.21}$$

Since the off-diagonal components of $g^{\mu\nu}$ are zero, we can expand Eq. (5.21) to obtain

$$g^{\rho\sigma} = g^{00} \Lambda^\rho_0 \Lambda^\sigma_0 + g^{ii} \Lambda^\rho_i \Lambda^\sigma_i . \tag{5.22}$$

Using $g^{00} = 1$ and $g^{ii} = -1$ and rearranging terms, we have

$$\begin{aligned} \Lambda_i^\rho \Lambda_i^\sigma &= \Lambda_0^\rho \Lambda_0^\sigma - g^{\rho\sigma} \\ &= \mathbf{u}^\rho \mathbf{u}^\sigma - g^{\rho\sigma} \ . \end{aligned} \tag{5.23}$$

In the last step we have used Eq. (5.20) to substitute $\Lambda_0^\rho = \mathbf{u}^\rho$ and $\Lambda_0^\sigma = \mathbf{u}^\sigma$. We can now apply the Lorentz transformation to the rest-frame value of the energy-momentum tensor to obtain the general form

$$\begin{aligned} T^{\rho\sigma} &= \Lambda_\mu^\rho \Lambda_\nu^\sigma T_R^{\mu\nu} \\ &= \Lambda_0^\rho \Lambda_0^\sigma \epsilon + \Lambda_i^\rho \Lambda_i^\sigma P \\ &= \mathbf{u}^\rho \mathbf{u}^\sigma \epsilon + (\mathbf{u}^\rho \mathbf{u}^\sigma - g^{\rho\sigma}) P \\ &= (\epsilon + P) \mathbf{u}^\rho \mathbf{u}^\sigma - g^{\rho\sigma} P \ . \end{aligned} \tag{5.24}$$

The energy-momentum tensor can also be derived from kinetic theory. For a detailed discussion of fluid mechanics and kinetic theory, see the book by Csernai [69]. Only a brief sketch is given here. The scalar function $N(\mathbf{x}, \mathbf{q})$ gives the distribution of fluid elements with four-momentum $\mathbf{q}^\mu$ at space-time points $\mathbf{x}^\mu$. Because N is a scalar, it can only depend on relativistic invariants, *i.e.* products of four-vectors such as $\mathbf{x}^2$, $\mathbf{x} \cdot \mathbf{q}$ or $\mathbf{q}^2$. (Note that $\mathbf{q}^\mu$ is the momentum of a fluid element and not that of a single particle.) Starting from $N(\mathbf{x}, \mathbf{q})$, the energy-momentum tensor of a fluid is defined as

$$T^{\mu\nu}(\mathbf{x}) = \int d^4 q \mathbf{q}^\mu \mathbf{q}^\nu N(\mathbf{x}, \mathbf{q}) \ . \tag{5.25}$$

In the rest frame of a perfect fluid with uniform temperature and no temperature gradients, $N(\mathbf{x}, \mathbf{q})$ is independent of $\mathbf{x} \cdot \mathbf{q}$. Then, when $\mu = \nu = 0$,

$$T^{00}(\mathbf{x}) = \int d^4 q q_0^2 N(\mathbf{x}, \mathbf{q}) \ . \tag{5.26}$$

To better see that Eq. (5.26) is really an energy density, we recall from Chapter 2 that

$$d^4 q = dq^2 \delta(\mathbf{q}^2 - M^2) \frac{d^3 q}{q_0} \ .$$

On mass shell, the integral over dq^2 vanishes, leaving

$$T^{00}(\mathbf{x}) = \int d^3q q_0 N(\mathbf{x}, \mathbf{q}) = \epsilon \tag{5.27}$$

where d^3q has units of GeV^3 or fm^{-3} in $\hbar = c = 1$ units. The pressure is defined from

$$\begin{aligned} T^{ij} &= \int d^4q q^i q^j N(\mathbf{x}, \mathbf{q}) \\ &= \frac{1}{3}\delta^{ij} \int d^4q |\vec{q}|^2 N(\mathbf{x}, \mathbf{q}) \end{aligned} \tag{5.28}$$

so that

$$P = T^{ii} \; . \tag{5.29}$$

Thus the rest frame form, $T_R^{\mu\nu}$, is seen to only be applicable to rotationally symmetric systems, independent of $\mathbf{x} \cdot \mathbf{q}$. To describe systems without rotational symmetry, three additional quantities are needed: the shear and bulk viscosities, η and ζ respectively, and the thermal conductivity, χ. See Ref. [70] for some more general calculations including the viscosities.

5.3 Hydrodynamic equations

We now calculate the hydrodynamic equations of motion for a perfect fluid [65]. The equations of motion arise from energy and momentum conservation as well as baryon number conservation,

$$\partial_\mu T^{\mu\nu} = 0 \; , \tag{5.30}$$

$$\partial_\mu \mathbf{J}^\mu = 0 \; . \tag{5.31}$$

In the rest frame, the derivative of the energy-momentum tensor,

$$\partial_\mu T_R^{\mu\nu} = 0 \; , \tag{5.32}$$

gives

$$\begin{aligned}\partial_0 T_R^{00} &= \frac{\partial \epsilon}{\partial t} = 0 \ , &(5.33)\\ \partial_i T_R^{ii} &= \frac{\partial P}{\partial x_i} = 0 \ . &(5.34)\end{aligned}$$

Thus the energy density is forced to be constant with time but could have spatial variations while the pressure is constant over all space (no pressure gradients) but could depend on time.

For a moving fluid, more work is needed. To obtain a scalar quantity from the requirement of energy-momentum conservation, we contract Eq. (5.30) with the velocity $\mathbf{u}_\nu$,

$$\mathbf{u}_\nu \, \partial_\mu T^{\mu\nu} = 0 \ , \tag{5.35}$$

obtaining

$$\mathbf{u}^\nu \partial_\nu \epsilon + (\epsilon + P) \partial_\nu \mathbf{u}^\nu = 0 \ . \tag{5.36}$$

EXAMPLE: Derive Eq. (5.36).
We begin by substituting the general form of the energy-momentum tensor, Eq. (5.24), for $T_{\mu\nu}$ in Eq. (5.35),

$$0 = \mathbf{u}_\nu \, \partial_\mu [(\epsilon + P)\mathbf{u}^\mu \mathbf{u}^\nu - g^{\mu\nu} P] \ . \tag{5.37}$$

Expanding, we have

$$\begin{aligned}0 &= \mathbf{u}_\nu \mathbf{u}^\mu \mathbf{u}^\nu \, \partial_\mu (\epsilon + P) + (\epsilon + P)(\mathbf{u}_\nu \mathbf{u}^\mu \, \partial_\mu \mathbf{u}^\nu \\ &\quad + \mathbf{u}_\nu \mathbf{u}^\nu \, \partial_\mu \mathbf{u}^\mu) - \mathbf{u}_\nu \, \partial^\nu P \ . \end{aligned}\tag{5.38}$$

Since $\mathbf{u}$ is a unit four-vector, $\mathbf{u}_\nu \mathbf{u}^\nu = \mathbf{u}^2 = 1$. In addition, this property also gives us $\mathbf{u}_\nu \partial_\mu \mathbf{u}^\nu = (1/2)\partial_\mu(\mathbf{u}_\nu \mathbf{u}^\nu) = 0$. The derivative is zero since $\mathbf{u}^2 = 1$. Using these expressions, Eq. (5.38) reduces to

$$0 = \mathbf{u}^\mu \, \partial_\mu (\epsilon + P) + (\epsilon + P) \, \partial_\mu \mathbf{u}^\mu - \mathbf{u}_\nu \, \partial^\nu P \ . \tag{5.39}$$

After canceling the $\mathbf{u}_\nu \partial^\nu P$ terms, we are left with Eq. (5.36).

Using Eq. (4.84) in Chapter 4,

$$\epsilon + P = Ts + \mu n_{\rm bar} \ ,$$

we can write Eq. (5.36) in terms of the entropy and baryon densities. When T, μ and P are held fixed, the derivative of the energy density is

$$\partial_\nu \epsilon = \partial_\nu s + \mu\, \partial_\nu n_{\rm bar} \; . \tag{5.40}$$

Replacing $\partial_\nu \epsilon$ by the right-hand side of Eq. (5.40) and $\epsilon + P$ by $Ts + \mu n_{\rm bar}$, Eq. (5.36) becomes

$$0 = T\mathbf{u}^\nu\, \partial_\nu s + \mu \mathbf{u}^\nu\, \partial_\nu n_{\rm bar} + (Ts + \mu n_{\rm bar})\, \partial_\nu \mathbf{u}^\nu \; . \tag{5.41}$$

Regrouping and rearranging terms, we have

$$\begin{aligned} 0 &= T(\mathbf{u}^\nu \partial_\nu s + s\partial_\nu \mathbf{u}^\nu) + \mu(\mathbf{u}^\nu \partial_\nu n_{\rm bar} + n_{\rm bar}\partial_\nu \mathbf{u}^\nu) \\ &= T\, \partial_\nu(s\mathbf{u}^\nu) + \mu\, \partial_\nu(n_{\rm bar}\mathbf{u}^\nu) \; . \end{aligned} \tag{5.42}$$

Since the particle current $\mathbf{J}^\nu$ is just $n_{\rm bar}\mathbf{u}^\nu$, the second term in Eq. (5.42) is zero by baryon number conservation. Thus we have

$$T\, \partial_\nu(s\mathbf{u}^\nu) = 0 \; , \tag{5.43}$$

entropy current conservation.

Another hydrodynamic equation can be derived by contracting $\partial_\mu T^{\mu\nu}$ with the tensor combination, $g_{\nu\lambda} - \mathbf{u}_\nu \mathbf{u}_\lambda$. In this case, the resulting equation will be a four-vector equation. We will find

$$0 = (\epsilon + P)\mathbf{u}^\mu \partial_\mu \mathbf{u}_\lambda - \partial_\lambda P + \mathbf{u}_\mu \mathbf{u}_\lambda \partial^\mu P \; . \tag{5.44}$$

EXAMPLE: Derive Eq. (5.44).
We begin with

$$0 = (g_{\nu\lambda} - \mathbf{u}_\nu \mathbf{u}_\lambda)\, \partial_\mu[(\epsilon + P)\mathbf{u}^\mu \mathbf{u}^\nu - g^{\mu\nu} P] \; . \tag{5.45}$$

Expanding the terms gives

$$\begin{aligned} 0 \;=\;& \mathbf{u}^\mu \mathbf{u}_\lambda\, \partial_\mu(\epsilon + P) + (\epsilon + P)(\mathbf{u}^\mu\, \partial_\mu \mathbf{u}_\lambda + \mathbf{u}_\lambda\, \partial_\mu \mathbf{u}^\mu) - \partial_\lambda P \\ & - (\epsilon + P)(\mathbf{u}_\lambda \mathbf{u}^\mu \mathbf{u}_\nu\, \partial_\mu \mathbf{u}^\nu + \mathbf{u}_\lambda\, (\mathbf{u}_\nu \mathbf{u}^\nu)\partial_\mu \mathbf{u}^\mu) \\ & - \mathbf{u}_\lambda \mathbf{u}^\mu (\mathbf{u}_\nu \mathbf{u}^\nu)\, \partial_\mu(\epsilon + P) + \mathbf{u}_\lambda \mathbf{u}^\mu\, \partial_\mu P \; . \end{aligned} \tag{5.46}$$

Making use of the relations $\mathbf{u}_\nu \mathbf{u}^\nu = 1$ and $\mathbf{u}_\nu \partial_\mu \mathbf{u}^\nu = 0$, the terms proportional to the derivative of $\epsilon + P$ cancel, as do the terms with the $\partial_\mu \mathbf{u}^\mu$ multiplying $\epsilon + P$ in Eq. (5.46). We are then left with Eq. (5.44).

When ϵ, s and $n_{\rm bar}$ are held fixed in Eq. (4.84), the derivative of the pressure is then

$$\partial_\lambda P = s\,\partial_\lambda T + n_{\rm bar}\,\partial_\lambda \mu \ . \tag{5.47}$$

Inserting this into Eq. (5.44), we have

$$\begin{aligned} 0 &= (sT + n_{\rm bar}\mu)\mathbf{u}^\mu\,\partial_\mu \mathbf{u}_\lambda - s\,\partial_\lambda T - n_{\rm bar}\,\partial_\lambda \mu \\ &\quad + s\mathbf{u}_\lambda \mathbf{u}^\mu\,\partial_\mu T + n_{\rm bar}\mathbf{u}_\lambda \mathbf{u}^\mu\,\partial_\mu \mu \ . \end{aligned} \tag{5.48}$$

We rewrite the equation collecting terms proportional to s and $n_{\rm bar}$ separately so that

$$\begin{aligned} 0 &= s(T\mathbf{u}^\mu\,\partial_\mu \mathbf{u}_\lambda - \partial_\lambda T + \mathbf{u}_\lambda \mathbf{u}^\mu\,\partial_\mu T) \\ &\quad + n_{\rm bar}(\mu\mathbf{u}^\mu\,\partial_\mu \mathbf{u}_\lambda - \partial_\lambda \mu + \mathbf{u}_\lambda \mathbf{u}^\mu\,\partial_\mu \mu) \\ &= s(\mathbf{u}^\mu\,\partial_\mu(\mathbf{u}_\lambda T) - \partial_\lambda T) + n_{\rm bar}(\mathbf{u}^\mu\,\partial_\mu(\mathbf{u}_\lambda \mu) - \partial_\lambda \mu) \ . \end{aligned} \tag{5.49}$$

If the net baryon density, $n_{\rm bar}$, is zero, the equation becomes much easier to handle. We are left with an equation for the temperature alone,

$$0 = \mathbf{u}^\mu\,\partial_\mu(T\mathbf{u}_\lambda) - \partial_\lambda T \ . \tag{5.50}$$

Returning to the more differential form of Eq. (5.52),

$$0 = T\mathbf{u}^\mu \partial_\mu \mathbf{u}_\lambda - \partial_\lambda T + \mathbf{u}^\mu \mathbf{u}_\lambda \partial_\mu T \ , \tag{5.51}$$

dividing both sides by T gives a differential equation in terms of $\ln T$,

$$0 = \mathbf{u}^\mu \partial_\mu \mathbf{u}_\lambda - \partial_\lambda \ln T + \mathbf{u}^\mu \mathbf{u}_\lambda \partial_\mu \ln T \ , \tag{5.52}$$

The entropy current conservation equation can also be rewritten in terms of temperature using Eq. (4.76), the speed of sound in the medium at zero chemical potential,

$$\partial_\nu \ln s = \frac{1}{c_s^2}\partial_\nu \ln T \ .$$

We then have

$$\begin{aligned} 0 &= \mathbf{u}^\mu \partial_\mu s + s\partial_\mu \mathbf{u}^\mu \\ &= \partial_\mu \mathbf{u}^\mu + \mathbf{u}^\mu \, \partial_\mu \ln s \\ &= \partial_\mu \mathbf{u}^\mu + \frac{1}{c_s^2}\mathbf{u}^\mu \, \partial_\mu \ln T \, . \end{aligned} \tag{5.53}$$

Equations (5.52) and (5.53) give the hydrodynamic equations in terms of a single fundamental thermodynamic quantity. The other thermodynamic quantities, $n_{\rm bar}$, ϵ, P and s are all functions of T and chemical potential μ. We will return to the formulation in Eqs. (5.52) and (5.53) later. We first proceed with solutions to Eqs. (5.36) and (5.44).

In heavy-ion collisions, as described in Chapter 3, the nuclei are accelerated toward each other in the center-of-mass frame of the system. Before the collision, all the motion is longitudinal. Most of the motion after the collision may still be assumed to be in the longitudinal direction. Thus the longitudinal coordinate is taken to be along the z axis. Some radial component is to be expected. Thus we discuss the expansion in a cylindrical coordinate system. The azimuthal angular dependence drops out because the system is rotationally symmetric. The four-vector velocity, normalized to unity, is then

$$\mathbf{u}^\mu = \frac{1}{\sqrt{1 - v_r^2 - v_z^2}}(1, v_z, v_r, 0) \tag{5.54}$$

where v_r and v_z are the longitudinal and radial components and the last component, v_ϕ, is zero. In general, $v_r < v_z$. We start with the assumption that $v_r \ll v_z$ so that the radial motion is negligible. Then we can write the velocity four-vector as

$$\mathbf{u}^\mu = (\cosh\theta, \sinh\theta, 0, 0) \tag{5.55}$$

where θ is the fluid velocity. In cylindrical coordinates, the differential four-vector, ∂_μ, is

$$\partial_\mu = \left(\frac{\partial}{\partial t}, -\frac{\partial}{\partial z}, -\frac{1}{r}\frac{\partial}{\partial r}, -\frac{\partial}{\partial \phi}\right) \, . \tag{5.56}$$

Only the first two components will be finite in the case of longitudinal expansion alone.

We now write the hydrodynamic equation in cylindrical coordinates. Note that

$$\begin{aligned}\mathbf{u}^\mu\partial_\mu &= \cosh\theta\frac{\partial}{\partial t}+\sinh\theta\frac{\partial}{\partial z}\,,\\ \partial_\mu\mathbf{u}^\mu &= \sinh\theta\frac{\partial\theta}{\partial t}+\cosh\theta\frac{\partial\theta}{\partial z}\,.\end{aligned}$$

Thus Eq. (5.36) becomes

$$\begin{aligned}0 &= \cosh\theta\frac{\partial\epsilon}{\partial t}+\sinh\theta\frac{\partial\epsilon}{\partial z}\\ &\quad+(\epsilon+P)\left(\sinh\theta\frac{\partial\theta}{\partial t}+\cosh\theta\frac{\partial\theta}{\partial z}\right)\,.\end{aligned}\tag{5.57}$$

We can show that Eq. (5.44) becomes

$$\begin{aligned}0 &= \sinh\theta\frac{\partial P}{\partial t}+\cosh\theta\frac{\partial P}{\partial z}\\ &\quad+(\epsilon+P)\left(\cosh\theta\frac{\partial\theta}{\partial t}+\sinh\theta\frac{\partial\theta}{\partial z}\right)\,.\end{aligned}\tag{5.58}$$

These two equations describe the longitudinal expansion of a gas.

EXAMPLE: Show that Eq. (5.58) comes from Eq. (5.44).

There are actually two equations that arise from Eq. (5.44), both leading to the same result, as we now show. The first comes from setting $\lambda = 0$ while the second arises from $\lambda = 1$. Note that $u_{\lambda=0} = \cosh\theta$ and $u_{\lambda=1} = \sinh\theta$. When $\lambda = 0$, we have

$$\begin{aligned}0 &= -\partial_0 P+\mathbf{u}_\mu u_0\partial^\mu P+(\epsilon+P)\mathbf{u}^\mu\partial_\mu u_0\\ &= -\frac{\partial P}{\partial t}+\cosh\theta\left(\cosh\theta\frac{\partial P}{\partial t}+\sinh\theta\frac{\partial P}{\partial z}\right)\\ &\quad+(\epsilon+P)\left(\cosh\theta\frac{\partial\cosh\theta}{\partial t}+\sinh\theta\frac{\partial\cosh\theta}{\partial z}\right)\\ &= (\cosh^2\theta-1)\frac{\partial P}{\partial t}++\cosh\theta\sinh\theta\frac{\partial P}{\partial z}\\ &\quad+(\epsilon+P)\left(\cosh\theta\sinh\theta\frac{\partial\theta}{\partial t}+\sinh^2\theta\frac{\partial\theta}{\partial z}\right)\end{aligned}\tag{5.59}$$

where we have used $u_0 = \cosh\theta$. Using the identity $\cosh^2\theta - 1 = \sinh^2\theta$ and dividing by $\sinh\theta$, we are left with Eq. (5.58). Likewise for $\lambda = 1$,

$$\begin{aligned}
0 &= -\partial_1 P + \mathbf{u}_\mu u_1 \partial^\mu P + (\epsilon + P)\mathbf{u}^\mu \partial_\mu u_1 \\
&= \frac{\partial P}{\partial z} + \sinh\theta\left(\cosh\theta\frac{\partial P}{\partial t} + \sinh\theta\frac{\partial P}{\partial z}\right) \\
&\quad + (\epsilon + P)\left(\cosh\theta\frac{\partial \sinh\theta}{\partial t} + \sinh\theta\frac{\partial \sinh\theta}{\partial z}\right) \\
&= (\sinh^2\theta + 1)\frac{\partial P}{\partial z} + \sinh\theta\cosh\theta\frac{\partial P}{\partial t} \\
&\quad + (\epsilon + P)\left(\cosh^2\theta\frac{\partial \theta}{\partial t} + \sinh\theta\cosh\theta\frac{\partial \theta}{\partial z}\right) \qquad (5.60)
\end{aligned}$$

using $u_1 = \sinh\theta$. After replacing $\sinh^2\theta + 1$ by $\cosh^2\theta$ and dividing by the common factor $\cosh\theta$, we are again left with Eq. (5.58).

We can simplify the equations somewhat by using the light-cone variables of rapidity, y, and proper time, τ,

$$y = \frac{1}{2}\ln\frac{t+z}{t-z}\,, \qquad (5.61)$$

$$\tau = \sqrt{t^2 - z^2}\,. \qquad (5.62)$$

In Fig. 5.1, the light-cone diagram shows contours of constant proper time, the hyperbolae, while the lines are contours of constant space-time rapidity. At the light-cone axes, where $t = \pm z$, $y = \pm\infty$ and $\tau = 0$. The t and z coordinates as functions of τ and y are

$$t = \tau\cosh y\,, \qquad (5.63)$$

$$z = \tau\sinh y\,. \qquad (5.64)$$

We now transform Eqs. (5.57) and (5.58) to the (τ, y) coordinate system. We first have to write the partial derivatives with respect to t and z as derivatives with respect to τ and y.

EXAMPLE: Write $\partial/\partial t$ and $\partial/\partial z$ in terms of $\partial/\partial\tau$ and $\partial/\partial y$.

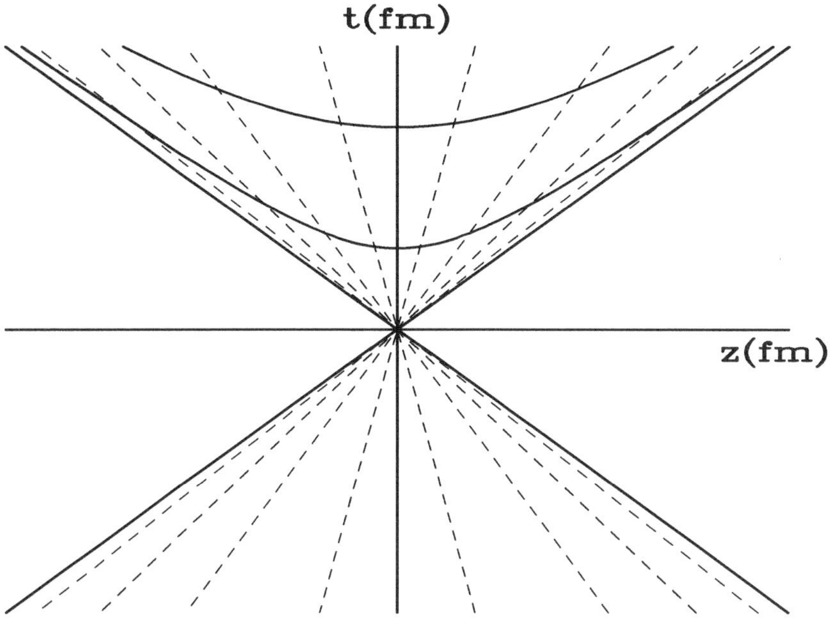

Figure 5.1: Light-cone diagram. The hyperbolae are contours of constant τ while the light dashed lines are contours of constant rapidity. The axes rotated 45° are the light-cone axes where $y = \pm\infty$.

We begin with

$$\frac{\partial}{\partial t} = \frac{\partial \tau}{\partial t}\frac{\partial}{\partial \tau} + \frac{\partial y}{\partial t}\frac{\partial}{\partial y} \tag{5.65}$$

$$\frac{\partial}{\partial z} = \frac{\partial \tau}{\partial z}\frac{\partial}{\partial \tau} + \frac{\partial y}{\partial z}\frac{\partial}{\partial y} \tag{5.66}$$

Using the definitions of y and τ in Eqs. (5.61) and (5.62), we have

$$\frac{\partial \tau}{\partial t} = \frac{t}{\tau} = \cosh y \, , \tag{5.67}$$

$$\frac{\partial y}{\partial t} = \frac{1}{2}\frac{t-z}{t+z}\left(\frac{t-z-(t+z)}{(t-z)^2}\right) = -\frac{z}{\tau^2} = -\frac{1}{\tau}\sinh y \, , \tag{5.68}$$

$$\frac{\partial \tau}{\partial z} = -\frac{z}{\tau} = -\sinh y \, , \tag{5.69}$$

$$\frac{\partial y}{\partial z} = \frac{1}{2}\frac{t-z}{t+z}\left(\frac{t-z+t+z}{(t-z)^2}\right) = \frac{t}{\tau^2} = \frac{1}{\tau}\cosh y \, . \tag{5.70}$$

Inserting the derivatives in Eqs. (5.67)-(5.70) into Eqs. (5.65) and (5.66), we finally have

$$\frac{\partial}{\partial t} = \cosh y \frac{\partial}{\partial \tau} - \frac{1}{\tau} \sinh y \frac{\partial}{\partial y} , \tag{5.71}$$

$$\frac{\partial}{\partial z} = -\sinh y \frac{\partial}{\partial \tau} + \frac{1}{\tau} \cosh y \frac{\partial}{\partial y} . \tag{5.72}$$

EXAMPLE: Write Eqs. (5.57) and (5.58) as functions of τ and y.
We begin with Eq. (5.57). Substituting Eqs. (5.71) and (5.72) for $\partial/\partial t$ and $\partial/\partial z$, we have

$$\begin{aligned} 0 = {} & \cosh\theta \left[\cosh y \frac{\partial \epsilon}{\partial \tau} - \frac{1}{\tau} \sinh y \frac{\partial \epsilon}{\partial y}\right] \\ & + \sinh\theta \left[-\sinh y \frac{\partial \epsilon}{\partial \tau} + \frac{1}{\tau} \cosh y \frac{\partial \epsilon}{\partial y}\right] \\ & + (\epsilon + P)\Big(\sinh\theta \left[\cosh y \frac{\partial \theta}{\partial \tau} - \frac{1}{\tau} \sinh y \frac{\partial \theta}{\partial y}\right] \\ & + \cosh\theta \left[-\sinh y \frac{\partial \theta}{\partial \tau} + \frac{1}{\tau} \cosh y \frac{\partial \theta}{\partial y}\right] \Big) . \end{aligned} \tag{5.73}$$

Regrouping terms to collect the derivatives with respect to τ and y, we obtain

$$\begin{aligned} 0 = {} & (\cosh\theta \cosh y - \sinh\theta \sinh y) \frac{\partial \epsilon}{\partial \tau} \\ & + \frac{1}{\tau} (\sinh\theta \cosh y - \cosh\theta \sinh y) \frac{\partial \epsilon}{\partial y} \\ & + (\epsilon + P)\Big([\sinh\theta \cosh y - \cosh\theta \sinh y] \frac{\partial \theta}{\partial \tau} \\ & + \frac{1}{\tau} [\cosh\theta \cosh y - \sinh\theta \sinh y] \frac{\partial \theta}{\partial y} \Big) . \end{aligned} \tag{5.74}$$

Recall the subtraction rules for hyperbolic functions,

$$\cosh(\theta - y) = \cosh\theta \cosh y - \sinh\theta \sinh y , \tag{5.75}$$

$$\sinh(\theta - y) = \sinh\theta \cosh y - \cosh\theta \sinh y . \tag{5.76}$$

With these, we find

$$\begin{aligned} 0 &= \cosh(\theta - y)\frac{\partial\epsilon}{\partial\tau} + \frac{1}{\tau}\sinh(\theta - y)\frac{\partial\epsilon}{\partial y} \\ &\quad + (\epsilon + P)\left(\sinh(\theta - y)\frac{\partial\theta}{\partial\tau} + \frac{1}{\tau}\cosh(\theta - y)\frac{\partial\theta}{\partial y}\right) . \end{aligned}$$

Finally, dividing by $\cosh(\theta - y)$ and multiplying by τ, we are left with

$$\begin{aligned} 0 &= \tau\frac{\partial\epsilon}{\partial\tau} + \tanh(\theta - y)\frac{\partial\epsilon}{\partial y} \\ &\quad + (\epsilon + P)\left(\tau\tanh(\theta - y)\frac{\partial\theta}{\partial\tau} + \frac{\partial\theta}{\partial y}\right) . \end{aligned} \tag{5.77}$$

Likewise for Eq. (5.58), we have

$$\begin{aligned} 0 &= \sinh\theta\left[\cosh y\frac{\partial P}{\partial\tau} - \frac{1}{\tau}\sinh y\frac{\partial P}{\partial y}\right] \\ &\quad + \cosh\theta\left[-\sinh y\frac{\partial P}{\partial\tau} + \frac{1}{\tau}\cosh y\frac{\partial P}{\partial y}\right] \\ &\quad + (\epsilon + P)\left(\cosh\theta\left[\cosh y\frac{\partial\theta}{\partial\tau} - \frac{1}{\tau}\sinh y\frac{\partial\theta}{\partial y}\right]\right. \\ &\quad \left. + \sinh\theta\left[-\sinh y\frac{\partial\theta}{\partial\tau} + \frac{1}{\tau}\cosh y\frac{\partial\theta}{\partial y}\right]\right) \\ &= \sinh(\theta - y)\frac{\partial P}{\partial\tau} + \frac{1}{\tau}\cosh(\theta - y)\frac{\partial P}{\partial y} \\ &\quad + (\epsilon + P)\left(\cosh(\theta - y)\frac{\partial\theta}{\partial\tau} + \frac{1}{\tau}\sinh(\theta - y)\frac{\partial\theta}{\partial y}\right) \\ &= \tau\tanh(\theta - y)\frac{\partial P}{\partial\tau} + \frac{\partial P}{\partial y} \\ &\quad + (\epsilon + P)\left(\tau\frac{\partial\theta}{\partial\tau} + \tanh(\theta - y)\frac{\partial\theta}{\partial y}\right) . \end{aligned} \tag{5.78}$$

We have so far not included baryon number conservation which gives us another equation. Starting from

$$\partial^\mu(n_{\text{bar}}\mathbf{u}_\mu) = 0 \; , \tag{5.79}$$

we have

$$\begin{aligned} 0 &= \mathbf{u}_\mu \partial^\mu n_{\rm bar} + n_{\rm bar} \partial^\mu \mathbf{u}_\mu \\ &= \cosh\theta \frac{\partial n_{\rm bar}}{\partial t} + \sinh\theta \frac{\partial n_{\rm bar}}{\partial z} \\ &\quad + n_{\rm bar} \left(\sinh\theta \frac{\partial \theta}{\partial t} + \cosh\theta \frac{\partial \theta}{\partial z} \right) . \end{aligned} \tag{5.80}$$

This last equation is identical to Eq. (5.57) with ϵ and $\epsilon + P$ replaced by $n_{\rm bar}$ so that

$$\begin{aligned} 0 &= \tau \frac{\partial n_{\rm bar}}{\partial \tau} + \tanh(\theta - y) \frac{\partial n_{\rm bar}}{\partial y} \\ &\quad + n_{\rm bar} \left(\tau(\tanh(\theta - y) \frac{\partial \theta}{\partial \tau} + \frac{\partial \theta}{\partial y} \right) . \end{aligned} \tag{5.81}$$

Note also that the entropy current conservation equation would be the same as Eq. (5.81) above with $n_{\rm bar}$ replaced by s.

5.4 Solutions to the hydrodynamic equations: longitudinal expansion

While we have three equations, we have four unknowns: energy density, ϵ, pressure, P, fluid velocity, θ, and baryon density, $n_{\rm bar}$. Thus, to solve the equations, we need a fourth equation, an equation of state relating ϵ and P. As we saw in Chapter 4, for a free gas of massless quarks and gluons,

$$P = \frac{1}{3}\epsilon . \tag{5.82}$$

We also related P and ϵ through the speed of sound in the system,

$$\frac{\partial P}{\partial \epsilon} = c_s^2 = \frac{1}{3} , \tag{5.83}$$

where the last equality only holds for the free gas of massless particles in Eq. (5.82). We can equally well assume that c_s^2 is not a fixed constant but depends on the global variables of the system while remaining independent of space and time. In that case, Eqs. (5.77) and (5.78) can be solved for the two variables, ϵ (or P) and θ. Then θ can be used to obtain $n_{\rm bar}$.

5.4.1 Initial conditions

The initial conditions provide the main physics input to the solutions of the hydrodynamical equations. In Chapter 3, we talked about the stopping power of the collision. There is one solution to the hydrodynamical evolution that begins from a system at rest, complete stopping of the initial nuclei,

$$\theta(t = 0, \vec{x}) = 0 \; . \qquad (5.84)$$

This is known as the Landau initial condition [71]. It is the initial condition that produces the highest densities, akin to those given at the beginning of this chapter.

Landau's model [71] built on Fermi's idea [72] to use hydrodynamic methods to study hadronic collisions (applied, at that time, to pp and pA collisions). Fermi assumed that, when two hadrons collide, the collision energy is released into a very small volume in the center of mass. The energy distribution in a small volume can be treated statistically without actually knowing the nuclear interaction. The volume is contracted in the direction of motion, as discussed in the introduction. Fermi's final assumption, that the final-state particles are formed instantaneously and immediately leave the collision volume without further interaction, is not justified. First, since hadrons interact strongly, it is unlikely that they would leave the volume without interacting with any of the other particles around them. In addition, particle production is not instantaneous. Finally, the system should expand and the number of produced particles is only 'frozen in' after interactions cease. Landau pointed out that relativistic hydrodynamics could be used to study the expansion because thermodynamics and hydrodynamics share the same range of validity, $\lambda \ll L$.

Landau assumed that after the collision $\lambda \ll L$ and statistical equilibrium is reached. The system then expands, governed by the relativistic hydrodynamics of an ideal fluid, as in our approach here. As the system expands, the mean-free path also increases until $\lambda \sim L$. At this point, the interaction cross section between the particles is decreased so that they effectively no longer interact. The system breaks up at a temperature similar to the mass of the pion, $T \sim m_\pi$, and the particles stream away from each other.

While Landau's description of the production, expansion and break up of the system is still that used today, the assumption that the system starts from rest is generally too extreme, especially if the initial nucleons are also expected to be stopped. The Gaussian distribution of produced particles [73], however, is similar to the multiplicity distributions seen at the CERN SPS. Even though the rapidity distributions are similar, other observables do not agree well with this assumption.

At the other extreme, one can assume scaling, *e.g.* that the initial system after the ions interact looks the same, regardless of the rapidity,

$$\theta(y, \tau_0) = y \ . \tag{5.85}$$

This is known as the Bjorken initial condition [74]. Bjorken's approach is very similar to Landau's. Only the initial boundary condition differs. He assumed that, at sufficiently high energy, the rapidity distribution has a central plateau. Thus the space-time evolution of the system is essentially the same in all frames near the center-of-mass frame. The plateau implies that particle production is frame independent. This assumption followed from the empirical evidence of experiments at the CERN S$p\overline{p}$S collider [75] and CERN ISR data from pp, $p\alpha$ and $\alpha\alpha$ collisions [76]. (Recall that an α particle is a ^{4}He nucleus.) In addition, the hot central region of the plateau has, on both sides, fragmentation regions which are the repository of the projectile and target baryon numbers. The assumption of a plateau is also a bit problematic for low energy AA collisions since it implies $\overline{p}/p \sim 1$ in the central region, a condition not yet satisfied at RHIC in $\sqrt{S_{NN}} = 200$ GeV collisions.

A schematic diagram of the multiplicity production in heavy-ion collisions in the Bjorken picture is shown in Fig. 5.2. The fragmentation regions, labeled $y_{\rm proj}$ and $y_{\rm targ}$, are localized around their initial rapidities, assuming the valence quarks do not stop but pass through each other. The hot central region is centered around $y_{\rm cm}$. A symmetric AA collision is shown. The central plateau is shown to extend from $y_{\rm cm}$ to near the fragmentation regions.

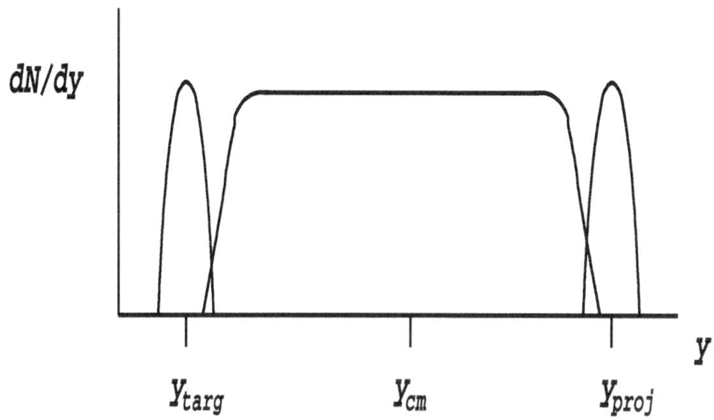

Figure 5.2: Schematic illustration of the Bjorken scaling picture. The projectile and target fragmentation regions are indicated, in addition to the plateau centered around $y_{\rm cm}$.

The condition in Eq. (5.85) also holds at later times so that $\partial\theta/\partial y = 1$, $\partial\theta/\partial\tau = 0$ and $\tanh(\theta - y) = 0$. Then Eqs. (5.77), (5.78) and (5.81) reduce to

$$0 = \tau\frac{\partial\epsilon}{\partial\tau} + \epsilon + P \ , \tag{5.86}$$

$$0 = \frac{\partial P}{\partial y} \ , \tag{5.87}$$

$$0 = \tau\frac{\partial n_{\rm bar}}{\partial\tau} + n_{\rm bar} \ . \tag{5.88}$$

We begin with the solution to the equation for baryon density conservation, Eq. (5.88), since its solution is rather straightforward. We have

$$n_{\rm bar}(\tau) = n_{\rm bar}(\tau_0)\frac{\tau_0}{\tau} \tag{5.89}$$

where τ_0 is the initial proper time at which the system reaches local equilibrium and the hydrodynamic solution is therefore valid.

EXAMPLE: Integrate Eq. (5.88) to obtain Eq. (5.89).
Rearranging terms in Eq. (5.88), we obtain

$$\begin{aligned} \frac{\partial n_{\rm bar}}{n_{\rm bar}} &= -\frac{\partial \tau}{\tau} , \\ \partial \ln n_{\rm bar} &= -\partial \ln \tau . \end{aligned}$$

We integrate the right-hand side from τ_0 to later time τ and the left-hand side from the baryon density at τ_0, $n_{\rm bar}(\tau_0)$ to $n_{\rm bar}(\tau)$ so that

$$\begin{aligned} \ln n_{\rm bar}\Big|_{n_{\rm bar}(\tau_0)}^{n_{\rm bar}(\tau)} &= -\ln \tau\Big|_{\tau_0}^{\tau} , \\ \ln\left(\frac{n_{\rm bar}(\tau)}{n_{\rm bar}(\tau_0)}\right) &= -\ln\left(\frac{\tau}{\tau_0}\right) . \end{aligned}$$

We exponentiate both sides to remove the logarithms and invert the τ ratio due to the negative sign, leaving us with Eq. (5.89).

The entropy density equation, the same form as the baryon density equation, thus has the solution

$$s(\tau) = s(\tau_0)\frac{\tau_0}{\tau} . \tag{5.90}$$

We now turn to the temperature evolution in this picture. In the (τ, y) coordinate system, Eq. (5.53) is

$$0 = c_s^2 + \tau\frac{\partial \ln T}{\partial \tau} . \tag{5.91}$$

After rearrangement of terms,

$$\partial \ln T = -c_s^2 \partial \ln \tau , \tag{5.92}$$

and the solution of the temperature equation is

$$T(\tau) = T(\tau_0)\left(\frac{\tau_0}{\tau}\right)^{c_s^2} . \tag{5.93}$$

Since $c_s^2 < 1$, the temperature drops more slowly than the entropy and baryon number densities.

We now solve for the energy density and pressure. Recall that, in this case, we need to include the bag pressure with the free gas quantities,

$$\epsilon = \epsilon_{\rm qgp} + B \ , \tag{5.94}$$
$$P = P_{\rm qgp} - B \ , \tag{5.95}$$

as in Chapter 4. If, in the quark-gluon plasma,

$$P_{\rm qgp} = c_s^2 \epsilon_{\rm qgp} \ , \tag{5.96}$$

then

$$\begin{aligned} P &= c_s^2 \epsilon_{\rm qgp} - B = c_s^2(\epsilon - B) - B \\ &= c_s^2 \epsilon - (1 + c_s^2) B \ . \end{aligned} \tag{5.97}$$

For the pressure to then be positive, the energy density must be larger than a minimum value,

$$\epsilon > \frac{1 + c_s^2}{c_s^2} B \ . \tag{5.98}$$

If $c_s^2 = 1/3$ and $B^{1/4} = 154$ MeV, $\epsilon > 4B = 4B/(\hbar c)^3 = 280$ MeV/fm^3. Inserting Eq. (5.97) into Eq. (5.86), we obtain

$$\begin{aligned} 0 &= \tau \frac{\partial \epsilon}{\partial \tau} + (1 + c_s^2)(\epsilon - B) \\ &= \tau \frac{\partial(\epsilon - B)}{\partial \tau} + (1 + c_s^2)(\epsilon - B) \ . \end{aligned} \tag{5.99}$$

Note that we can subtract B in the derivative with respect to τ since B is constant. The solution to this equation is then similar to those for entropy and baryon density,

$$\begin{aligned} \partial \ln(\epsilon - B) &= -(1 + c_s^2) \partial \ln \tau \ , \\ \epsilon(\tau) - B &= [\epsilon(\tau_0) - B] \left(\frac{\tau_0}{\tau} \right)^{1 + c_s^2} \ . \end{aligned} \tag{5.100}$$

5.4.2 Relation to the final state

We now discuss how these solutions can be related to physical observables. The only observable results come from the final state after expansion, cooling and breakup. The observables are the final-state particle rapidity and transverse momentum, p_T, distributions.

As shown in Eq. (4.182) of Chapter 4, the number density and entropy density of an ideal gas of gluons and two light quark-antiquark flavors are proportional to each other,

$$s = \xi n \ , \tag{5.101}$$

where $\xi = 3.92$. Recall that here the number density is the total density rather than the baryon density alone. The rapidity distribution is obtained by integrating the particle density over the space and time coordinates where

$$d^4x = \tau d\tau dy d^2x \ . \tag{5.102}$$

EXAMPLE: Show Eq. (5.102).
In general, $d^4x = dtdxdydz = dtdzd^2x$ where d^2x is the transverse area. We have related (t, z) to (τ, y) through

$$t = \tau \cosh y \ , \qquad z = \tau \sinh y \ . \tag{5.103}$$

Then the Jacobian $dtdz$ is related to $d\tau dy$ by

$$dt\, dz = \begin{pmatrix} \cosh y & \tau \sinh y \\ \sinh y & \tau \cosh y \end{pmatrix} d\tau\, dy = \tau d\tau\, dy \ . \tag{5.104}$$

At fixed, final, proper time τ_f, the rapidity distribution, dN/dy, is

$$\frac{dN}{dy} = \int d^2x\, \tau_f n(\tau_f, y, x) \ . \tag{5.105}$$

Using Eq. (4.182) to relate the number density to the entropy density, we have

$$\frac{dN}{dy} = \int d^2x \frac{\tau_f}{\xi} s(\tau_f, y, x) \ . \tag{5.106}$$

We can then relate the final entropy density to the initial entropy density through Eq. (5.90) so that

$$\frac{dN}{dy} = \int d^2x \frac{\tau_0}{\xi} s(\tau_0, y, x) \ . \tag{5.107}$$

Thus the final-state particle distribution is only a function of the initial entropy density. If we also assume that there is no transverse expansion, we can integrate over d^2x to obtain

$$\frac{dN}{dy} = \frac{\pi R_A^2 \tau_0}{\xi} s(\tau_0, y) \tag{5.108}$$

where πR_A^2 is the transverse area of the nucleus in an AA collision. If there is no net baryon number in the system, we can relate the initial entropy density to the initial energy density by

$$s(\tau_0)T(\tau_0) = \epsilon(\tau_0) + P(\tau_0) = [\epsilon(\tau_0) - B](1 + c_s^2) \ . \tag{5.109}$$

EXAMPLE: Using Eqs. (5.108) and (4.216), calculate the energy density at time τ_0 from Eq. (5.109).

Equation (4.216) tells us that for $\mu = 0$, the difference between the energy density of the quark-gluon plasma, $\epsilon_{QGP} \equiv \epsilon(\tau_0)$ here, and the vacuum energy density, B, is proportional to $T(\tau_0)^4$. We represent the proportionality factor by λ so that $\epsilon(\tau_0) - B = \lambda T^4(\tau_0)$. Since the equation of state of a massless quark-gluon gas is $\epsilon = 3P$ with $c_s^2 = 1/3$, the energy density is proportional to T^4. If we assume a different equation of state, the energy density is proportional to T^{1+1/c_s^2}. This gives

$$T(\tau_0) = \left[\frac{\epsilon(\tau_0) - B}{\lambda}\right]^{c_s^2/(1+c_s^2)} \ . \tag{5.110}$$

Inserting this value for the temperature into the left-hand side of Eq. (5.109), the entropy density is

$$s(\tau_0) = \left[\frac{\epsilon(\tau_0) - B}{\lambda}\right]^{1/(1+c_s^2)} \lambda(1 + c_s^2) \ . \tag{5.111}$$

Rearranging terms to solve for the initial energy density, we find

$$\begin{aligned} \epsilon(\tau_0) &= B + \lambda \left[\frac{s(\tau_0)}{\lambda(1 + c_s^2)}\right]^{1+c_s^2} \\ &= B + \lambda \left[\frac{\xi}{\pi R_A^2 \tau_0 \lambda (1 + c_s^2)} \frac{dN}{dy}\right]^{1+c_s^2} \ . \end{aligned} \tag{5.112}$$

Table 5.1: The initial energy densities in a quark-gluon plasma for given values of bag constant and multiplicity.

dN/dy	$\epsilon(\tau_0)$ (GeV^4)	$\epsilon(\tau_0)$ GeV/fm^3
	$B^{1/4} = 145$ MeV	
4	4.67×10^{-4}	0.06
60	0.00137	0.18
600	0.020	2.65
2000	0.0996	12.97
	$B^{1/4} = 235$ MeV	
4	0.003	0.4
60	0.00398	0.518
600	0.023	2.99
2000	0.102	13.31

In the last step we have replaced the initial entropy density by the final-state multiplicity from Eq. (5.108).

We now calculate the initial energy density for several values of dN/dy with $c_s^2 = 1/3$, $\lambda = 37\pi^2/30$ and $\xi = 3.92$. We will assume $\tau_0 = 1$ fm and $R_A = 6.6$ fm, the radius of lead nucleus. We check the importance of the bag constant, B, by taking two different values, $B^{1/4} = 145$ MeV, as in the original formulation of the MIT bag model, and 235 MeV, the value we obtained in Chapter 4 with $T_c = 170$ MeV. The results are shown in Table 5.1. The multiplicities range from a low value compatible with pp collisions up to $dN/dy = 2000$, a moderate value for heavy-ion collisions at the LHC. Note that for low multiplicities, the energy density is very sensitive to the value of the bag constant while, at high multiplicities, the second term in Eq. (5.112) is so large that the bag constant only has a small effect on the energy density. We also note that the energy density of a smaller system with the same final-state multiplicity would be considerably higher. A shorter initial value of τ_0 would increase $\epsilon(\tau_0)$ since the initial time appears in the denominator.

Table 5.2: The initial energy densities in an ideal gas of massless pions for several values of the final-state multiplicity.

dN/dy	$\epsilon(\tau_0)$ (GeV^4)	$\epsilon(\tau_0)$ GeV/fm^3
4	5.16×10^{-5}	0.0067
60	0.00191	0.249
600	0.041	5.35
2000	0.205	26.63

If no quark-gluon plasma is formed and the system started out as an ideal gas of massless pions, the initial energy density would be significantly different, as shown in Table 5.2. In this case, $\lambda = 3\pi^2/30$, $\xi = 3.6$ and the bag constant is absent. We see that for low values of the multiplicity, the absence of the bag constant makes a large difference. The energy density is about a factor of two lower in the hadron gas with $dN/dy = 4$. As the multiplicity increases, however, the energy density becomes larger in the hadron gas due to the relative values of λ. In the quark-gluon plasma $\lambda \sim 12.2$ while in the hadron gas it is ~ 1. It is clear that these higher densities are not really consistent with a hadronic system since they are considerably higher than those for both a quark-gluon plasma and also than the energy density in the center of the nucleon, calculated at the beginning of Chapter 4.

The initial temperature of the system, $T(\tau_0)$, can be obtained from the value of the initial energy density calculated in Tables 5.1 and 5.2. Thus entropy current conservation provides a direct connection between the observed multiplicity and the initial temperature. This connection is exploited to set the initial conditions for evolution, usually with the assumption that $\tau_0 = 1$ fm. The 1 fm value is obtained if one assumes that the mean transverse momentum is approximately that of the internal transverse momentum of nucleons in a nucleus, $p_T \sim 200$ MeV. Thus, from the uncertainty principle,

$$p_T \tau_0 \sim \hbar c \,, \tag{5.113}$$

we have $\tau_0 \sim 1$ fm since $\hbar c \sim 200$ MeV fm.

Other, shorter, equilibration times can be inferred if one instead assumes that, at high center-of-mass energies such as at RHIC and the LHC particle production is dominated by hard interactions that can be calculated perturbatively. The minimum p_T for perturbative $2 \to 2$ parton scattering is ~ 2 GeV. These relatively low p_T partons are referred to as 'minijets' in the literature [77] since high p_T partons form 'jets'. The resulting multiplicity distributions of the produced partons are very sensitive to the low momentum fraction, x, behavior of the parton distribution functions. A leading order calculation of the multiplicity resulted in the estimate of ~ 5000 particles per unit rapidity [78] at midrapidity for $\sqrt{S_{NN}} = 5.5$ TeV, the Pb+Pb collision energy at the LHC before any higher-order corrections were taken into account. Most of the multiplicity was due to gluon production through the $gg \to gg$ process. Assuming that the matter equilibrated with this multiplicity, the initial time is $\tau_0 \sim \hbar c/p_T \sim 0.1$ fm for $p_T = 2$ GeV. The initial temperature calculated for these values of dN/dy and τ_0 was of order 1 GeV, considerably higher than most of the values we have discussed. The RHIC multiplicities calculated this way were somewhat lower than dynamically based phenomenological models (compare the RHIC results in Ref. [78] with those listed in Ref. [79]).

One difficulty with assuming an arbitrary fixed minimum p_T for this minijet production is that, for $p_T \sim 2$ GeV at the LHC, the initial parton may scatter more than once as it passes through the nucleus [80]. However, for perturbative QCD to be applicable, the parton density must be dilute [17]. If the transverse density of partons becomes equal to the transverse area of the target, the parton density is said to 'saturate'. Fixing the minimum p_T to the point of saturation results in a higher minimum p_T at the LHC, $p_T \sim 2.2$ GeV, while the minimum p_T at RHIC drops to ~ 1.2 GeV [81]. The RHIC multiplicity in the saturation approach agreed rather well with the data [79] while the predicted LHC multiplicity was reduced. Other models of the initial state, such as shadowing and the color-glass condensate mentioned in Chapter 2 also lead to reduced multiplicities, on the order of 2000 at midrapidity, resulting in lower initial temperatures. With shadowing parameterizations alone, however, there is still the problem of setting the initial scale. In the color-glass condensate approach the scale is automatically the saturation scale at the collision energy. Given these uncertainties, it is difficult to accurately predict the final-state multiplicity for a new energy regime and much model tuning typically occurs after the first data are available.

5.4.3 Space-time history of an ideal quark-gluon plasma

We now work our way through the longitudinal expansion of a quark-gluon plasma from formation to freezeout, assuming a first order phase transition with a mixed phase. If a quark-gluon plasma is produced at proper time τ_0 with initial temperature $T(\tau_0)$, assuming $\mu = 0$, the initial entropy density is

$$s(\tau_0) = \frac{4g_{\rm qgp}\pi^2}{90}T^4(\tau_0) \ . \tag{5.114}$$

We have not specified the number of degrees of freedom. This way, if the initial temperature is high enough for the strange quark to be an active degree of freedom, it can also be included. From the initial time and temperature, the system expands and cools down. When the system has cooled to T_c, it is no longer in the quark-gluon plasma phase but is entering the mixed phase. The time at which this occurs, τ_m, depends on $T(\tau_0)$ and τ_0, giving, for $c_s^2 = 1/3$,

$$\tau_m = \tau_0 \left(\frac{T(\tau_0)}{T_c}\right)^3 \tag{5.115}$$

using Eq. (5.93).

The system remains in the mixed phase until it has totally changed from a quark-gluon plasma to a hadron gas. The number of degrees of freedom is reduced from the quark-gluon plasma value to the hadronic value. Recall that entropy current conservation gives us

$$s(\tau_0)\tau_0 = s(\tau)\tau \ . \tag{5.116}$$

This relation holds throughout the expansion, even when the temperature is constant since $T = T_c$ throughout the mixed phase. The entropy density at the beginning of the mixed phase, when quarks and gluons are still the only relevant degrees of freedom, is

$$s_Q(\tau_m) = \frac{4g_{\rm qgp}\pi^2}{90} T_c^3 \ . \tag{5.117}$$

At the end of the mixed phase, the relevant degrees of freedom are only hadrons so that the entropy density when $\tau = \tau_H$, the time at which the mixed phase ends and the system has only hadronic degrees of freedom, is

$$s_H(\tau_H) = \frac{4g_\pi\pi^2}{90} T_c^3 \ . \tag{5.118}$$

The time at the end of the mixed phase, τ_H, can be found from the relation

$$s_Q(\tau_m)\tau_m = s_H(\tau_H)\tau_H \tag{5.119}$$

so that

$$\tau_H = \tau_m \frac{s_{\rm qgp}(\tau_m)}{s_H(\tau_H)} = \tau_m \frac{g_{\rm qgp}}{g_\pi} \ . \tag{5.120}$$

From this expression it is clear that the duration of the mixed phase is very sensitive to the number of degrees of freedom in each phase. When the ratio of the degrees of freedom is large, as that between an ideal quark-gluon plasma with $2-3$ active light quark flavors and a massless pion gas, the mixed phase can be of long duration. If the ratio is close to unity, the mixed phase will be short.

It is not clear how the transition occurs. One speculation is that clumps of hadrons, or bubbles, appear in the plasma. These bubbles then expand, taking up more of the volume of the system until the whole system has converted to hadrons. Another possibility is for the system to form hadrons only at the edges, 'boiling off' hadrons until the edges reach the center.

In any case, as long as the system remains in the mixed phase, the total entropy density will be a combination of the quark-gluon and hadron gas contributions,

$$s(\tau) = \lambda(\tau)s_Q(\tau_m) + [1-\lambda(\tau)]s_H(\tau_H) \ . \tag{5.121}$$

By construction, $\lambda(\tau_m) = 1$ and $\lambda(\tau_H) = 0$. To construct the time dependence of $\lambda(\tau)$, we rely on entropy current conservation, noting that

$$s(\tau)\tau = s(\tau_0)\tau_0 = s_Q(\tau_m)\tau_m = s_H(\tau_H)\tau_H \ . \tag{5.122}$$

From Eq. (5.121), we have

$$\lambda(\tau) = \frac{s(\tau) - s_H(\tau_H)}{s_Q(\tau_m) - s_H(\tau_H)} = \frac{[s(\tau)/s_H(\tau_H)] - 1}{g_{\rm qgp}/g_\pi - 1} \tag{5.123}$$

where we used Eq. (5.120) for the last equality. To introduce τ explicitly, we apply Eq. (5.122) to obtain

$$\lambda(\tau) = \frac{(\tau_0/\tau)[s(\tau_0)/s_H(\tau_H)] - 1}{g_{\rm qgp}/g_\pi - 1} \ . \tag{5.124}$$

Finally, since $s \propto T^3$, we obtain an expression in terms of the initial conditions and the degrees of freedom,

$$\lambda(\tau) = \frac{(\tau_0/\tau)(g_{\rm qgp}/g_\pi)(T(\tau_0)/T_c)^3 - 1}{g_{\rm qgp}/g_\pi - 1} \tag{5.125}$$

when $c_s^2 = 1/3$.

EXAMPLE: Find $\lambda(\tau)$ for a two-flavor quark-gluon plasma and a massless pion gas.
For a two-flavor quark-gluon plasma, the effective number of degrees of freedom are

$$g_{\rm qgp} = g_g + \frac{7}{8}(g_f + g_{\overline{f}}) = 2 \times 8 + \frac{7}{8} \times 2 \times 2 \times 3 \times 2 = 37 \ .$$

We have an extra factor of two relative to the 12 quark degrees of freedom included in Chapter 4 since we need to explicitly include both quarks and antiquarks in the effective number of degrees of freedom. The massless pion gas has $g_\pi = 3$ so that $g_{\rm qgp}/g_\pi = 37/3$. The denominator of Eq. (5.125) is then 34/3. Removing a factor of 1/3 gives

$$\lambda(\tau) = \frac{37}{34}\frac{\tau_0}{\tau}\left(\frac{T(\tau_0)}{T_c}\right)^3 - \frac{3}{34} \ . \tag{5.126}$$

After the mixed phase ends, the system is a hadron gas until freezeout at temperature $T = T(\tau_{\rm fo})$. At freezeout, the system has become so dilute that the mean-free path, λ, is larger than L, the system size, and the application of hydrodynamics to the system is no longer valid. The particles no longer interact but simply move away from each other on linear trajectories. This is the thermal freezeout, when elastic scattering stops. The time at which this occurs is

$$\tau_{\rm fo} = \tau_H \left(\frac{T_c}{T(\tau_{\rm fo})} \right)^3 . \tag{5.127}$$

There is also chemical freezeout, the time at which inelastic scattering stops and the particle identities are set until they decay. The two times are not necessarily the same. Chemical freezeout is typically assumed to happen first since more energy is generally needed for inelastic processes that can change quark flavor and thus hadron identity.

The evolution of the system as just described goes from an initial quark-gluon plasma to freezeout. Such evolution should be typical of RHIC collisions where the initial temperature should be well above T_c. However, at lower energies the system may initially be in either a mixed phase if $T(\tau_0) \sim T_c$ or in a hadron gas if $T(\tau_{\rm fo}) < T < T_c$. If we begin in a hadron gas,

$$s(\tau_0) = \frac{4 g_\pi \pi^2}{90} T^3(\tau_0) , \tag{5.128}$$

$$\tau_{\rm fo} = \tau_0 \left(\frac{T(\tau_0)}{T(\tau_{\rm fo})} \right)^3 . \tag{5.129}$$

Figures 5.3 and 5.4 show the change of temperature with time assuming $T_c = 170$ MeV, $T(\tau_{\rm fo}) = 140$ MeV and $\tau_0 = 1$ fm. In Fig. 5.3, the system begins in a quark-gluon plasma with $T(\tau_0) = 250$ MeV. A logarithmic scale has been used on the proper time axis to make the lifetime of the plasma phase more obvious. The system spends ~ 36 fm in the mixed phase, leading to $\tau_{\rm fo} = 70.2$ fm. Such a long lifetime is rather unrealistic, indicating that the longitudinal scaling expansion is too naive. A transverse expansion, which would cool the system more rapidly, reducing the lifetime, is also necessary. On the other hand, in Fig. 5.4, $T(\tau_0) = 150$ MeV. If $T(\tau_{\rm fo}) = 140$ MeV here also, the lifetime of the system is quite short, $\tau_{\rm fo} \sim 1.2$ fm. Note the short, linear scale on the x-axis of Fig. 5.4.

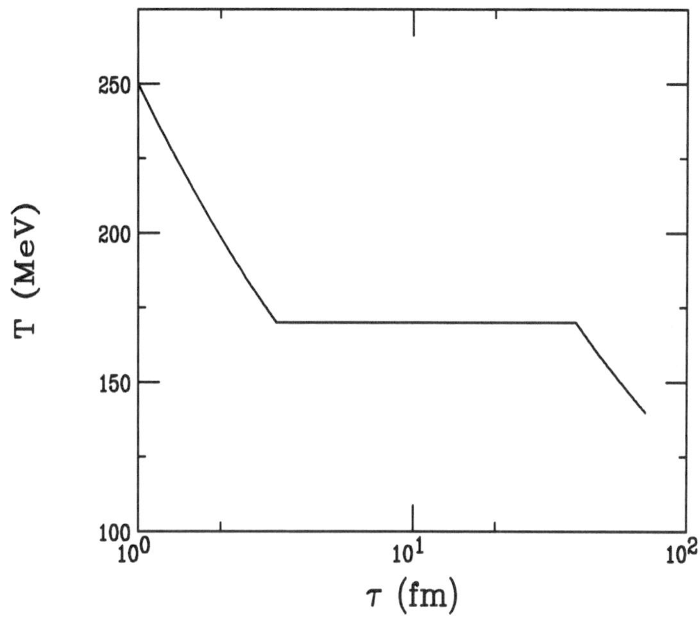

Figure 5.3: Temperature evolution of a system starting in the quark-gluon plasma phase with $T(\tau_0) = 250$ MeV.

Figures 5.5 and 5.6 show the proper time contours for the systems described above with $T(\tau_0) = 250$ MeV and 150 MeV representing initial quark-gluon plasma and hadron gas production respectively. The lifetime of the system is so long relative to the equilibration time, τ_0, for the system with $T(\tau_0) = 250$ MeV that the contours representing times τ_0 and τ_m, the beginning of the mixed phase, are indistinguishable.

5.4.4 Modified longitudinal scaling solutions

Before looking into the transverse expansion, we first consider a perturbation on the longitudinal solution. Instead of only $\theta = y$, as in

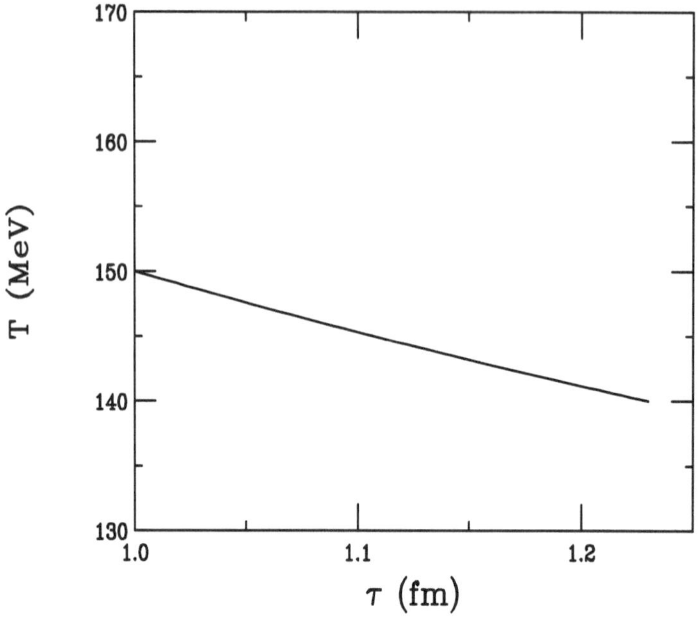

Figure 5.4: Temperature evolution of a system starting in the hadron gas phase with $T(\tau_0) = 150$ MeV.

Eq. (5.85), we consider

$$\theta(y,\tau) = y + \theta' \tag{5.130}$$

where $y \gg \theta'$ so that $\tanh(\theta - y) \approx \theta - y = \theta'$. Then Eqs. (5.77) and (5.78) become

$$0 = \tau\frac{\partial\epsilon}{\partial\tau} + \theta'\frac{\partial\epsilon}{\partial y} + \epsilon + P\,, \tag{5.131}$$

$$0 = \tau\theta'\frac{\partial P}{\partial\tau} + \frac{\partial P}{\partial y} + (\epsilon + P)\theta'\,. \tag{5.132}$$

If we assume that $\theta'\partial\epsilon/\partial y \sim 0$ and $\tau\theta'\partial P/\partial\tau \sim 0$, then Eq. (5.86) is unchanged while Eq. (5.87) is

$$\frac{\partial P}{\partial y} = -(\epsilon + P)\theta'\,. \tag{5.133}$$

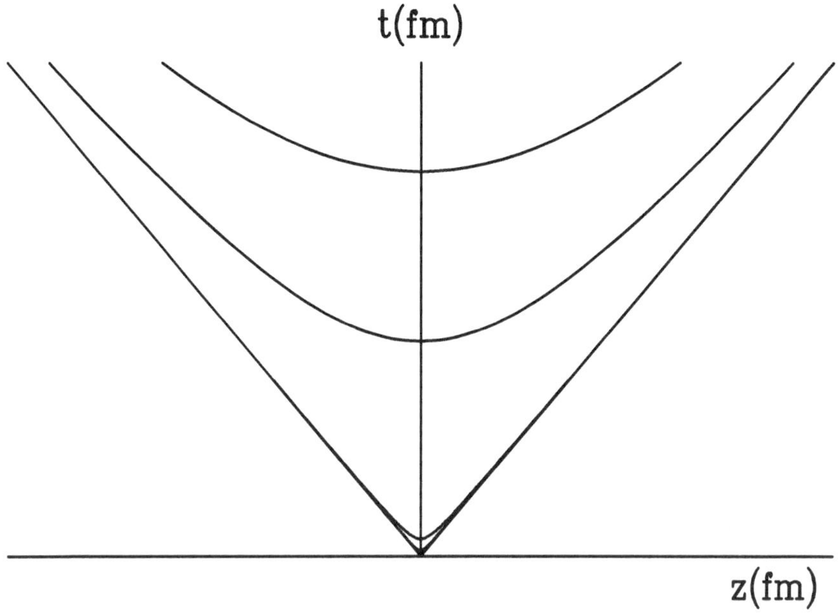

Figure 5.5: Proper time contours for a system starting in a quark-gluon plasma with $T(\tau_0) = 250$ MeV. The contours for $\tau_0 = 1$ fm and the beginning of the mixed phase at $\tau_m = 3.2$ fm are indistinguishable. The intermediate contour shows the end of the mixed phase at $\tau_H = 39.2$ fm while the top contour is at system freezeout, $\tau_{\rm fo} = 70.2$ fm. Only the $t > 0$ half plane is shown.

We can rewrite the derivative of the pressure with respect to the fluid rapidity as

$$\frac{\partial P}{\partial y} = \frac{\partial P}{\partial \epsilon}\frac{\partial \epsilon}{\partial y} = c_s^2 \frac{\partial(\epsilon - B)}{\partial y} \ . \tag{5.134}$$

Adding Eqs. (5.94) and (5.95) and using Eq. (5.96), we have

$$\epsilon + P = (\epsilon - B)(1 + c_s^2) \ . \tag{5.135}$$

Putting these relations into Eq. (5.133), we obtain

$$\partial \ln(\epsilon - B) = -\theta' \left(\frac{1 + c_s^2}{c_s^2} \right) \partial y \ . \tag{5.136}$$

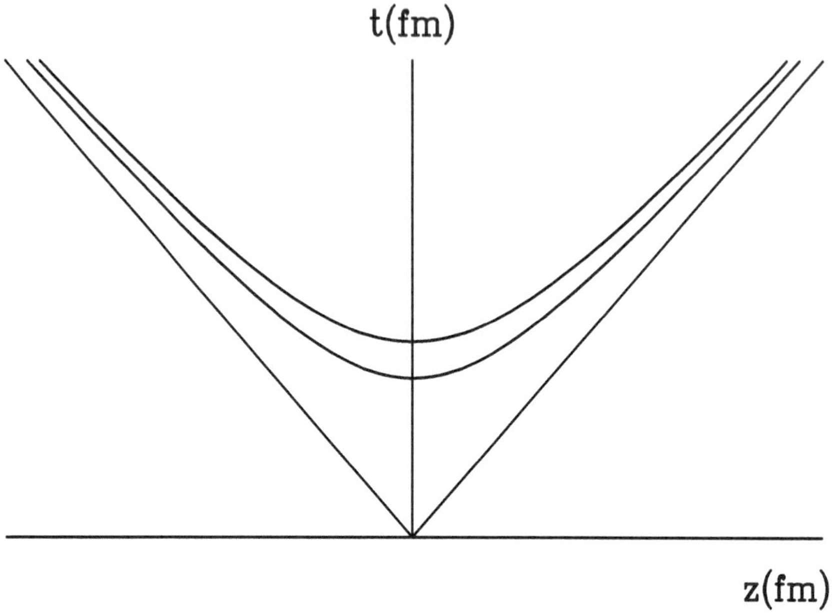

Figure 5.6: Contours of proper time for a system starting in a hadron gas at $T(\tau_0) = 150$ MeV. The lower hyperbola is at $\tau_0 = 1$ fm while the upper is at freezeout, $\tau = 1.23$ fm. Only the $t > 0$ half plane is shown.

The solution is

$$\epsilon(\tau) - B = [\epsilon(\tau_0) - B] \exp[-(1 + c_s^2) y \theta' / c_s^2] \ , \tag{5.137}$$

different from that of Eq. (5.100). The previous solution depended only on time and not on rapidity. Our solution here is independent of time but is limited in its extent in rapidity space.

Another interesting solution of the longitudinal expansion scenario is found when searching for solutions independent of proper time. Equations (5.77) and (5.78) become

$$0 = \tanh(\theta - y) \frac{\partial \epsilon}{\partial y} + (\epsilon + P) \frac{\partial \theta}{\partial y} \ , \tag{5.138}$$

$$0 = \frac{\partial P}{\partial y} + (\epsilon + P) \tanh(\theta - y) \frac{\partial \theta}{\partial y} \ . \tag{5.139}$$

We multiply Eq. (5.138) by $\tanh(\theta - y)$ and subtract Eq. (5.139) to obtain

$$0 = \tanh^2(\theta - y)\frac{\partial \epsilon}{\partial y} - \frac{\partial P}{\partial y} \tag{5.140}$$

since the terms proportional to $\epsilon + P$ cancel upon subtraction. Replacing $\partial P/\partial y$ by $c_s^2 \partial\epsilon/\partial y$, we have

$$0 = [c_s^2 - \tanh^2(\theta - y)]\frac{\partial \epsilon}{\partial y} \; . \tag{5.141}$$

If $\partial\epsilon/\partial y \neq 0$, then

$$c_s^2 = \tanh^2(\theta - y) \; . \tag{5.142}$$

We can solve this equation for the fluid velocity θ, obtaining

$$\theta = y \pm \tanh^{-1} c_s = y \pm y_s \; . \tag{5.143}$$

In this solution, a rarefaction wave propagates into the fluid, reducing the density of the medium.

5.5 Solutions to the hydrodynamic equations: transverse (radial) expansion

We now turn to expansion in the radial direction, the transverse direction in cylindrical coordinates. If $v_r \neq 0$, we write the velocity four-vector as

$$\mathbf{u}^\mu = \frac{1}{\sqrt{1 - v_z^2 - v_r^2}}(1, v_z, v_r, 0) \; . \tag{5.144}$$

The hydrodynamic equations can be solved either at $z = 0$ or $r = 0$ and the solutions boosted to finite values of r and z. The scaling solution, $s\tau = s(\tau_0)\tau_0$, tells us that the transverse motion at any z is related to the motion at $z = 0$ Lorentz boosted in the z direction.

We thus write the entropy and temperature equations, Eqs. (5.43) and (5.50) respectively, in vector notation for more transparency. The entropy current conservation is then

$$0 = \frac{\partial}{\partial t}(s\gamma) + \vec{\nabla}(s\gamma\vec{v}) \ . \tag{5.145}$$

The temperature equation will become

$$\frac{\partial}{\partial t}(T\gamma\vec{v}) + \vec{\nabla}(T\gamma) = \vec{v} \times \vec{\nabla} \times (T\gamma\vec{v}) \ . \tag{5.146}$$

EXAMPLE: Show Eq. (5.146) can be obtained from Eq. (5.50).
Equation (5.146) is most easily derived taking the spatial part of $\mathbf{u}_\lambda$ and ∂_λ. Taking $\mathbf{u} = \gamma(1, \vec{v})$, we can write

$$\mathbf{u}^\mu \partial_\mu = \gamma\frac{\partial}{\partial t} - \gamma\vec{v}\cdot\vec{\nabla} \ , \tag{5.147}$$

giving us

$$0 = \gamma\frac{\partial}{\partial t}(T\gamma\vec{v}) + \gamma\vec{v}\cdot\vec{\nabla}(T\gamma\vec{v}) + \nabla T \ . \tag{5.148}$$

The middle term can be rewritten using the property of vector algebra,

$$\vec{a} \times \vec{b} \times \vec{c} = \vec{b}(\vec{a}\cdot\vec{c}) - \vec{c}(\vec{a}\cdot\vec{b})$$

so that

$$\vec{v} \times \vec{\nabla} \times (T\gamma\vec{v}) = \vec{\nabla}(T\gamma v^2) - \vec{v}\cdot\vec{\nabla}(T\gamma\vec{v}) \ . \tag{5.149}$$

Inserting Eq. (5.149) into Eq. (5.148), after rearranging terms we obtain

$$\vec{\nabla}T + \gamma\frac{\partial}{\partial t}(T\gamma\vec{v}) + \gamma\vec{\nabla}(T\gamma v^2) = \gamma\vec{v} \times \vec{\nabla} \times (T\gamma\vec{v}) \ . \tag{5.150}$$

Note that

$$\gamma\vec{\nabla}(T\gamma v^2) = \gamma^2 v^2\vec{\nabla}T + \gamma T\vec{\nabla}(\gamma v^2) \ . \tag{5.151}$$

Recall that $\gamma^2 v^2 = \gamma^2 - 1$ so that

$$\vec{\nabla}T + \gamma^2 v^2\vec{\nabla}T = \gamma^2\vec{\nabla}T \ . \tag{5.152}$$

We can also let

$$\gamma T\vec{\nabla}(\gamma v^2) = \frac{1}{2}T\vec{\nabla}(\gamma^2 v^2) \ . \tag{5.153}$$

Equation (5.148) can then be written as

$$\gamma^2\vec{\nabla}T + \frac{1}{2}T\vec{\nabla}(\gamma^2 v^2) + \gamma\frac{\partial}{\partial t}(T\gamma\vec{v}) = \gamma\vec{v}\times\vec{\nabla}\times(T\gamma\vec{v}) \ . \tag{5.154}$$

We now rewrite the first term in Eq. (5.154) as

$$\gamma^2\vec{\nabla}T = \gamma\vec{\nabla}(T\gamma) - \gamma T\vec{\nabla}\gamma = \gamma\vec{\nabla}(T\gamma) - \frac{1}{2}T\vec{\nabla}\gamma^2 \ . \tag{5.155}$$

Combining the term proportional to $\vec{\nabla}\gamma^2$ with the second term in Eq. (5.154), we have

$$\frac{1}{2}T\vec{\nabla}(\gamma^2 v^2 - \gamma^2) = \frac{1}{2}T\vec{\nabla}(1) = 0 \ . \tag{5.156}$$

We are left with

$$\frac{\partial}{\partial t}(T\gamma\vec{v}) + \gamma\vec{\nabla}(T\gamma) = \gamma\vec{v}\times\vec{\nabla}\times(T\gamma\vec{v}) \ . \tag{5.157}$$

Finally, we drop the extra factor of γ in Eq. (5.157) to obtain Eq. (5.146). We note that in the case of symmetric motion, as studied here, the cross terms on the right-hand side of Eq. (5.146) will vanish.

We now turn to the specific case where the velocity four-vector is given by Eq. (5.144). The entropy current conservation equation is then

$$0 = \frac{\partial}{\partial t}(s\gamma) + \frac{1}{r}\frac{\partial}{\partial r}(rs\gamma v_r) + \frac{\partial}{\partial z}(s\gamma v_z) \ . \tag{5.158}$$

We do not immediately take $z \to 0$ but wait until after we have expanded the derivatives. We then let $z = 0$ and drop terms linear in v_z. Thus $\partial v_z/\partial t = 1/t$ but *e.g.* $\partial s/\partial z = 0$ since z is fixed. Equation (5.158) becomes

$$\begin{aligned} 0 &= \frac{\partial}{\partial t}(s\gamma) + \frac{1}{r}(s\gamma v_r) + \frac{\partial}{\partial r}(s\gamma v_r) + \frac{1}{t}(s\gamma) + v_z\frac{\partial}{\partial z}(s\gamma) \\ &= \frac{\partial}{\partial t}(s\gamma) + \frac{\partial}{\partial r}(s\gamma v_r) + s\gamma(\frac{v_r}{r} + \frac{1}{t}) \ . \end{aligned} \tag{5.159}$$

At $z = 0$, the temperature equation becomes

$$0 = \frac{\partial}{\partial t}(T\gamma v_r) + \frac{\partial}{\partial r}(T\gamma) \; . \tag{5.160}$$

The radial velocity v_r is related to the transverse rapidity variable as

$$\alpha = \tanh^{-1} v_r \tag{5.161}$$

so that

$$v_r = \tanh\alpha \; , \qquad \gamma = \cosh\alpha \; , \qquad \gamma v_r = \sinh\alpha \; . \tag{5.162}$$

By substitution we can show that

$$0 = \frac{\partial}{\partial t}(rts\cosh\alpha) + \frac{\partial}{\partial r}(rts\sinh\alpha) \; , \tag{5.163}$$

$$0 = \frac{\partial}{\partial t}(T\sinh\alpha) + \frac{\partial}{\partial r}(T\cosh\alpha) \; . \tag{5.164}$$

EXAMPLE: Obtain Eqs. (5.163) and (5.164) from Eqs. (5.159) and (5.160).
We start with the entropy equation. Using the values for v_r and γ in Eq. (5.162), we have

$$0 = \frac{\partial}{\partial t}(s\cosh\alpha) + \frac{1}{t}(s\cosh\alpha) + \frac{\partial}{\partial r}(s\sinh\alpha) + \frac{1}{r}(s\sinh\alpha) \; . \tag{5.165}$$

We now multiply both sides by r and t so that

$$0 = rt\frac{\partial}{\partial t}(s\cosh\alpha) + rs\cosh\alpha + rt\frac{\partial}{\partial r}(s\sinh\alpha) + ts\sinh\alpha \; . \tag{5.166}$$

Note, *e.g.* that

$$\partial/\partial t(rts\cosh\alpha) = rt\partial/\partial t(s\cosh\alpha) + rs\cosh\alpha \tag{5.167}$$

since $\partial(rt)/\partial t = r$. With this, rearranging terms simply gives us Eq. (5.163). The factor of rt in Eq. (5.163) is a phase space factor for cylindrical geometry. Equation (5.164) is obtained simply by substitution.

To numerically evaluate the hydrodynamic equations, it is more convenient to write them in the form

$$0 = \frac{\partial \ln s}{\partial t} + v_r \frac{\partial \ln s}{\partial r} + v_r \frac{\partial \alpha}{\partial t} + \frac{\partial \alpha}{\partial r} + \frac{1}{t} + \frac{v_r}{r} , \tag{5.168}$$

$$0 = v_r \frac{\partial \ln T}{\partial t} + \frac{\partial \ln T}{\partial r} + \frac{\partial \alpha}{\partial t} + v_r \frac{\partial \alpha}{\partial r} . \tag{5.169}$$

EXAMPLE: Derive Eqs. (5.168) and (5.169).
To obtain Eqs. (5.168) and (5.169), we must expand all the derivatives of Eqs. (5.163) and (5.164). In the case of Eq. (5.163), we have

$$\begin{aligned} 0 &= rs \cosh\alpha + rts \sinh\alpha \frac{\partial \alpha}{\partial t} + rt \cosh\alpha \frac{\partial s}{\partial t} \\ &\quad + rts \cosh\alpha \frac{\partial \alpha}{\partial r} + rt \sinh\alpha \frac{\partial s}{\partial r} + st \sinh\alpha . \end{aligned} \tag{5.170}$$

We next divide both sides by rst and rewrite $\partial s/s$ as $\partial \ln s$ to obtain

$$\begin{aligned} 0 &= \frac{1}{t} \cosh\alpha + \sinh\alpha \frac{\partial \alpha}{\partial t} + \cosh\alpha \frac{\partial \ln s}{\partial t} \\ &\quad + \cosh\alpha \frac{\partial \alpha}{\partial r} + \sinh\alpha \frac{\partial \ln s}{\partial r} + \frac{1}{r} \sinh\alpha . \end{aligned} \tag{5.171}$$

Finally, both sides are divided by $\cosh\alpha$ and $\tanh\alpha$ is replaced by v_r to obtain Eq. (5.168).

The derivation of the temperature result is simpler but follows the same lines. We first expand all the derivatives to find

$$0 = T \cosh\alpha \frac{\partial \alpha}{\partial t} + \sinh\alpha \frac{\partial T}{\partial t} + T \sinh\alpha \frac{\partial \alpha}{\partial r} + \cosh\alpha \frac{\partial T}{\partial r} . \tag{5.172}$$

Dividing both sides by $T \cosh\alpha$, replacing $\partial T/T$ by $\partial \ln T$ and $\tanh\alpha$ by v_r, we are left with Eq. (5.169).

We can recast Eqs. (5.168) and (5.169) still further by recalling that $\partial \ln T / \partial \ln s = c_s^2$. Then a temperature dependent potential can be introduced, defined as

$$\partial \phi = c_s \partial \ln s = \frac{1}{c_2} \partial \ln T . \tag{5.173}$$

We can then make the replacements $\partial \ln T = c_s \partial\phi$ and $\partial \ln s = \partial\phi / c_s$. The two equations then become

$$0 = \frac{\partial\phi}{\partial t} + v_r c_s \frac{\partial\alpha}{\partial t} + v_r \frac{\partial\phi}{\partial r} + c_s \frac{\partial\alpha}{\partial r} + c_s \left(\frac{1}{t} + \frac{v_r}{r} \right) , \quad (5.174)$$

$$0 = v_r c_s \frac{\partial\phi}{\partial t} + \frac{\partial\alpha}{\partial t} + c_s \frac{\partial\phi}{\partial r} + v_r \frac{\partial\alpha}{\partial r} . \quad (5.175)$$

Adding Eqs. (5.174) and (5.175) gives

$$\begin{aligned} 0 &= (1 + v_r c_s) \frac{\partial(\phi+\alpha)}{\partial t} + (v_r + c_s) \frac{\partial(\phi+\alpha)}{\partial r} \\ &\quad + c_s \left(\frac{1}{t} + \frac{v_r}{r} \right) \end{aligned} \quad (5.176)$$

while subtracting Eq. (5.175) from Eq. (5.174) gives

$$\begin{aligned} 0 &= (1 - v_r c_s) \frac{\partial(\phi+\alpha)}{\partial t} + (v_r - c_s) \frac{\partial(\phi+\alpha)}{\partial r} \\ &\quad + c_s \left(\frac{1}{t} + \frac{v_r}{r} \right) . \end{aligned} \quad (5.177)$$

We now change variables to

$$\ln a_\pm = \phi \pm \alpha \quad (5.178)$$

so that

$$\partial \ln a_\pm = \frac{\partial a_\pm}{a_\pm} . \quad (5.179)$$

After dividing both sides of Eqs. (5.174) and (5.175) by $1 \pm v_r c_s$ and multiplying by $a_\pm$, the combined result is

$$0 = \frac{\partial a_\pm}{\partial t} + \frac{v_r \pm c_s}{1 \pm v_r c_s} \frac{\partial a_\pm}{\partial r} + \frac{c_s}{1 \pm v_r c_s} \left(\frac{1}{t} + \frac{v_r}{r} \right) . \quad (5.180)$$

The new variable $a_\pm$ is better behaved as $T \to 0$ since $\phi \propto \ln T \to -\infty$ while, instead, $a_\pm \to 0$. The potential ϕ and the radial velocity v_r can be defined in terms of $a_\pm$ as

$$\phi = \frac{1}{2} \ln(a_+ a_-) , \quad (5.181)$$

$$v_r = \frac{a_+ - a_-}{a_+ - a_-} = \frac{\exp(\alpha) + \exp(-\alpha)}{\exp(\alpha) - \exp(-\alpha)} = \tanh\alpha . \quad (5.182)$$

Equation (5.180) has been written in such a way as to make it easier to obtain a numerical solution. The solution is obtained using the method of characteristics, often applied to hydrodynamic calculations. This method is a way of solving systems of partial differential equations (PDEs). For a first order PDE, a characteristic is a straight line in phase space where the PDE is an ordinary differential equation, a function of only one independent variable and derivatives only with respect to that variable. A two-dimensional PDE can be written in the form

$$F(x, y, u, u_x, u_y) = 0 \tag{5.183}$$

where x and y are independent variables, u is the solution and u_x, u_y are partial derivatives of u with respect to x and y. Curves of constant F, called level curves, are used to determine u. If u is smooth and differentiable, this procedure is exact. The level curves do not intersect but the characteristic curves sometimes do, resulting in multi-valued solutions. Discontinuities in the form of shock waves (compression waves) exist in multi-valued solutions. If the characteristics do not cover the full domain of the PDE, rarefactions (reductions in density and pressure due to the passage of a sound wave through the medium) may develop.

The wave equation

$$0 = c\frac{\partial u}{\partial x} + \frac{\partial u}{\partial t} \tag{5.184}$$

describes the one-dimensional motion of a wave with no modification in shape. Any curve in the x, t plane can be expressed as $x \equiv x(r)$ and $t \equiv t(r)$ where r is a distance along a characteristic curve starting from $x = x_0$ at $t = 0$ when $r = 0$. If the ordinary differential equation determined by the characteristics is a solution, then u is known all along the characteristic curve at $x = x_0$. Choosing a different value of x_0 would give another solution for u. Thus

$$u(x, t) = u(x(r), t(r)) \ , \tag{5.185}$$

making u an indirect function of r. Then

$$\frac{\partial u}{\partial r} = \frac{\partial x}{\partial r}\frac{\partial u}{\partial x} + \frac{\partial t}{\partial r}\frac{\partial u}{\partial t} \ . \tag{5.186}$$

Starting from Eq. (5.184), we have $\partial u/\partial r = 0$ as long as $\partial x/\partial r = c$ and $\partial t/\partial r = 1$. Then $t = r$ and $t = 0$ at $r = 0$. We can immediately also see that $x = cr + x_0 = ct + x_0$ where x_0 is an integration constant.

Equation (5.180) has an inhomogeneous term, linear in $a_\pm$,

$$\frac{c_s a_\pm}{1 \pm v_r c_s}\left(\frac{1}{t} + \frac{v_r}{r}\right) . \tag{5.187}$$

It is easier to first find the solution for an example without an inhomogeneous term. We can calculate the solution for a one-dimensional expansion in the transverse direction (with transverse coordinate x) of a semi-infinite slab of matter from $x = -\infty$ to $x = R > 0$. The expansion occurs at $t > 0$. At $x < R$, $T = T_0$ and $v_x = 0$. A simple solution may be obtained but first we have to go back to Eqs. (5.43) and (5.50) to write them in characteristic form, finding

$$\frac{\partial a_\pm}{\partial t} + \frac{v_x \pm c_s}{1 \pm v_x c_s}\frac{\partial a_\pm}{\partial x} = 0 . \tag{5.188}$$

EXAMPLE: Show that Eq. (5.188) follows from Eqs. (5.43) and (5.50). In this case, only the x coordinate is used so that $\mathbf{u} = \gamma(1, v_x)$ and $\vec{\nabla} = \hat{x}\partial/\partial x$. Thus the entropy and temperature equations become

$$0 = \frac{\partial}{\partial t}(s\gamma) + \frac{\partial}{\partial x}(s\gamma v_x) , \tag{5.189}$$

$$0 = \frac{\partial}{\partial t}(T\gamma v_x) + \frac{\partial}{\partial x}(T\gamma) . \tag{5.190}$$

We now let $v_x = \tanh\theta$ so that $\gamma = \cosh\theta$ and $\gamma v_x = \sinh\theta$ so that the entropy equation is

$$0 = \frac{\partial}{\partial t}(s\cosh\theta) + \frac{\partial}{\partial x}(s\sinh\theta) \tag{5.191}$$

$$= s\sinh\theta\frac{\partial\theta}{\partial t} + \cosh\theta\frac{\partial s}{\partial t} + s\cosh\theta\frac{\partial\theta}{\partial x} + \sinh\theta\frac{\partial s}{\partial x} \tag{5.192}$$

$$= \frac{\partial \ln s}{\partial t} + v_x\frac{\partial\theta}{\partial t} + v_x\frac{\partial \ln s}{\partial x} + \frac{\partial\theta}{\partial x} \tag{5.193}$$

while the temperature equation is

$$
\begin{aligned}
0 &= \frac{\partial}{\partial t}(T\sinh\theta) + \frac{\partial}{\partial x}(T\cosh\theta) && (5.194)\\
&= T\cosh\theta\frac{\partial\theta}{\partial t} + \sinh\theta\frac{\partial T}{\partial t} + T\sinh\theta\frac{\partial\theta}{\partial x} + \cosh\theta\frac{\partial T}{\partial x} && (5.195)\\
&= v_x\frac{\partial \ln T}{\partial t} + \frac{\partial\theta}{\partial t} + \frac{\partial \ln T}{\partial x} + v_x\frac{\partial\theta}{\partial x} \; . && (5.196)
\end{aligned}
$$

Using the definition of $\partial\phi$ in Eq. (5.173), we obtain

$$
\begin{aligned}
0 &= \frac{\partial\phi}{\partial t} + v_x c_s\frac{\partial\theta}{\partial t} + v_x\frac{\partial\phi}{\partial x} + c_s\frac{\partial\theta}{\partial x} \; , && (5.197)\\
0 &= v_x c_s\frac{\partial\phi}{\partial t} + \frac{\partial\theta}{\partial t} + c_s\frac{\partial\phi}{\partial x} + v_x\frac{\partial\theta}{\partial x} \; . && (5.198)
\end{aligned}
$$

Adding and subtracting the two equations gives a combined result of

$$
0 = (1 \pm v_x c_s)\frac{\partial}{\partial t}(\phi \pm \theta) + (v_x \pm c_s)\frac{\partial}{\partial x}(\phi \pm \theta) \; . \tag{5.199}
$$

We again replace $\phi \pm \theta$ by $\ln a_\pm$. After dividing by $1 \pm v_x c_s$, we obtain the result in Eqs. (5.188).

We can now recast Eq. (5.188) in terms of Eq. (5.186) to find the solution for $a_\pm$ and the temperature dependence of the expansion. In the case of longitudinal expansion alone, the expansion is governed by a single set of characteristics, with the negative sign, while, for expansion in more than one dimension, the characteristic equations are coupled [82]. Now $a_+ \equiv 1$ so that

$$
\alpha = -\phi = \tanh^{-1} v_x = \frac{1}{2}\ln\left(\frac{1+v_x}{1-v_x}\right) \; . \tag{5.200}
$$

We must now solve for a_-. Applying Eq. (5.186), $\partial t/\partial r = 1$ while

$$
\frac{\partial x}{\partial r} = \frac{v_x - c_s}{1 - v_x c_s} = C \; . \tag{5.201}
$$

The characteristic line is then

$$
x(t, x_0) = Cr + x_0 = \frac{v_x - c_s}{1 - v_x c_s}t + x_0 \; . \tag{5.202}
$$

The function a_- is constant along this line so that

$$a_-(x,t) = a_-(x_0, t=0) = a_-\left(x - \frac{v_x - c_s}{1 - v_x c_s} t, t = 0\right) . \qquad (5.203)$$

The fluid velocity depends only on the temperature since $\phi \sim \ln T$. The region of nonzero temperature is near the edge of the system, $x \sim R$, at $t = 0$ so that, at later times, a point with temperature $T < T_0$ is

$$x(T) = \frac{v_x - c_s}{1 - v_x c_s} t + R . \qquad (5.204)$$

The coefficient of t above is the velocity of sound relative to the medium. At $T = T_0$, $v_x = 0$ and $x = R - c_s t$. Thus a rarefaction wave propagates into the medium. Solving for v_x when $R + t \geq x \geq R - c_s t$ (the upper limit on x when $v_x = 1 + c_s$) we obtain

$$v_x = \frac{x - R + c_s t}{t + c_s(x - R)} . \qquad (5.205)$$

Note that to insure a positive velocity, for $x < R - c_s t$ we take $v_x = 0$. We can obtain the value of α from the expression for v_x,

$$\alpha = \frac{1}{2} \ln \left[\frac{1 + c_s}{1 - c_s} \frac{t + x - R}{t - x + R} \right] . \qquad (5.206)$$

If c_s is constant, we can solve for the temperature,

$$\begin{aligned} \partial\phi(T) &= \frac{1}{c_s} \partial \ln T , \\ \phi &= \frac{1}{c_s} \ln \left(\frac{T}{T_0} \right) = -\alpha , \end{aligned} \qquad (5.207)$$

giving us

$$\begin{aligned} T &= T_0 \exp(-c_s \alpha) \\ &= T_0 \left(\frac{1 - c_s}{1 + c_s} \frac{t - x + R}{t + x - R} \right)^{c_s/2} \end{aligned} \qquad (5.208)$$

using the result for α in Eq. (5.206). The temperature is equal to T_0 when the expression in brackets in Eq. (5.208) is equal to unity or

$x = R - c_s t$. This solution to the time dependence of the temperature can be compared to that of Bjorken scaling [74],

$$T = T(\tau_0)\left(\frac{\tau_0}{\tau}\right)^{c_s^2} . \tag{5.209}$$

Baym *et al.* [82] calculated solutions to the hydrodynamic equations in one dimension as well as in spherical and cylindrical coordinates. The characteristic curves used for their calculations are given in Eq. (5.188) for a one-dimensional expansion and in Eq. (5.180) for cylindrical geometry. We will show their results in the one-dimensional and cylindrical cases as well as briefly mention their results in spherical coordinates since they are a good illustration of the basic effects.

In the case of longitudinal and spherical expansions, the hydrodynamic equations are invariant under scale transformations so that the results can be presented as functions of r/R and t/R where r is the spatial coordinate and R is the system size. The temperature profile as a function of r/R is shown in Fig. 5.7 for several t/R contours. At early times, $t/R < 1/c_s$, the system can be described by the solutions found using the method of characteristics as discussed above. At later times, there is approximate scaling at relatively low values of r/R. The kinks in the temperature profiles at $r/R \sim 2$ and 3 correspond to $t/R = 4$ and 5 occur at the boundary of the two regions. There is a stationary point where the $t/R \leq 3$ curves all cross. At this point, $v_x(T) = c_s$ and

$$T_{\rm stat} = T_0\left(\frac{1 - c_s}{1 + c_s}\right)^{c_s/2} = 0.68 T_0 . \tag{5.210}$$

After $T < T_{\rm stat}$, the stationary point no longer exists. Note that it is only at $t/R \sim 5$ that the temperature has dropped to half its initial value at $r/R \leq 1$. If $R \sim 6.6$ fm, the radius of a lead nucleus, then $t \sim 33$ fm, a rather long time.

Figure 5.8 shows individual isotherms (curves of constant temperature) in the t, r plane. They are labeled as T/T_0. The linear parts emanating from $r/R = 1$ correspond to the characteristic solutions while the hyperbolic-like parts resemble the Bjorken scaling results in Figs. 5.3 and 5.4.

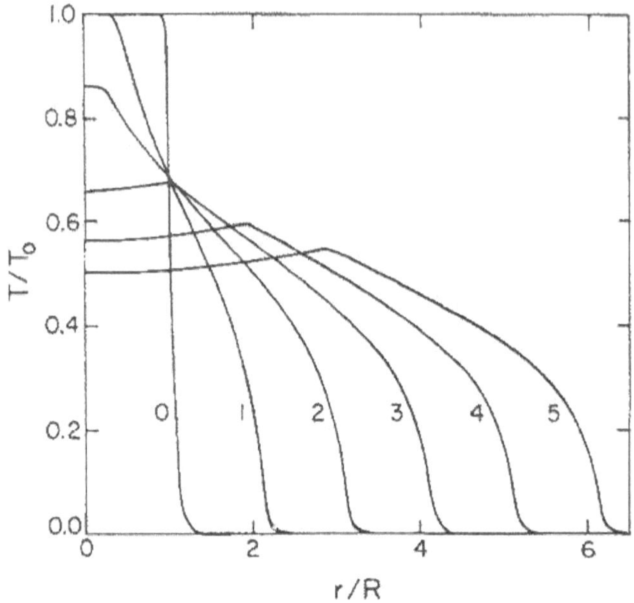

Figure 5.7: Temperature (T/T_0) profiles are shown as a function of r/R for a one-dimensional expansion at successive values of relative time, t/R. Note that the expansion begins at $t = 0$. Reprinted from Ref. [82] with permission from Elsevier.

A spherically expanding system cools much more rapidly. The temperature decreases to half its value already at $t/R < 2$, about a third the time of the one-dimensional expansion [82]. The cooling is typically too fast for a scaling regime to develop.

In cylindrical coordinates, Eq. (5.180), there is an inhomogeneous term. The two parts of this term are due to the two-dimensional character of the expansion and the coupling between the transverse and longitudinal motions. There is thus a difference in the longitudinal and transverse coordinates and the hydrodynamic equations are no longer invariant under scale transformations. There are, instead, two scales: the radius of the initial distribution, R, and the initial time at which the hydrodynamic evolution begins, t_0. The results for the temperature evolution shown in Fig. 5.9 are for $t_0 = 1$ fm and $R = 7$

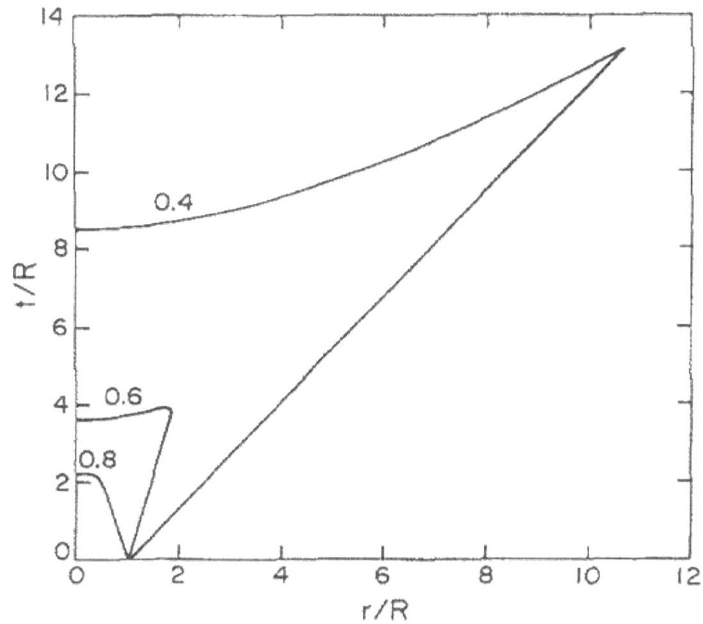

Figure 5.8: Isotherms in the $r/R, t/R$ plane for a one-dimensional expansion. The label on each isotherm is the value of T/T_0. Reprinted from Ref. [82] with permission from Elsevier.

fm. Increasing t_0 decreases the cooling due to the longitudinal part of the expansion. Isotherms in the transverse direction are shown in Fig. 5.10. The horizontal portions of the isotherm indicate uniform cooling.

5.6 Observable consequences

Before ending this chapter, we briefly touch on some experimental observables related to the equation of state and the expansion of the system. The first is the transverse momentum or transverse mass distribution of produced particles. The slope of the distribution includes contributions from the freezeout temperature and the radial expansion velocity. The second is the azimuthal distribution. Pressure gradients

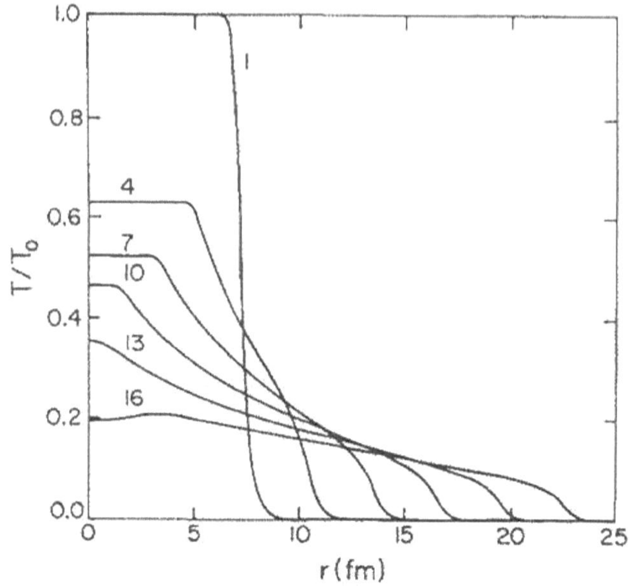

Figure 5.9: Temperature (T/T_0) profiles are shown as a function of r for a cylindrical expansion coupled to a longitudinal expansion. The curves are labeled by the time (in fm) that has elapsed since the start of the evolution. The initial time here is assumed to be $t_0 = 1$ fm. Reprinted from Ref. [82] with permission from Elsevier.

within the participant matter develop into an azimuthally anisotropic flow pattern which is discernible in the final state.

5.6.1 Particle spectra

One way to quantify the expansion (fluid) velocity of the system produced in heavy-ion collisions is to study the distributions of the produced particles, particularly the transverse momentum distribution. The invariant rate, $EdN/d^3p = dN/\pi dp_T^2 dy$, can be parameterized in terms of the multiplicity in a given rapidity range, dN/dy, and the Maxwell-Boltzmann distribution, $\exp(-E/T)$ in the rest frame of the particle and $\exp[-m_T \cosh(y - \eta)/T]$ where η is the fluid rapidity in

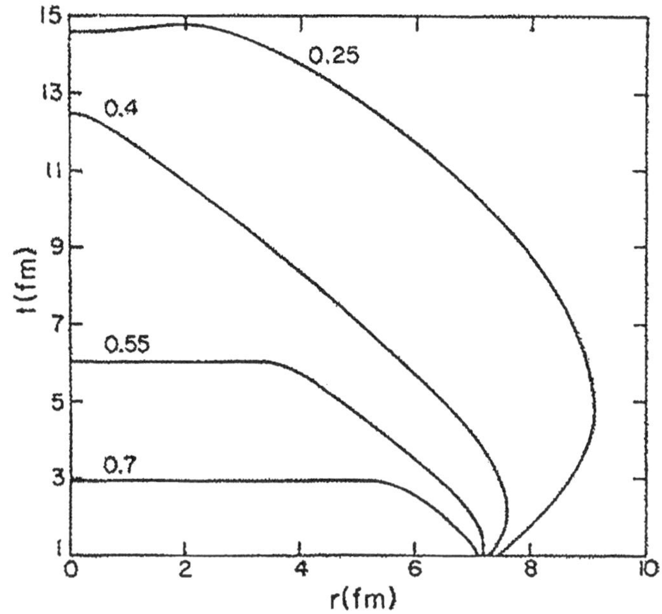

Figure 5.10: Isotherms the r, t plane for a cylindrical expansion coupled to a longitudinal expansion. The initial time is assumed to be $t_0 = 1$ fm. The label on each isotherm is the value of T/T_0. Reprinted from Ref. [82] with permission from Elsevier.

the frame of the fluid. Thus the expression

$$\frac{dN}{2\pi p_T dp_T dy} \propto \frac{dN}{dy} \exp[m_T \cosh(y - \eta)/T] , \tag{5.211}$$

is fit to data. The final-state distribution is governed by the temperature at which the system freezes out and the particles no longer interact. If the values of T and η agree for different particle species, *e.g.* pions, kaons, protons and antiprotons, this indicates a global freezeout temperature and fluid velocity. When no expansion is assumed, a global temperature cannot necessarily be identified for all particle species.

We note that this type of fitting procedure will give a meaningful result for the temperature only at the low p_T end of the spectrum since low p_T particles are more likely to be in thermal equilibrium. High

p_T particles are more likely to have been produced in hard scatterings, governed by perturbative QCD. However, it must be remarked that while high p_T particles may not come to thermal equilibrium, they are still affected by the surrounding medium.

To further quantify the effects of the hot, dense matter on the transverse momentum distributions, a reference spectrum at the same energy, within the same rapidity range, would be useful. Data from pp collisions are often used as reference spectra since they contain no nuclear effects. At high p_T, perturbative QCD can be used to calculate the pp result rather accurately, making it a useful reference distribution. The ratio of the AA result to the reference pp result at the same energy is denoted R_{AA},

$$R_{AA}(p_T) = \frac{d\sigma_{AA}/dp_T}{\langle \sigma_{inel} T_{AA} \rangle d\sigma_{pp}/dp_T} \, . \tag{5.212}$$

The factor multiplying the pp cross section in Eq. (5.212) is the average number of nucleon-nucleon collisions, see Chapter 3, and is used to obtain the AA cross section per nucleon. At high p_T in AA collisions, if the particles are not in equilibrium and production is dominated by gluons, which eliminates the dependence on the proton and neutron number, the AA result would be the pp result scaled up by a factor of A^2 since the integral of the nuclear overlap function, $T_{AB}(b)$, over impact parameter is AB or $R_{AA} = 1$.

Figure 5.11 shows both the per nucleon p_T distributions in pp and AA collisions assuming that the AA result, in addition to pp-like production in the first collisions, includes particles produced in the thermal system with some 'thermal' slope and the resulting R_{AA} as a function of p_T. The left-hand side of Fig. 5.11 is a schematic illustration of two p_T distributions, one in pp and the other in AA collisions, at the same energy. The AA per nucleon yield is higher due to additional particle production in the hot medium produced in the collision. However, these secondary particles have, on average, rather low p_T so that the yield is only important for p_T less than a few GeV. At higher p_T, the per nucleon yields would be indistinguishable in the two cases. The right-hand side shows the R_{AA} resulting from forming the ratio of the two distributions. The line at $R_{AA} = 1$ indicates what the ratio would be

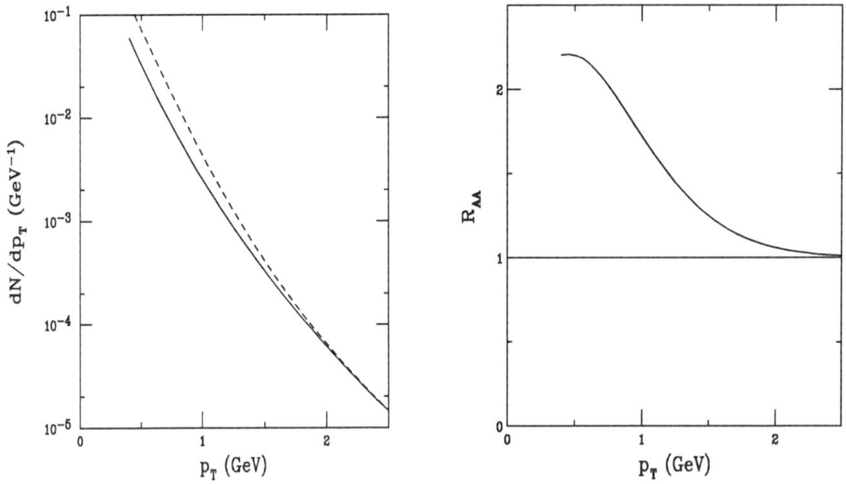

Figure 5.11: Left-hand side: The p_T distributions in pp (solid) and AA (dashed) collisions. Right-hand side: The ratio R_{AA} as a function of p_T due to hydrodynamic effects alone.

if AA collisions were a simple superposition of pp collisions. (In Chapter 7, we discuss thermal dilepton production in a more quantitative way.)

It would be nice if something so obvious as an enhanced yield due to thermal production could be observed in the p_T distributions alone. However, this is not the case. Indeed, in pA interactions, broadening of the p_T distributions relative to the pp distributions has been observed. Such initial-state effects must be taken into account when assessing AA data. However, the RHIC data indicate that something more than hydrodynamic expansion and initial-state effects are affecting the p_T distributions since, in fact, $R_{AA} < 1$ over a large p_T range and the AA yield is suppressed relative to the pp result. Measurements of R_{AA} in d+Au collisions at RHIC showed that, while some initial-state effects may be present, they are not strong enough to cause the suppression seen in Au+Au collisions up to $p_T \geq 10$ GeV which must then be attributed to final-state effects. At such high p_T, any thermal production is negligible. The high p_T results may be due to energy loss by partons moving through matter, see *e.g.* Ref. [83].

The transverse momentum distributions should also depend on centrality since more central collisions are more likely to come to thermal equilibrium. Thus in addition to comparing with a pp reference process, the centrality of the collision can be studied by taking the ratio of the results in central collisions to those in more peripheral collisions, especially useful if no pp reference data is available. This central to peripheral ratio is known as R_{CP} where C is for central and P is for peripheral. The denominator is typically the most peripheral bin with meaningful statistics. Measurements of R_{CP} indicate that it is also less than unity. As the numerator of R_{CP} goes from the most central bin to more peripheral bins, R_{CP} increases toward unity. See the reports from the RHIC experiments published in Ref. [83] for more details.

5.6.2 Azimuthally anisotropic collective flow

In the hydrodynamic expansion following a heavy-ion collision, the matter develops a flow pattern. The flow pattern is related to the equation of state of the system through the dependence of the pressure (or energy per particle) on temperature and density. The phenomenon of this collective flow has been investigated over a wide range of energies, from tens of MeV per nucleon to the RHIC regime, see Ref. [84] for a review of the data prior to the startup of RHIC.

To study flow in experiments, it is important to determine the reaction plane, a reference plane, of the collision. In Cartesian coordinates, the $\hat{z}$ unit vector is in the beam direction. The $\hat{x}$ unit vector lies in the direction of the impact parameter vector and forms the reaction plane with the $\hat{z}$ vector. The $\hat{y}$ unit vector is normal to the reaction plane. Particle rapidity, p_T and azimuthal angle with respect to the reaction plane can be determined from its motion. The azimuthal distribution of emitted particles in a given rapidity range can be characterized using the Fourier expansion,

$$\frac{d^2N}{dp_T d\phi dy} = \Sigma_{n=0}^{\infty} v_n(p_T)\cos(n\phi) \tag{5.213}$$

where ϕ is the azimuthal angle with respect to the reaction plane. The v_0 coefficient describes the overall yield of particles in the given rapidity range. It is typically taken to be an overall normalization parameter allowing all the higher order coefficients to be given as a percent. The v_1 coefficient gives the strength of the directed flow in the reaction plane with preferred emission at $\phi = 0$ in the forward hemisphere and $\phi = \pi$ in the backward hemisphere. In the case of symmetric AA collisions, v_1 must be zero around midrapidity. The coefficient v_2 most closely describes the elliptic flow. For collisions with $\sqrt{S_{NN}} < 1.5$ GeV, the elliptic flow corresponds to an out-of-plane squeeze-out with maximum emission at $\phi = \pi/2$ and $3\pi/2$. When $\sqrt{S_{NN}} > 1.5$ GeV, the elliptic flow is oriented in the reaction plane with maximum emission at $\phi = 0$ and π [85]. At RHIC, v_2 has been measured as a function of p_T, see *e.g.* Ref. [86]. The effects are in rather good agreement with hydrodynamic calculations as long as the transverse momentum is not so large since high p_T particles are not expected to fully thermalize. As a function of rapidity, v_2 is observed to be largest at midrapidity and decreasing at forward and backward rapidity [87]. The higher order Fourier moments (v_3 and above) can help match the simplistic Fourier expansion to the shape of the measured experimental distribution.

The most flow is generated from collisions that are not fully central but still have significant overlap, see Fig. 5.12. The shaded overlap region formed by the participants is initially an ellipsoid shaped somewhat like an almond with maximum compression along the broad flat sides. As the system expands and cools, its coordinate-space shape becomes more spherical. However, the momentum-space distributions retain the memory of the early pressure anisotropies (note that the coordinate space distributions are not directly experimentally observable). The initial azimuthal anisotropy, with arrows indicating the direction of particle motion, is shown in Fig. 5.13. Indeed, the strongest elliptic flow effects in Ref. [86] are seen for the $45-85\%$ most central (more peripheral) collisions, the weakest for the $0-11\%$ most central events. Meson flow (*e.g.* π and K) is stronger than baryon flow (*e.g.* p and Λ).

Comparing different collision systems as a function of centrality is a good way to study the effect of the initial coordinate-space azimuthal

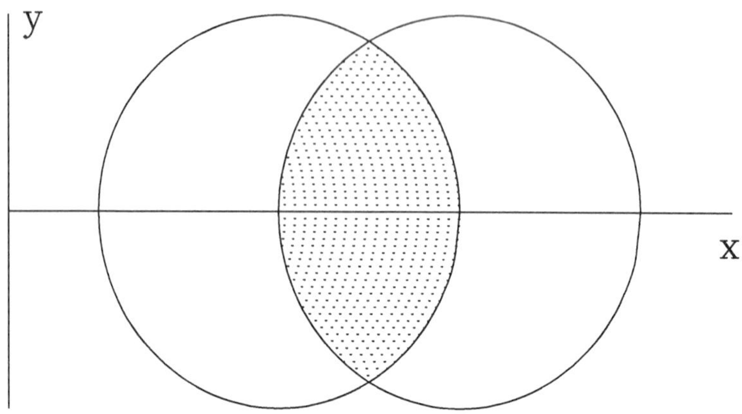

Figure 5.12: A semi-central collision of two equal size nuclei is shown in the transverse plane. The x and y unit vectors are labeled. The z axis (the beam axis) goes into the page. The shaded overlap region indicates the nucleon participants in the collision.

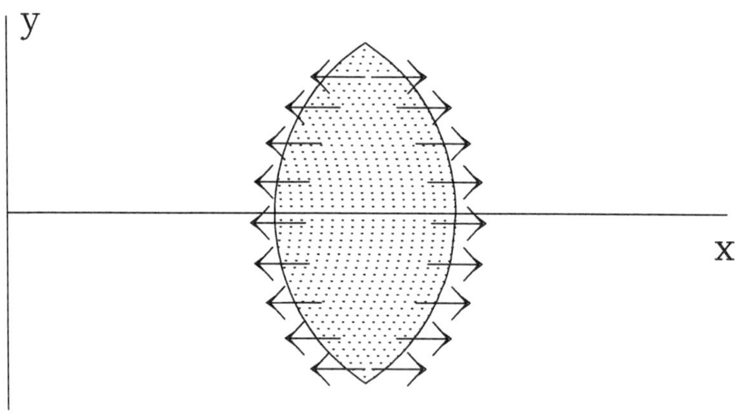

Figure 5.13: The ellipsoid of hot participant matter left behind at midrapidity after the spectators have departed. The arrows indicate the direction of the expansion. The x and y unit vectors are labeled. The z axis (the beam axis) goes into the page.

anisotropy on the observed momentum-space elliptic flow. The PHOBOS collaboration looked for similarity in Cu+Cu and Au+Au elliptic flow when $N_{\rm part} = 100$ in both systems [88]. In Cu+Cu collisions, $N_{\rm part} = 100$ is more central than in Au+Au collisions so that the participant matter should be have a smaller anisotropy in Cu+Cu collisions. The two systems are compared through the eccentricity of the coordinate-space ellipsoidal region formed by the participants. The eccentricity is defined as

$$\mathcal{E} = \frac{\sigma_y^2 - \sigma_x^2}{\sigma_y^2 + \sigma_x^2} \tag{5.214}$$

where σ_x and σ_y are the widths of the particle distributions in the $\hat{x}$ and $\hat{y}$ directions. In the most central collisions, $\sigma_y^2 \sim \sigma_x^2$ in a symmetric AA collision, resulting in a small eccentricity, while the eccentricity is larger in more peripheral collisions. The eccentricity can be positive or negative, depending on the orientation of the ellipse. The definition of the reaction plane requires that $\hat{y}$ is the long axis of the ellipse. However, orientations where $\sigma_x^2 > \sigma_y^2$ are due to fluctuations in the positions of the participant nucleons [88]. These fluctuations seem to be more important in smaller systems.

Exercises

1. Show the steps in the derivation of Eq. (5.91) from Eq. (5.53).

2. Given $T_c = 170$ MeV and $T_{\text{fo}} = 100$ MeV, calculate the time evolution of the temperature for the following cases:

   - $\tau_0 = 0.1$ fm, $T(\tau_0) = 1000$ MeV;
   - $\tau_0 = 0.5$ fm, $T(\tau_0) = 550$ MeV;
   - $\tau_0 = 1$ fm, $T(\tau_0) = 250$ MeV;
   - $\tau_0 = 1$ fm, $T(\tau_0) = 150$ MeV.

   Plot the temperature as a function of proper time.

3. For each case above, calculate the final-state multiplicity, dN/dy, corresponding to the values of τ_0 and $T(\tau_0)$.

Chapter 6

Lattice gauge theory

6.1 Introduction

The estimates of the critical temperature for the phase transition given in Chapters 4 and 5 are ad hoc, based on limiting physical behavior such as massless pions, zero baryon number, and so on. Even the order of the QCD phase transition was unknown in the early days of heavy-ion collisions. A method of reliably calculating the thermodynamic quantities in the nonperturbative regime of QCD, necessary to determine the order of the phase transition and the critical temperature, is needed. Lattice gauge theory [89, 90, 91] has proven to be a useful method for doing these calculations, especially as computing power has continued to improve. In lattice QCD, the thermodynamic partition function is related to the path integral formulation of quantum mechanics. In Euclidean space, with imaginary time, the action, S_E, is related to the Hamiltonian operator in the partition function and the imaginary time is related to the inverse temperature.

In this chapter, to motivate the lattice formulation of QCD at finite temperature, we will first describe some important symmetries in field theory that lead to the QCD Lagrangian. We thus discuss field theory in terms of the Lagrangian and motivate how the invariance of QCD under rotations in color space leads to the QCD Lagrangian. We then introduce the path integral and show how the path integral method is related to the partition function. We will then go on to describe how the QCD Lagrangian is discretized to a lattice of finite space-time points. We will discuss, rather generally, how fermion (quark) fields are handled and then describe some of the basic finite temperature lattice results.

6.2 Symmetries and the Lagrangian

An important concept in theoretical physics is that interactions are dictated by symmetries of the theory. The invariance of a theory under transformations at any point in space-time (local symmetries) is intimately connected to conserved quantities. The invariance of a physical system under certain symmetry transformations implies conservation laws, see also Refs. [4, 92]. Before we delve into lattice gauge theory as applied to finite temperature QCD, it is worth spending some time on this connection.

In the previous chapters, we have used these principles without explicit reference to these symmetries. For example, in classical physics, we can relate invariance in time ($t \rightarrow t+\Delta t$), space ($x_i \rightarrow x_i+\Delta x_i$) and rotation ($r_i \rightarrow R_{ij}r_j$) to energy ($dE/dt = 0$), momentum ($dp_i/dt = 0$) and angular momentum ($dJ_i/dt = 0$) conservation. Quantum mechanical operators are associated with observables. In the Heisenberg picture of quantum mechanics, the time evolution of an operator is governed by its commutator with the Hamiltonian, the energy operator, $H = T+V$ where T is the kinetic energy and V is the potential energy,

$$\frac{d\mathcal{O}}{dt} = i[H,\mathcal{O}] \ . \tag{6.1}$$

The conservation law is the equivalent of saying that the operator, $\mathcal{O}$, commutes with the Hamiltonian,

$$[H,\mathcal{O}] = 0 \ . \tag{6.2}$$

Angular momentum conservation, $dJ_i/dt = 0$, is the equivalent of the statement $[J_i, H] = 0$. Another way of making the same statement involves group theory and states that the Hamiltonian is invariant under the rotation group O(3) with generators J_1, J_2 and J_3 which satisfy the commutation relation

$$[J_i, J_j] = i\epsilon_{ijk} J_k \ . \tag{6.3}$$

6.2.1 The Lagrangian

In classical mechanics, the single particle equations of motion can be derived from Lagrange's equation,

$$\frac{d}{dt}\left(\frac{\partial L}{\partial x_i'}\right) - \frac{\partial L}{\partial x_i} = 0 \tag{6.4}$$

where x_i is the spatial coordinate, t is the time, and x_i' is the time derivative, dx_i/dt. The Lagrangian is the difference between the kinetic and potential energies,

$$L = T - V \ . \tag{6.5}$$

It is easy to apply the formalism of a discrete system with coordinates $x_i(t)$ to a system with continuously varying coordinates, $\phi(\mathbf{x})$, where $\mathbf{x}$ is a four-vector in space-time. Then

$$L(x_i, x_i', t) \rightarrow \mathcal{L}(\phi, \partial_\mu \phi, \mathbf{x}_\mu) \tag{6.6}$$

and the Lagrange equation becomes

$$\partial_\mu \left(\frac{\partial \mathcal{L}}{\partial(\partial_\mu \phi)}\right) - \frac{\partial \mathcal{L}}{\partial \phi} = 0 \ . \tag{6.7}$$

Here $\mathcal{L}$ is the Lagrange density (although it is often referred to as the Lagrangian itself) with

$$L = \int d^3x \mathcal{L} \ . \tag{6.8}$$

Thus, instead of writing down a wave equation to start with, here we begin instead with a Lagrangian.

If we take a Lagrange density for fermions,

$$\mathcal{L} = i\overline{\psi}\gamma^\mu \partial_\mu \psi - m\overline{\psi}\psi \ , \tag{6.9}$$

and regard ψ and $\overline{\psi}$ as independent variables, we can derive the Dirac equations for both the fermion field, ψ, and its adjoint, $\overline{\psi}$, from the Lagrangian.

EXAMPLE: Verify this.
We begin with the adjoint equation, obtained by taking the derivatives with respect to the fermion field, ψ. We have

$$\begin{aligned} \frac{\partial \mathcal{L}}{\partial \psi} &= -m\overline{\psi} , \\ \frac{\partial \mathcal{L}}{\partial(\partial_\mu \psi)} &= i\overline{\psi}\gamma^\mu , \\ \partial_\mu \frac{\partial \mathcal{L}}{\partial(\partial_\mu \psi)} &= i\partial_\mu \overline{\psi}\gamma^\mu . \end{aligned}$$

The Lagrange equation for the fermion field is the Dirac equation for the adjoint field,

$$i\partial_\mu \overline{\psi}\gamma^\mu + m\overline{\psi} = 0 . \tag{6.10}$$

The Lagrange equation of motion for the ψ field is found by taking the derivatives with respect to the adjoint field, $\overline{\psi}$. The derivative with respect to $\overline{\psi}$ is rather trivial,

$$\frac{\partial \mathcal{L}}{\partial \overline{\psi}} = -m\psi . \tag{6.11}$$

To take the derivative with respect to $\partial_\mu \overline{\psi}$, we need to move the derivative from ψ to $\overline{\psi}$. We can replace $\overline{\psi}\gamma^\mu \partial_\mu \psi$ by $\partial_\mu(\overline{\psi}\gamma^\mu\psi) - (\partial_\mu\overline{\psi})\gamma^\mu\psi$. The argument of the total derivative, $\overline{\psi}\gamma^\mu\psi$, is the conserved current, $\mathbf{j}^\mu = \overline{\psi}\gamma^\mu\psi$. Since $\partial_\mu \mathbf{j}^\mu = 0$, the Lagrange density becomes

$$\mathcal{L} = -i(\partial_\mu\overline{\psi})\gamma^\mu\psi - m\overline{\psi}\psi \tag{6.12}$$

where now

$$\begin{aligned} \frac{\partial \mathcal{L}}{\partial(\partial_\mu \overline{\psi})} &= -i\gamma^\mu\psi , \\ \partial_\mu \frac{\partial \mathcal{L}}{\partial(\partial_\mu \overline{\psi})} &= -i\gamma^\mu\partial_\mu\psi . \end{aligned}$$

The Lagrange equation for $\overline{\psi}$ is finally

$$i\gamma^{\mu}\partial_{\mu}\psi - m\psi = 0 \ . \tag{6.13}$$

In field theory, the Maxwell equations for electromagnetism can be written in terms of the gauge field, $\mathbf{A}_{\mu}$,

$$\partial_{\mu}F^{\mu\nu} = \mathbf{j}^{\nu} \tag{6.14}$$

where $F^{\mu\nu} = \partial^{\mu}\mathbf{A}^{\nu} - \partial^{\nu}\mathbf{A}^{\mu}$.

EXAMPLE: Show that the Lagrange density,

$$\mathcal{L} = -\frac{1}{4}F_{\mu\nu}F^{\mu\nu} - \mathbf{j}^{\mu}\mathbf{A}_{\mu} \ , \tag{6.15}$$

leads to Eq. (6.14) above. Then show that the current is conserved, $\partial_{\nu}\mathbf{j}^{\nu} = 0$.
Using the definition of the field strength tensor, $F_{\mu\nu}$, we can write the Lagrange density as

$$\begin{aligned}
\mathcal{L} &= -\frac{1}{4}(\partial_{\mu}\mathbf{A}_{\nu} - \partial_{\nu}\mathbf{A}_{\mu})(\partial^{\mu}\mathbf{A}^{\nu} - \partial^{\nu}\mathbf{A}^{\mu}) - \mathbf{j}^{\mu}\mathbf{A}_{\mu} \\
&= -\frac{1}{4}g^{\lambda\alpha}g^{\sigma\beta}(\partial_{\alpha}\mathbf{A}_{\beta} - \partial_{\beta}\mathbf{A}_{\alpha})(\partial_{\lambda}\mathbf{A}_{\sigma} - \partial_{\sigma}\mathbf{A}_{\lambda}) - \mathbf{j}^{\mu}\mathbf{A}_{\mu} \ .
\end{aligned} \tag{6.16}$$

The derivative of $\mathcal{L}$ with respect to the field $\mathbf{A}_{\mu}$ is

$$\frac{\partial\mathcal{L}}{\partial\mathbf{A}_{\mu}} = -\mathbf{j}^{\mu} \ . \tag{6.17}$$

The derivative with respect to $\partial_{\nu}\mathbf{A}_{\mu}$ is a bit more complicated since we need to retain all possible terms and watch the indices. We have

$$\begin{aligned}
\frac{\partial\mathcal{L}}{\partial(\partial_{\nu}\mathbf{A}_{\mu})} &= -\frac{1}{4}g^{\lambda\alpha}g^{\sigma\beta}[\delta^{\alpha}_{\nu}\delta^{\beta}_{\mu}F_{\lambda\sigma} - \delta^{\beta}_{\nu}\delta^{\alpha}_{\mu}F_{\lambda\sigma} + \delta^{\lambda}_{\nu}\delta^{\sigma}_{\mu}F_{\alpha\beta} - \delta^{\sigma}_{\nu}\delta^{\lambda}_{\mu}F_{\alpha\beta}] \\
&= -\frac{1}{4}[g^{\lambda\nu}g^{\sigma\mu}F_{\lambda\sigma} - g^{\lambda\mu}g^{\sigma\nu}F_{\lambda\sigma} + g^{\nu\alpha}g^{\mu\beta}F_{\alpha\beta} - g^{\mu\alpha}g^{\nu\beta}F_{\alpha\beta}] \\
&= -\frac{1}{4}[F^{\nu\mu} - F^{\mu\nu} + F^{\nu\mu} - F^{\mu\nu}] \\
&= F^{\mu\nu}
\end{aligned} \tag{6.18}$$

since $F^{\nu\mu} = -F^{\mu\nu}$. Finally then

$$\partial_\mu F^{\mu\nu} = \mathbf{j}^\nu \ . \tag{6.19}$$

Taking the derivative of both sides with respect to ∂_ν, we find

$$\partial_\nu \mathbf{j}^\nu = \partial_\nu \partial_\mu F^{\mu\nu} = \partial_\nu \partial_\mu (\partial^\mu \mathbf{A}^\nu - \partial^\nu \mathbf{A}^\mu) \ . \tag{6.20}$$

After rearranging terms so that terms with the same indices are grouped together, we get

$$\partial_\nu \mathbf{j}^\nu = \partial_\mu \partial^\mu (\partial^\nu \mathbf{A}_\nu) - \partial^\nu \partial_\nu (\partial^\mu \mathbf{A}_\mu) = 0 \ . \tag{6.21}$$

Thus the current $\mathbf{j}^\nu$ is conserved.

If we add a mass term, $(1/2)m^2\mathbf{A}_\mu\mathbf{A}^\mu$, to the Lagrange density in Eq. (6.15),

$$\mathcal{L} = -\frac{1}{4}F_{\mu\nu}F^{\mu\nu} + \frac{1}{2}m^2\mathbf{A}_\mu\mathbf{A}^\mu - \mathbf{j}^\mu\mathbf{A}_\mu \ , \tag{6.22}$$

the derivative $\partial\mathcal{L}/\partial\mathbf{A}_\mu$ has a new contribution, $m^2\mathbf{A}^\mu$, since the derivative acts on both $\mathbf{A}_\mu$ terms, canceling the 1/2. The Lagrange equation is then

$$\partial_\mu F^{\mu\nu} + m^2\mathbf{A}^\nu = \mathbf{j}^\nu \ . \tag{6.23}$$

We can rewrite the derivative term as

$$\partial_\mu F^{\mu\nu} = \partial_\mu(\partial^\mu \mathbf{A}^\nu - \partial^\nu \mathbf{A}^\mu) = \Box^2\mathbf{A}^\nu - \partial^\nu(\partial_\mu \mathbf{A}^\mu) \ . \tag{6.24}$$

Since the electromagnetic fields are invariant under Lorentz transformations,

$$\mathbf{A}_\mu \to \mathbf{A}'_\mu = \mathbf{A}_\mu + \partial_\mu \chi \ , \tag{6.25}$$

we can choose $\partial_\mu \mathbf{A}^\mu = 0$ so that the field equation becomes

$$\Box^2\mathbf{A}^\mu + m^2\mathbf{A}_\mu = \mathbf{j}^\mu \tag{6.26}$$

and the term proportional to m^2 would correspond to a photon mass in quantum electrodynamics (QED). As we will discuss later, such a term would break gauge invariance so that the photon must be massless.

To relate the Lagrangian approach to the single particle equations used in Chapter 2 to discuss interactions, the terms in the Lagrangian are associated with propagators and vertex factors. The propagators arise from terms in the Lagrangian quadratic in the fields such as $i\overline{\psi}\gamma^{\mu}\partial_{\mu}\psi$. The interaction terms are associated with vertices. In QED, the electron-photon interaction is $ie\overline{\psi}\gamma^{\mu}\psi\mathbf{A}_{\mu}$, coupling the electron current, $\mathbf{j}^{\mu} = -e\overline{\psi}\gamma^{\mu}\psi$ to the photon field $\mathbf{A}_{\mu}$ with vertex factor $ie\gamma^{\mu}$.

6.2.2 Symmetries and gauge invariance

In field theory, Noether's theorem states that for a system described by a Lagrangian, any continuous symmetry transformation which leaves the action invariant implies a conserved current,

$$\partial_{\mu}\mathbf{j}^{\mu} = 0 . \tag{6.27}$$

The charge associated with this current,

$$Q = \int d^3x\, j^0(x) . \tag{6.28}$$

is a constant of motion (conserved),

$$\frac{dQ}{dt} = 0 , \tag{6.29}$$

because $\vec{\nabla}\cdot\vec{j} \to 0$ at $x \to \infty$ so that

$$\frac{dQ}{dt} = \int d^3x\, \partial_0 j^0 \approx \int d^3x\, \partial_{\mu}\mathbf{j}^{\mu} = 0 . \tag{6.30}$$

If the Lagrangian is invariant under transformation in some symmetry group, G, the conserved charges, Q, are the generators of the symmetry group. Thus if $\mathcal{L}(\phi)$ is invariant under the transformation

$$\phi_i(x) \to \phi_i'(x) = \phi_i(x) + \delta\phi_i(x) = \phi_i(x) + i\alpha^a t^a_{ij}\phi_j(x) \tag{6.31}$$

where α is a space-time independent small parameter and the t's are matrices satisfying

$$[t^a, t^b] = ic_{abc}t^c . \tag{6.32}$$

If the transformation is the same at every space-time point, as it is if α is a constant, the symmetry is 'global' and we refer to 'global gauge invariance'. Thus the Dirac Lagrange density,

$$\mathcal{L} = i\overline{\psi}\gamma^{\mu}\partial_{\mu}\psi - m\overline{\psi}\psi \ , \tag{6.33}$$

is invariant under the transformation

$$\psi(x) \to \psi'(x) = \exp(i\alpha)\psi(x) \ . \tag{6.34}$$

EXAMPLE: Show that the Dirac Lagrange density is globally gauge invariant.
Since α is constant, we have

$$\partial_{\mu}\psi \to \partial_{\mu}\psi' = \exp(i\alpha)\partial_{\mu}\psi \ . \tag{6.35}$$

We also have to obtain the transformation of the adjoint field,

$$\begin{aligned} \overline{\psi} = \psi^{\dagger}\gamma^0 \to (\psi')^{\dagger}\gamma^0 &= [\exp(i\alpha)\psi]^{\dagger}\gamma^0 \\ &= \psi^{\dagger}\gamma^0\exp(-i\alpha) \\ &= \overline{\psi}\exp(-i\alpha) \ . \end{aligned} \tag{6.36}$$

Finally,

$$\begin{aligned} \mathcal{L} \to \mathcal{L}' &= i\overline{\psi}\exp(-i\alpha)\gamma^{\mu}\exp(i\alpha)\partial_{\mu}\psi - m\overline{\psi}\exp(-i\alpha)\exp(i\alpha)\psi \\ &= \mathcal{L} \ . \end{aligned} \tag{6.37}$$

The parameter α can run over all real numbers. Thus the transformations $U(\alpha) = \exp(i\alpha)$ form the unitary Abelian (commuting) U(1) group. Because of Noether's theorem, indexNoether's theorem U(1) invariance implies a conserved current. If we let $\psi \to \psi + \delta\psi = (1+i\alpha)\psi$, requiring $\delta\mathcal{L} = 0$ leads to

$$0 = \frac{\partial\mathcal{L}}{\partial\psi}\delta\psi + \frac{\partial\mathcal{L}}{\partial(\partial_{\mu}\psi)}\delta(\partial_{\mu}\psi) + \delta\overline{\psi}\frac{\partial\mathcal{L}}{\partial\overline{\psi}} + \delta(\partial_{\mu}\overline{\psi})\frac{\partial\mathcal{L}}{\partial(\partial_{\mu}\overline{\psi})} \ . \tag{6.38}$$

Replacing $\delta\psi$ with $i\alpha\psi$ and $\delta\overline{\psi}$ with $-i\alpha\overline{\psi}$, we have

$$0 = i\alpha\left(\frac{\partial\mathcal{L}}{\partial\psi}\psi + \frac{\partial\mathcal{L}}{\partial(\partial_{\mu}\psi)}\partial_{\mu}\psi\right) - i\alpha\left(\overline{\psi}\frac{\partial\mathcal{L}}{\partial\overline{\psi}} + \partial_{\mu}\overline{\psi}\frac{\partial\mathcal{L}}{\partial(\partial_{\mu}\overline{\psi})}\right) . \tag{6.39}$$

We can rewrite the derivative terms as

$$\begin{aligned} \frac{\partial \mathcal{L}}{\partial(\partial_\mu \psi)}\partial_\mu \psi &= \partial_\mu\left(\frac{\partial \mathcal{L}}{\partial(\partial_\mu \psi)}\psi\right) - \partial_\mu\left(\frac{\partial \mathcal{L}}{\partial(\partial_\mu \psi)}\right)\psi \,, \\ \partial_\mu \overline{\psi}\frac{\partial \mathcal{L}}{\partial(\partial_\mu \overline{\psi})} &= \partial_\mu\left(\overline{\psi}\frac{\partial \mathcal{L}}{\partial(\partial_\mu \overline{\psi})}\right) - \overline{\psi}\partial_\mu\left(\frac{\partial \mathcal{L}}{\partial(\partial_\mu \overline{\psi})}\right) \end{aligned} \tag{6.40}$$

so that

$$\begin{aligned} 0 &= i\alpha\left(\frac{\partial \mathcal{L}}{\partial \psi} - \partial_\mu\left(\frac{\partial \mathcal{L}}{\partial(\partial_\mu \psi)}\right)\right)\psi + i\alpha\partial_\mu\left(\frac{\partial \mathcal{L}}{\partial(\partial_\mu \psi)}\psi\right) \\ &\quad - i\alpha\overline{\psi}\left(\frac{\partial \mathcal{L}}{\partial \overline{\psi}} - \partial_\mu\left(\frac{\partial \mathcal{L}}{\partial(\partial_\mu \overline{\psi})}\right)\right) - i\alpha\partial_\mu\left(\overline{\psi}\frac{\partial \mathcal{L}}{\partial(\partial_\mu \overline{\psi})}\right) . \end{aligned} \tag{6.41}$$

The first and third terms on the right-hand side of Eq. (6.41) are simply the Euler-Lagrange equations for the ψ and $\overline{\psi}$ fields and are zero. Combining the remaining two terms gives us

$$0 = i\alpha\partial_\mu\left(\frac{\partial \mathcal{L}}{\partial(\partial_\mu \psi)}\psi - \overline{\psi}\frac{\partial \mathcal{L}}{\partial(\partial_\mu \overline{\psi})}\right) \tag{6.42}$$

$$= \partial_\mu \mathbf{j}^\mu \,. \tag{6.43}$$

where

$$\mathbf{j}^\mu = \frac{ie}{2}\left(\frac{\partial \mathcal{L}}{\partial(\partial_\mu \psi)}\psi - \overline{\psi}\frac{\partial \mathcal{L}}{\partial(\partial_\mu \overline{\psi})}\right) , \tag{6.44}$$

normalized so that $\mathbf{j}^\mu$ is the electromagnetic charge current density for an electron.

EXAMPLE: Let

$$\mathcal{L} = (\partial_\mu \phi^*)(\partial^\mu \phi) - \mu^2 \phi^* \phi - \lambda(\phi^* \phi)^2 \tag{6.45}$$

where $\phi = (1/\sqrt{2})(\phi_1 + i\phi_2)$ *and* $\phi^* = (1/\sqrt{2})(\phi_1 - i\phi_2)$. *Show that, under* $U(1)$ *transformations,*

$$\mathbf{j}^\mu = -ie(\phi^* \partial_\mu \phi - \phi \partial_\mu \phi^*) \,. \tag{6.46}$$

We first note that the Lagrange density can be written in terms of the real fields, ϕ_1 and ϕ_2, using

$$\begin{aligned}
\partial_\mu\phi &= \frac{1}{\sqrt{2}}(\partial_\mu\phi_1 + i\partial_\mu\phi_2) \ , \\
\partial_\mu\phi^* &= \frac{1}{\sqrt{2}}(\partial_\mu\phi_1 - i\partial_\mu\phi_2) \ , \\
\phi^*\phi &= \frac{1}{2}(\phi_1^2 + \phi_2^2) \ .
\end{aligned} \tag{6.47}$$

Then

$$\mathcal{L} = \frac{1}{2}[(\partial_\mu\phi_1)^2 + (\partial^\mu\phi_2)^2] - \frac{1}{2}\mu^2(\phi_1^2+\phi_2^2) - \frac{1}{4}\lambda(\phi_1^2+\phi_2^2)^2 \ . \tag{6.48}$$

We can show this Lagrange density to be invariant under the transformation

$$\begin{aligned}
\phi_1 &\rightarrow \phi_1' = \phi_1\cos\alpha - \phi_2\sin\alpha \ , \\
\phi_2 &\rightarrow \phi_2' = \phi_1\sin\alpha + \phi_2\cos\alpha \ .
\end{aligned} \tag{6.49}$$

We then have

$$(\phi_1')^2 + (\phi_2')^2 = \phi_1^2 + \phi_2^2 \tag{6.50}$$

using $\sin^2\alpha + \cos^2\alpha = 1$ and canceling the cross terms. Similarly, the derivatives transform as

$$\begin{aligned}
\partial_\mu\phi_1' &= (\cos\alpha)\partial_\mu\phi_1 - (\sin\alpha)\partial_\mu\phi_2 \ , \\
\partial_\mu\phi_2' &= (\sin\alpha)\partial_\mu\phi_1 + (\cos\alpha)\partial_\mu\phi_2 \ .
\end{aligned} \tag{6.51}$$

so that

$$(\partial_\mu\phi_1')^2 + (\partial_\mu\phi_2')^2 = (\partial_\mu\phi_1)^2 + (\partial_\mu\phi_2)^2 \ . \tag{6.52}$$

If $\alpha \ll 1$, $\cos\alpha \sim 1$ and $\sin\alpha \sim \alpha$. It is easy to see that the transformation is equivalent to rotations in the (ϕ_1, ϕ_2) plane since

$$\begin{aligned}
\phi_1' &= \phi_1 - \alpha\phi_2 \ , \\
\phi_2' &= \phi_2 + \alpha\phi_1 \ .
\end{aligned} \tag{6.53}$$

This can be written more generally as

$$\phi_i' = \phi_i + i\alpha t_{ij}\phi_j \tag{6.54}$$

where

$$t = \begin{pmatrix} 0 & i \\ -i & 0 \end{pmatrix} \tag{6.55}$$

is the rotation matrix. The complex field thus transforms as

$$\begin{aligned} \phi' &= \frac{1}{\sqrt{2}}(\phi_1' + i\phi_2') \\ &= \frac{1}{\sqrt{2}}[\phi_1 \cos\alpha - \phi_2 \sin\alpha + i(\phi_1 \sin\alpha + \phi_2 \cos\alpha)] \\ &= \frac{1}{\sqrt{2}}[\phi_1 + i\phi_2](\cos\alpha + i \sin\alpha) \\ &= \phi \exp(i\alpha) \ . \end{aligned} \tag{6.56}$$

We differentiate the Lagrange density in terms of the complex fields, as in the case of the ψ and $\overline{\psi}$ fields above, to find

$$\begin{aligned} 0 &= \frac{\partial \mathcal{L}}{\partial \phi}\delta\phi + \frac{\partial \mathcal{L}}{\partial(\partial_\mu \phi)}\delta(\partial_\mu \phi) + \delta\phi^* \frac{\partial \mathcal{L}}{\partial \phi^*} + \delta(\partial_\mu \phi^*)\frac{\partial \mathcal{L}}{\partial(\partial_\mu \phi^*)} \\ &= i\alpha\partial_\mu \left(\frac{\partial \mathcal{L}}{\partial(\partial_\mu \phi)}\phi - \phi^* \frac{\partial \mathcal{L}}{\partial(\partial_\mu \phi^*)} \right) \end{aligned} \tag{6.57}$$

where

$$\begin{aligned} \mathbf{j}^\mu &= i\left(\frac{\partial \mathcal{L}}{\partial(\partial_\mu \phi)}\phi - \phi^* \frac{\partial \mathcal{L}}{\partial(\partial_\mu \phi^*)} \right) \\ &= -i[\phi^*(\partial^\mu \phi) - \phi(\partial^\mu \phi^*)] \ . \end{aligned} \tag{6.58}$$

If instead, we write the current density in terms of the real fields, ϕ_1 and ϕ_2, we can show that

$$\mathbf{j}^\mu = \phi_1(\partial^\mu \phi_2) - \phi_2(\partial^\mu \phi_1) \ . \tag{6.59}$$

This is left as an exercise.

We have now thoroughly discussed the case where α is constant, the case of 'global' gauge invariance. More generally, if α is no longer a constant but depends on space-time, we have 'local' gauge invariance if the Lagrangian is invariant under this type of transformation.

If we look back at the Dirac Lagrange density, it is *not* locally gauge invariant as written. Here, with

$$\begin{aligned}\psi' &= \exp[i\alpha(x)]\psi \\ \overline{\psi}' &= \exp[-i\alpha(x)]\overline{\psi}\end{aligned} \tag{6.60}$$

the mass term is invariant since

$$m\overline{\psi}'\psi' = m\overline{\psi}\exp[-i\alpha(x)]\exp[i\alpha(x)]\psi = m\overline{\psi}\psi \ . \tag{6.61}$$

However, the derivative term breaks the invariance because

$$\partial_\mu\psi' = \exp[i\alpha(x)]\partial_\mu\psi + i\exp[i\alpha(x)]\psi\partial_\mu\alpha(x) \ . \tag{6.62}$$

For the Lagrangian to be locally gauge invariant, we have to replace the derivative by a covariant derivative that remains invariant under transformation,

$$(\mathbf{D}_\mu\psi)' = \exp[i\alpha(x)]\mathbf{D}_\mu\psi \ . \tag{6.63}$$

We introduce the vector gauge field, $\mathbf{A}_\mu$, so that the covariant derivative is defined as

$$\mathbf{D}_\mu = \partial_\mu - ie\mathbf{A}_\mu \ . \tag{6.64}$$

If $\mathbf{A}_\mu$ transforms as

$$\mathbf{A}'_\mu = \mathbf{A}_\mu + \frac{1}{e}\partial_\mu\alpha(x) \ , \tag{6.65}$$

where now the Dirac Lagrange density is

$$\mathcal{L} = i\overline{\psi}\gamma^\mu\mathbf{D}_\mu\psi - m\overline{\psi}\psi \ , \tag{6.66}$$

local gauge invariance is restored since

$$\begin{aligned}(\mathbf{D}_\mu\psi)' &= \exp[i\alpha(x)][\partial_\mu\psi + i\psi\partial_\mu\alpha(x) - ie\mathbf{A}_\mu - i\psi\partial_\mu\alpha(x)] \\ &= \exp[i\alpha(x)]\mathbf{D}_\mu\psi \ .\end{aligned} \tag{6.67}$$

The full locally gauge invariant Lagrange density for quantum electrodynamics, QED, also includes the kinetic energy term for the gauge field, $-(1/4)F_{\mu\nu}F^{\mu\nu}$, so that

$$\mathcal{L} = i\overline{\psi}\gamma^\mu\partial_\mu\psi + e\overline{\psi}\gamma^\mu\mathbf{A}_\mu\psi - m\overline{\psi}\psi - \frac{1}{4}F_{\mu\nu}F^{\mu\nu} \ . \tag{6.68}$$

Note that there is no photon mass term as in Eq. (6.26) since such a term would break local gauge invariance, giving a contribution proportional to $(\partial_\mu \alpha)^2$. Thus the photon, the gauge particle in QED, is massless.

Armed with all this, we can turn to the QCD Lagrange density. Things will be a little more complicated because the quarks carry color. The U(1) gauge group has to be replaced by the color group SU(3). In this case, we have transformations of the type

$$\psi \rightarrow \psi' = U\psi = \exp[i\alpha_a(x)\lambda_a/2]\psi \tag{6.69}$$

where ψ is now a three-dimensional vector in color space and U is a 3×3 matrix. In matrix algebra, $U^\dagger U = 1$, implying $(\det U)^*(\det U) = 1$ so that $\det U = \exp(i\phi)$. The SU(3) group restricts $\det U$ to be 1. For $U^\dagger = U^{-1}$, the infinitesimal transformation of U^{-1} is $1 - i\alpha_a(x)\lambda_a/2$ while that of $U^\dagger$ is $1 - i\alpha_a^*(x)\lambda_a^\dagger/2$, telling us that $\alpha_a(x)\lambda_a/2 = \alpha_a^*(x)\lambda_a^\dagger/2$. Since the SU(3) generators, $\lambda_a/2$, do not commute,

$$\left[\frac{\lambda_a}{2}, \frac{\lambda_b}{2}\right] = i f_{abc}\frac{\lambda_c}{2} \ , \tag{6.70}$$

where the f_{abc}'s are the structure constants of the group, the SU(3) group is non-Abelian.

The requirement that the QCD Lagrange density is locally gauge invariant implies that the color gauge field,

$$\mathbf{A}_\mu = \frac{\lambda_a}{2}\mathbf{A}_\mu^a \ , \tag{6.71}$$

transforms as

$$\mathbf{A}_\mu^a \rightarrow \mathbf{A}_\mu^a - \frac{1}{g}\partial_\mu\alpha_a - f_{abc}\alpha_b\mathbf{A}_\mu^c \tag{6.72}$$

where g is the QCD coupling constant. The eight color fields are those of the gluon, the QCD gauge particle. Since the theory is non-Abelian, the gluons of QCD can couple to each other while the Abelian photon fields of QED cannot. The QCD Lagrange density is

$$\mathcal{L} = i\overline{\psi}\gamma^\mu\partial_\mu\psi - g\overline{\psi}\gamma^\mu\psi\mathbf{A}_\mu^a - m\overline{\psi}\psi - \frac{1}{4}F_{\mu\nu}^a F_a^{\mu\nu} \tag{6.73}$$

where, as we discuss in more detail later,

$$\begin{aligned}
F_{\mu\nu} &= \frac{i}{g}[\mathbf{D}_\mu, \mathbf{D}_\nu] \\
&= \partial_\mu \mathbf{A}_\nu - \partial_\nu \mathbf{A}_\mu - ig[\mathbf{A}_\mu, \mathbf{A}_\nu] \\
F^a_{\mu\nu} &= \partial_\mu \mathbf{A}^a_\nu - \partial_\nu \mathbf{A}^a_\mu + g f_{abc} \mathbf{A}^b_\mu \mathbf{A}^c_\nu \; .
\end{aligned} \tag{6.74}$$

A sum over index a is implied in Eq. (6.73). Local gauge invariance implies that the gluon, like the photon of QED, is massless. We will discuss local gauge invariance in QCD in more detail when we describe its formulation on the lattice. We will also discuss other symmetries broken by the QCD Lagrangian.

6.3 Basics of lattice gauge theory

Lattice gauge theory, discussed in this chapter, is a way to probe nonperturbative aspects of QCD. As we discussed previously, QCD is asymptotically free at short distances where perturbative calculations are possible. However, when Feynman diagrams of numerous processes are evaluated, the large momentum scales, $p \to \infty$, corresponding to short distances, $r \to 0$, may cause divergences in the momentum integrations. These types of divergences are called "ultraviolet divergences", corresponding to the ultraviolet or small wavelength part of th electromagnetic spectrum. In perturbation theory, ultraviolet divergences are handled using renormalization group methods. In nonperturbative QCD, the perturbative expansion is no longer valid and another way to handle the divergences is required.

Lattice gauge theory reformulates QCD on a lattice of discrete space-time points. The spacing between lattice points provides a finite distance scale. This lattice spacing gives a minimum distance and thus also a maximum momentum scale which acts as a momentum cutoff in the integrations. Thus terms that would become infinite remain finite. Of course, in this manner, the result for any quantity calculated on the lattice depends on the lattice spacing, a. The way to remove this dependence is to perform the same calculation multiple times with different, decreasing, a until the scaling regime is reached. The lattice spacing is related to the coupling constant. Thus when the lattice results as a function of a agree with the behavior of the coupling constant as a function of scale Q^2, the scaling regime is reached. In the scaling regime, the lattice spacing can be related to a physical scale.

In the remainder of this chapter, we describe how the QCD Lagrangian is put on the lattice. We will then discuss the lattice QCD results relevant to heavy-ion collisions such as studies of the QCD phase transition. To perform these studies, we need the partition function which can be related to the path-integral formulation of quantum mechanics, as we first show.

6.3.1 The path integral and the partition function

The partition function developed in Chapter 4 at finite temperature,

$$Z = \mathrm{Tr}\exp[-H/T] \ , \tag{6.75}$$

can be shown to be related to the expectation value of the time evolution operator, $\exp(iHt)$ in imaginary time. In this section, we show how this happens.

The quantum mechanical description of the relationship between the partition function and the evolution of a system comes about in the path integral formulation developed by Feynman. The path integral approach generalizes the classical action. In classical terms, there is a well-defined way of moving from point A at location x_a and time t_a to point B at x_b and t_b. Quantum mechanically, however, there are an infinite number of ways to get from A to B.

Feynman proposed several postulates for the path-integral formalism, see Ref. [93] for a complete study. The probability is given by the absolute square of a complex amplitude. The amplitude of an event is obtained by summing all paths that include that event. The of a particular history, or space-time path, is proportional to $\exp[iS(L)]$ where $S(L)$ is the action (time integral of the Lagrangian) for that path. The proportionality factor is determined by normalizing to the square root of the total probability so that the the total probability from all normalized amplitudes is unity.

The probability amplitude is integrated over all possible histories, no matter how indirect or absurd, see Fig. 6.1 for some indirect histories. The path integral assigns equal amplitudes to all histories, no

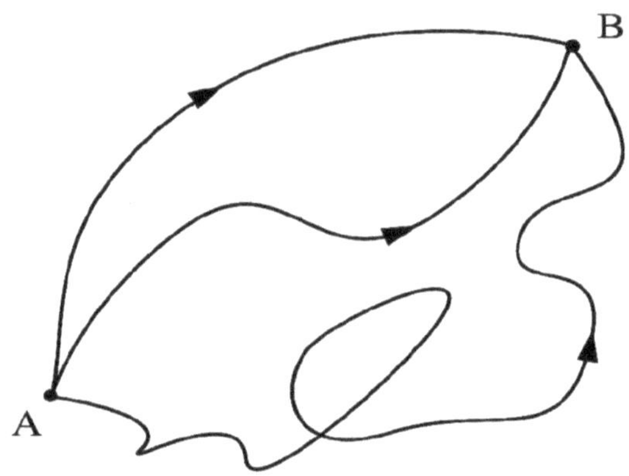

Figure 6.1: Three paths contributing to the quantum-mechanical amplitude for a particle to move from A to B [94].

matter how unlikely they may seem. Only the phase differs. The more absurd paths will be suppressed.

In the limit where the action is large, the path integral is dominated by solutions that are stationary points of the action since amplitudes with similar histories will constructively interfere. On the other hand, the complex phases for histories far from the stationary points will vary rapidly, leading to destructive interference. Thus the most significant histories are the classical solutions to the Euler-Lagrange equation, recovering classical mechanics.

The path is described as $x(t)$ for $t_A < t < t_B$. As already stated, there are infinitely many possible paths between the the starting and ending points with a corresponding value of the action,

$$S_M[L(x,x',t)] = \int_{t_A}^{t_B} dtL(x,x',t) \tag{6.76}$$

where $L = T - V$ is the Lagrangian. The subscript M on the action denotes that the action is in Minkowski space where the metric tensor is as described in Chapter 1,

$$g_{\mu\nu} = \begin{pmatrix} 1 & 0 & 0 & 0 \\ 0 & -1 & 0 & 0 \\ 0 & 0 & -1 & 0 \\ 0 & 0 & 0 & -1 \end{pmatrix} . \tag{6.77}$$

We make the distinction because we will soon rotate into Euclidean space with a different metric tensor.

The classical trajectory minimizes the Minkowski action. This is known as the principle of least action. In quantum mechanics, however, all paths may contribute. They just have to satisfy the criteria of starting at (x_A, t_A) and ending at (x_B, t_B). Figure 6.1 shows three such possible paths.

As stated above, the amplitude is given by summing all paths that satisfy the criteria and is proportional to the weight, $\exp[iS_M]$, for that path,

$$\sum_{\text{all paths}} \exp[iS_M] . \tag{6.78}$$

Quantum mechanically, this is just

$$\langle x_B(t_B)| \exp[-H(t_B - t_B)]|x_A(t_A)\rangle \tag{6.79}$$

so that

$$\langle x_b(t_b)| \exp[-H(t_b - t_a)]|x_a(t_a)\rangle = \sum_{\text{all paths}} \exp[iS_M] . \tag{6.80}$$

To sum over paths, the spatial coordinates of points between A and B are treated as variables. The time difference, $t_B - t_A$, is divided into n discrete intervals: $t_0, t_1, \cdots t_n$ where $t_A = t_0$ and $t_B = t_n$. Thus the space points, x_1, which pass through the first intermediate time point, t_1, can vary from $-\infty$ to $+\infty$. All paths through t_1 can be written as an integral instead of a sum,

$$\sum_{\text{all paths}} \exp[iS_M] \longrightarrow \int_{-\infty}^{\infty} dx_1 \exp[iS_M(x_1)] \; . \tag{6.81}$$

This "time slicing" can be done for each intermediate point, giving

$$\lim_{\substack{\Delta t \to 0, n \to \infty \\ n\Delta t = t_B - t_A}} \int_{-\infty}^{\infty} dx_1 \cdots \int_{-\infty}^{\infty} dx_{n-1} \exp[iS_M(x_1, \cdots, x_{n-1})] \; . \tag{6.82}$$

An example is shown in Fig. 6.2. The amplitude is then

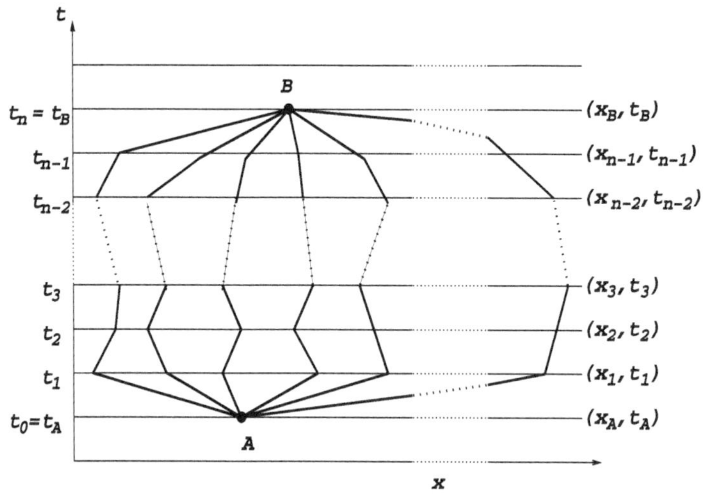

Figure 6.2: Several paths between (x_A, t_A) and (x_B, t_B). The time slices are shown as horizontal lines. The dots represent intermediate space-time points along a particular path.

$$\langle x_B | \exp[-iH(t_B - t_A)] | x_A \rangle = \int_{-\infty}^{\infty} \prod_{i=1}^{n-1} dx_i \exp[iS_M(x_1, \cdots, x_{n-1})] \; . \tag{6.83}$$

There is no integral over x_0 and x_n since these are the known values, x_A and x_B respectively.

We now relate the amplitude to the partition function,

$$Z = \mathrm{Tr}[\exp(-H/T)] = \sum_{x_A} \langle x_A| \exp[-H/T] |x_A\rangle \ . \tag{6.84}$$

On the left-hand side of Eq. (6.83), the operator is $\exp[-iH(t_B - t_A)]$ while in Eq. (6.84) it is $\exp[-H/T]$. The two operators become identical if we go to imaginary time, $t = -i\tau$, and let $\tau_A = 0$, $\tau_B = 1/T$ so that the inverse temperature in the partition function is equivalent to the imaginary time interval in the amplitude, $-i(t_B - t_A) = -\tau_B = -1/T$. Next, the initial and final states are identical in the partition function while this is not a requirement in the calculation of the quantum-mechanical amplitude. To make the equivalence of the initial and final states a requirement in the amplitude, periodic boundary conditions are imposed so that $x_A = x_B$ or $x(\tau_A) = x(\tau_B)$. Lastly, while the amplitude involves only a single term, the partition function is a summation over all states x_A. To remove this last distinction, the partition function can be obtained from the amplitude by summing over $x_A \equiv x_n$. Then, after imposing periodic boundary conditions and summing over x_A, the left-hand side of Eq. (6.83) becomes equivalent to the partition function. The right-hand side of Eq. (6.83) now includes an additional integral over x_n,

$$Z = \int dx_n \prod_{i=1}^{n-1} dx_i \exp[iS_M|_{t=-i\tau}] = \int \prod_{i=1}^{n} dx_i \exp[iS_M|_{t=-i\tau}] \ . \tag{6.85}$$

Finally, we introduce the notation

$$[dx] \equiv \prod_{i=1}^{n} dx_i \tag{6.86}$$

so that

$$Z = \int [dx] \exp[iS_M|_{t=-i\tau}] \ . \tag{6.87}$$

We next have to write the Minkowski-space action in the imaginary time formalism. In general, for a single particle in Minkowski space,

$$L_M(x, x') = \frac{1}{2}m(x')^2 - V(x) \tag{6.88}$$

where $x' = dx/dt$. When $t = -i\tau$, $dt = -id\tau$ and $d/dt = id/d\tau$. Thus the kinetic energy term in the Lagrangian is

$$\frac{1}{2}m(x')^2 = \frac{1}{2}m\left(i\frac{dx}{d\tau}\right)^2 = -\frac{1}{2}m\left(\frac{dx}{d\tau}\right)^2 \tag{6.89}$$

and

$$\begin{aligned} L_M|_{t=-i\tau} &= -\frac{1}{2}m\left(\frac{dx}{d\tau}\right)^2 - V(x) \\ &= -\left(\frac{1}{2}m\left(\frac{dx}{d\tau}\right)^2 + V(x)\right) \\ &= -(T(x) + V(x)) \ . \end{aligned} \tag{6.90}$$

Note that the Hamiltonian operator is $T(x)+V(x)$ so that $L_M|_{t=-i\tau} = -H$. The action is then

$$S_M|_{t=-i\tau} = i\int_{\tau_A}^{\tau_B} d\tau\left[\frac{1}{2}m\left(\frac{dx}{d\tau}\right)^2 + V(x)\right] \tag{6.91}$$

where the negative signs in $L_M|_{t=-i\tau}$ and the differential ($dt = -id\tau$) cancel. We have $iS_M|_{t=-i\tau}$ in the partition function which we can redefine as the Euclidean action [95],

$$-S_E = iS_M|_{t=-i\tau} = -\int_{\tau_A}^{\tau_B} d\tau\left[\frac{1}{2}m\left(\frac{dx}{d\tau}\right)^2 + V(x)\right] \ . \tag{6.92}$$

Note that, in the Minkowski action, the sign of the potential is negative while, in the Euclidean action, the sign is positive. Finally then

$$Z = \int [dx] \exp[-S_E(x)] \ . \tag{6.93}$$

EXAMPLE: Write the simple harmonic oscillator in Euclidean space. The Hamiltonian operator for a particle of unit mass, $m \equiv 1$, is

$$\hat{H} = \frac{1}{2}(\hat{p}^2 + \omega^2 \hat{x}^2) \tag{6.94}$$

where $\hat{x}$ and $\hat{p}$ are operators such that:

$$\begin{array}{ll} \hat{x}|x\rangle = x|x\rangle \ , & \hat{p}|p\rangle = p|p\rangle \ , \\ \langle x|x'\rangle = \delta(x - x') \ , & \langle p|p'\rangle = \delta(p - p') \ , \\ 1 = \int dx|x\rangle\langle x| \ , & 1 = \int dp|p\rangle\langle p| \ , \\ \langle x|p\rangle = \exp[ip \cdot x] \ , & \langle p'|x'\rangle = \exp[-ip' \cdot x'] \ . \end{array} \tag{6.95}$$

The partition function is

$$Z = \int dx \, \langle x| \exp[-\beta H]|x\rangle \ . \tag{6.96}$$

We divide the time interval β into n equal time steps of size ϵ where we will eventually take $\epsilon \to 0$. The time steps divide the spatial interval x into n equal steps, x_0, x_1, x_2, $\cdots$, x_n where $x_n = x_0 = x$. We insert the unit operator, *e.g.*, $1 = \int dx_1|x_1\rangle\langle x_1|$ at each step so that

$$\begin{aligned} Z &= \int dx \, \langle x|(\exp[-(\epsilon H)]^n|x\rangle \\ &= \int dx \prod_{i=1}^{n-1} dx_i \langle x| \exp[-\epsilon H]|x_1\rangle\langle x_1| \exp[-\epsilon H]|x_2\rangle \ldots \langle x_{n-1}| \exp[-\epsilon H]|x\rangle \\ &= \int \prod_{i=1}^{n} dx_i \, \langle x_{i-1}| \exp[-\epsilon H]|x_i\rangle \\ &= \int [dx]\langle x_{i-1}| \exp[-\epsilon H]|x_i\rangle \ . \end{aligned} \tag{6.97}$$

At each step we have

$$\begin{aligned} \Delta(-\epsilon H) &\equiv \langle x_{i-1}| \exp[-\epsilon H]|x_i\rangle \\ &= \left\langle x_{i-1} \middle| \exp\left\{-\frac{1}{2}\epsilon(\hat{p}^2 + \omega^2\hat{x}^2)\right\} \middle| x_i \right\rangle \ . \end{aligned} \tag{6.98}$$

The x operator is easily done, leaving

$$\Delta(-\epsilon H) = \exp\left[-\frac{\epsilon}{2}\omega^2 x_i^2\right]\left\langle x_{i-1} \middle| \exp\left\{-\frac{\epsilon}{2}\hat{p}^2\right\} \middle| x_i \right\rangle \ . \tag{6.99}$$

We then introduce the unit operator twice in p space, $1 = \int dpdp' |p\rangle\langle p|p'\rangle\langle p'|$ so that

$$\begin{aligned}\Delta(-\epsilon H) &= \exp\left[-\frac{\epsilon}{2}\omega^2 x_i^2\right] \\ &\times \int dp\, dp' \langle x_{i-1}| \exp[-\frac{\epsilon}{2}\hat{p}^2]|p\rangle\langle p|p'\rangle\langle p'|x_i\rangle\ . \qquad (6.100)\end{aligned}$$

Then $\exp[-(\epsilon/2)\hat{p}^2]|p\rangle = \exp[-(\epsilon/2)p^2]|p\rangle$ and, using $\langle p|p'\rangle = \delta(p-p')$, we have

$$\begin{aligned}\Delta(-\epsilon H) &= \exp\left[-\frac{\epsilon}{2}\omega^2 x_i^2\right] \\ &\times \int dp\, dp' \langle x_{i-1}|p\rangle\delta(p-p')\langle p'|x_i\rangle \exp[-\frac{\epsilon}{2}p^2]\ . \qquad (6.101)\end{aligned}$$

Next we use the definitions of $\langle x|p\rangle$ and $\langle p|x\rangle$ to obtain

$$\begin{aligned}\Delta(-\epsilon H) &= \exp\left[-\frac{\epsilon}{2}\omega^2 x_i^2\right] \int dp\, dp' \ \exp[ip\cdot x_{i-1} - ip'\cdot x_i] \\ &\times \delta(p-p') \exp\left[-\frac{\epsilon}{2}p^2\right]\ . \qquad (6.102)\end{aligned}$$

We use the delta function to do the integration over dp', leaving us with

$$\Delta(-\epsilon H) = \exp\left[-\frac{\epsilon}{2}\omega^2 x_i^2\right] \int dp \ \exp\left[-\frac{\epsilon}{2}p^2 + ip(x_{i-1} - x_i)\right] \qquad (6.103)$$

We then can complete the square in the exponent of the term inside the integral,

$$-\frac{1}{2}\epsilon p^2 + ip(x_{i-1} - x_i) = \frac{1}{2}\left(\frac{x_{i-1} - x_i}{\sqrt{\epsilon}} + i\sqrt{\epsilon}p\right)^2 - \frac{1}{2}\left(\frac{x_{i-1} - x_i}{\sqrt{\epsilon}}\right)^2 \qquad (6.104)$$

Then

$$\begin{aligned}\Delta(-\epsilon H) &= \exp\left[-\frac{1}{2}\left(\frac{(x_{i-1} - x_i)^2}{\epsilon} + \epsilon\omega^2 x_i^2\right)\right] \\ &\times \int dp \ \exp\left[\frac{1}{2}\left(\sqrt{\epsilon}p - \frac{i(x_i - x_i)}{\sqrt{\epsilon}}\right)^2\right]\ . \qquad (6.105)\end{aligned}$$

The integrand of dp is a Gaussian, giving a constant contribution to the partition function,

$$Z = \int \prod_{i=1}^{n} dx_i \exp\left[-\frac{1}{2}\left(\frac{(x_{i-1} - x_i)^2}{\epsilon} + \epsilon\omega^2 x_i^2 \right) \right] . \qquad (6.106)$$

The exponent can be written as the sum,

$$Z = \int \prod_{i=1}^{n} dx_i \exp\left[-\sum_{i=1}^{n} \left(\frac{1}{2\epsilon}(x_{i-1} - x_0)^2 + \frac{1}{2}\epsilon\omega^2 x_0^2 \right) \right] . \qquad (6.107)$$

Letting $\epsilon = d\tau$, the second term becomes $\int_0^\beta d\tau(1/2)\omega^2 x^2$ while, in the first term, $(x_{i-1} - x_i)^2 = (dx)^2$ and $1/\epsilon = 1/d\tau = d\tau/(d\tau)^2$ so that the first term becomes $\int_0^\beta d\tau(1/2)(dx/d\tau)^2$. We then have

$$\begin{aligned} Z &= \int \prod_{i=1}^{n} dx_i \exp\left[\int_0^\beta d\tau \left(\frac{1}{2}\left(\frac{dx}{d\tau}\right)^2 + \frac{1}{2}\omega^2 x^2 \right) \right] \\ &\equiv \sum_{\text{paths}} \exp[-S_E] , \end{aligned} \qquad (6.108)$$

as expected.

These single particle results can be generalized to a scalar field ϕ to find the amplitude at $\phi(x_B, t_B)$ from the initial state $\phi(x_A, t_A)$. In this case the Minkowski action is

$$S_M = \int_{t_A}^{t_B} dt \int dx \mathcal{L}_M[\phi(x,t), \partial_\mu\phi(x,t)] \qquad (6.109)$$

where $\mathcal{L}_M$ is the Lagrange density in Minkowski space and the Lagrangian is $\int dx\mathcal{L}_M$. In the diagonal basis, the field operator $\hat{\phi}(x,t)$ gives

$$\hat{\phi}(x_A, t_A)|\phi(x_A, t_A)\rangle = \phi(x_A, t_A)|\phi(x_A, t_A)\rangle . \qquad (6.110)$$

The state vector, after evolution from t_A to t_B, is

$$\exp[-iH(t_B - t_A)]|\phi(x_A, t_A)\rangle . \qquad (6.111)$$

The probability amplitude for the field is [95]

$$\langle\phi(x_B,t_B)|\exp[-iH(t_B-t_A)]|\phi(x_A,t_A)\rangle = \sum_{\text{all paths}} \exp[iS_M] \; . \quad (6.112)$$

The right-hand side can be discretized in the time coordinate into n_t time slices so that the field is $\phi(x, t_i)$ at all points x. The field can also be discretized in the x coordinate into n_x spatial slices so that for a given time slice, t_j, we have $\phi(x_1, t_j)$, $\phi(x_2, t_j) \cdots \phi(x_n, t_j)$. The path then specifies all field configurations $\phi(x, t)$ between t_A and t_B, as shown in Fig. 6.3. The sum over paths through (x_i, t_j) is

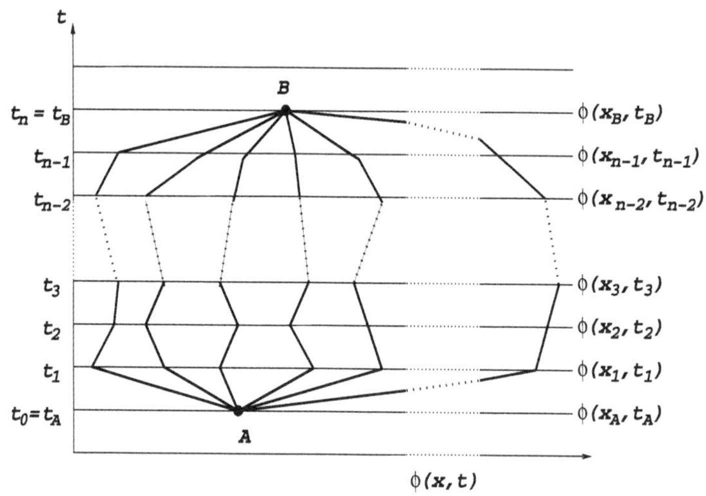

Figure 6.3: Several paths between $\phi(x_A, t_A)$ and $\phi(x_B, t_B)$. The time slices are shown as horizontal lines. The dots represent intermediate space-time points along a particular path.

$$\int d\phi(x_i, t_j) \exp[iS_M] \; . \quad (6.113)$$

To sum over all paths through t_j and all space points x_1 to xn_x, we have an n_x-dimensional integral,

$$\int \prod_{i=1}^{n_x} d\phi(x_i, t_j) \exp[iS_M] \; . \quad (6.114)$$

A similar argument can be made for each time slice so that including all time slices up to $n_t - 1$ (excluding the final time point, t_B) gives the final sum over paths with amplitude

$$\begin{aligned} &\langle \phi(x_B, t_B)| \exp[-iH(t_B - t_A)] |\phi(x_A, t_A)\rangle \\ &\quad = \int \prod_{j=1}^{n_t-1} \prod_{i=1}^{n_x} d\phi(x_i, t_j) \exp[iS_M] \ . \end{aligned} \tag{6.115}$$

We can go through the same arguments as before for the single particle case to relate the path integral for the field to the partition function. The action for the fields is

$$S_M = \int_{t_A}^{t_B} dt \int dx \mathcal{L}_M[\phi, \partial_\mu \phi] \ . \tag{6.116}$$

The Minkowski Lagrange density for a scalar field is

$$\begin{aligned} \mathcal{L}_M &= \frac{1}{2}\partial^\mu \phi \partial_\mu \phi - V(\phi) \\ &= \frac{1}{2}\left(\frac{\partial \phi}{\partial t}\right)^2 - \frac{1}{2}\left(\frac{\partial \phi}{\partial x}\right)^2 - V(\phi) \ . \end{aligned} \tag{6.117}$$

EXAMPLE: Show that the Lagrange density in Eq. (6.117) generates the Klein-Gordon equation when $V(\phi) = (1/2)m^2\phi^2$.
Recall that the equation of motion for the field is

$$\partial_\mu \left(\frac{\partial \mathcal{L}}{\partial(\partial_\mu \phi)}\right) - \frac{\partial \mathcal{L}}{\partial \phi} = 0 \ . \tag{6.118}$$

In the case of Eq. (6.117), $\partial\mathcal{L}/\partial(\partial_\mu\phi) = \partial^\mu \phi$ and $\partial\mathcal{L}/\partial\phi = -m^2\phi$. Thus the equation of motion is

$$\partial_\mu \partial^\mu \phi + m^2 \phi = 0 \ , \tag{6.119}$$

the Klein-Gordon equation of Chapter 2.

To write the Lagrange density in Euclidean space, we again make the replacement $t = -i\tau$. Then

$$\mathcal{L}_M|_{t=-i\tau} = -\frac{1}{2}\left(\frac{\partial \phi}{\partial \tau}\right)^2 - \frac{1}{2}\left(\frac{\partial \phi}{\partial x}\right)^2 - V(\phi) \ . \tag{6.120}$$

We can again define the Euclidean action relative to the Minkowski action by

$$
\begin{aligned}
iS_M|_{t=-i\tau} = -S_E &= -\int_{\tau_A}^{\tau_B} d\tau \int dx \left[\frac{1}{2}\left(\frac{\partial\phi}{\partial\tau}\right)^2 + \frac{1}{2}\left(\frac{\partial\phi}{\partial x}\right)^2 + V(\phi)\right] \\
&= -\int d\tau dx \left[\frac{1}{2}\delta^{\mu\nu}\partial_\mu\phi\partial_\nu\phi + V(\phi)\right] \\
&= -\int d\tau dx \mathcal{L}_E \qquad (6.121)
\end{aligned}
$$

where

$$\mathcal{L}_E = -\mathcal{L}_M|_{t=-i\tau} \qquad (6.122)$$

and $\delta^{\mu\nu}$ is the Euclidean metric tensor with $\delta^{\mu\nu} = 1$ when $\mu = \nu$ and 0 otherwise.

EXAMPLE: Show that $\mathcal{L}_E = -\mathcal{L}_M|_{t=-i\tau}$ also holds for the QCD Lagrange density.
The QCD Lagrangian in Minkowski space, Eq. (6.73), is

$$\mathcal{L}_M = i\overline{\psi}\gamma^\mu \mathbf{D}_\mu\psi - m\overline{\psi}\psi - \frac{1}{4}F_{\mu\nu}F^{\mu\nu} \qquad (6.123)$$

while, in Euclidean space we have,

$$\mathcal{L}_E = \overline{\psi}\gamma^\mu \mathbf{D}_\mu\psi + m\overline{\psi}\psi + \frac{1}{4}F_{\mu\nu}F^{\mu\nu} \qquad (6.124)$$

where $\mathbf{D}_\mu = \partial_\mu - ig\mathbf{A}_\mu$.

In imaginary time coordinates, the zeroth or time component of the derivative is $i\partial_\tau$ instead of ∂_t. To factor an i out of D_τ, we also need $A_t = iA_\tau$ in the gauge field so that $D_t = iD_\tau$. The space-time component of the field strength tensor, F_{tx}, is

$$F_{tx} = \partial_t A_x - \partial_x A_t - g[A_t, A_x] \ . \qquad (6.125)$$

Changing to imaginary time coordinates, we have

$$
\begin{aligned}
F_{\tau x} &= i\partial_\tau A_x - i\partial_x A_\tau - ig[A_\tau, A_x] \\
&= i(\partial_\tau A_x - \partial_x A_\tau - g[A_\tau, A_x]) \qquad (6.126)
\end{aligned}
$$

so that

$$F^a_{tx}F^{tx\,a} = -F^a_{\tau x}F^{\tau x\,a} \ . \tag{6.127}$$

The convolution of the field strength tensors is unchanged when both μ and ν are space coordinates.

The gamma matrices anticommute in Minkowski space,

$$\{\gamma^\mu, \gamma^\nu\} = 2g^{\mu\nu} \ . \tag{6.128}$$

In Euclidean space, we would similarly have

$$\{\gamma^\mu_E, \gamma^\nu_E\} = 2\delta^{\mu\nu} \ . \tag{6.129}$$

For this to be true, we need

$$\gamma^0 = \gamma^0_E \tag{6.130}$$

for the time coordinate while the anticommutation relations for the spatial gamma matrices tell us that

$$\begin{aligned} (\gamma^i)^2 &= -1 \ , \\ (\gamma^i_E)^2 &= 1 \end{aligned} \tag{6.131}$$

giving us the relation

$$\gamma^i = i\gamma^i_E \ . \tag{6.132}$$

Since $\overline{\psi} = \psi^\dagger\gamma^0$, it is unchanged when going from real time to imaginary time. Then

$$\begin{aligned} \mathcal{L}_M|_{t=-i\tau} &= i\overline{\psi}\gamma^0_E(iD_\tau)\psi - i\overline{\psi}(i\gamma^i_E)D_i - m\overline{\psi}\psi - \frac{1}{4}(-F^a_{\tau i}F^{\tau i\,a} - F^a_{ji}F^{ji\,a}) \\ &= -[\overline{\psi}\gamma^\mu_E\mathbf{D}_\mu\psi + m\overline{\psi}\psi + \frac{1}{4}F^a_{\mu\nu}F^{\mu\nu\,a}] \\ &= -\mathcal{L}_E \end{aligned} \tag{6.133}$$

A comparison of Minkowski and Euclidean-space quantities is given in Table 6.1.

Table 6.1: Comparison of Minkowski and Euclidean-space quantities. (Note that in the last column, the equality holds for the Lagrange density and the action when the Minkowski-space quantity is evaluated with $t = -i\tau$.) See also [95].

Quantity	Minkowski	Euclidean	Relation
coordinates, $\mathbf{x}^\mu$	(t, x, y, z)	(τ, x, y, z)	$t = -i\tau$
volume element, d^4x	$dt\, dx\, dy\, dz$	$d\tau\, dx\, dy\, dz$	$dt = -id\tau$
line element	$g_{\mu\nu} d\mathbf{x}^\mu\, d\mathbf{x}^\nu$	$\delta_{\mu\nu}\, d\mathbf{x}^\mu\, d\mathbf{x}^\nu$	
metric tensor	$g_{\mu\nu}$: $g_{tt} = 1, g_{xx} = g_{yy} = g_{zz} = -1$	$\delta_{\mu\nu}$: $\delta_{\tau\tau} = \delta_{xx} = \delta_{yy} = \delta_{zz} = 1$	
	$g_{ij} = 0,\ i \neq j$	$\delta_{ij} = 0,\ i \neq j$	
derivatives, ∂_μ	$(\partial_t, \partial_x, \partial_y, \partial_z)$	$(\partial_\tau, \partial_x, \partial_y, \partial_z)$	$\partial_t = i\partial_\tau$
gauge field, $\mathbf{A}_\mu$	(A_t, A_x, A_y, A_z)	(A_τ, A_x, A_y, A_z)	$A_t = iA_\tau$
gamma matrices, γ^μ	$(\gamma^0, \gamma^1, \gamma^2, \gamma^3)$	$(\gamma_E^0, \gamma_E^1, \gamma_E^2, \gamma_E^3)$	$\gamma^0 = \gamma_E^0$
	$\{\gamma^\mu, \gamma^\nu\} = 2g^{\mu\nu}$	$\{\gamma_E^\mu, \gamma_E^\nu\} = 2\delta^{\mu\nu}$	$\gamma^i = i\gamma_E^i$
scalar Lagrangian	$\frac{1}{2}[(\partial_t\phi)^2 - \lvert\vec{\nabla}\phi\rvert^2] - V(\phi)$	$\frac{1}{2}[(\partial_\tau\phi)^2 + \lvert\vec{\nabla}\phi\rvert^2] + V(\phi)$	$\mathcal{L}_M = -\mathcal{L}_E$
QCD Lagrangian	$i\overline{\psi}\gamma^\mu \mathbf{D}_\mu \psi - m\overline{\psi}\psi - \frac{1}{4}F_{\mu\nu}F^{\mu\nu}$	$\overline{\psi}\gamma_E^\mu \mathbf{D}_\mu \psi + m\overline{\psi}\psi + \frac{1}{4}F_{\mu\nu}F^{\mu\nu}$	$\mathcal{L}_M^{\text{QCD}} = -\mathcal{L}_E^{\text{QCD}}$
action	$S_M = \int dt\, dx\, \mathcal{L}_M$	$S_E = \int d\tau\, dx\, \mathcal{L}_E$	$iS_M = -S_E$

6.3.2 Putting QCD on the lattice

To discretize the QCD Lagrangian, a lattice of space-time points is set up with imaginary time coordinates. Each space-time point is given a label, *e.g.* n, in the d-dimensional coordinate system. There is a vector with length equal to the lattice spacing between the lattice points, labeled by *e.g.* μ, ν and κ in the x, y and z directions. The lattice points nearest to n in the x, y and z directions are $n+\mu$, $n+\nu$ and $n+\kappa$ respectively. The lattice spacing in all d dimensions is usually chosen to be the same, typically labeled a.

At each lattice point, the fermion fields (ψ and $\overline{\psi}$) and vector gauge fields ($\mathbf{A}_\mu$) need to be specified. The partition function for the QCD Lagrangian involves integration over ψ, $\overline{\psi}$ and $\mathbf{U}_\mu$, related to the gauge field, $\mathbf{A}_\mu$, at each space-time point,

$$\begin{aligned} Z &= \int \prod_n d\overline{\psi}(n) d\psi(n) d\mathbf{U}_\mu(n) \exp[-S_E] \\ &= \int [d\overline{\psi}][d\psi][dU] \exp[-S_E] \,. \end{aligned} \tag{6.134}$$

The fermion fields can be integrated separately and obey different rules for integration and differentiation, as we will show. That leaves only the vector gauge field $\mathbf{A}_\mu$ to specify the field configuration. The gauge field $\mathbf{A}_\mu$ interacts with the fermion field since it couples to the fermion current by the interaction term $\overline{\psi}\gamma_E^\mu \mathbf{A}^\mu \psi$.

In QED, there is only one fermion field per point while, in QCD, there are three for the color degrees of freedom. The fermion fields are represented by three-dimensional vectors in color space. The gauge fields specify the rotation of the fermion field in color space. The interaction Lagrangian should be independent of the local frame in color space. Thus the QCD Lagrangian should be locally gauge invariant, as we discussed previously. Since QCD possesses SU(3) color symmetry, the gauge fields can be expanded in terms of the SU(3) generators.

An element of the SU(3) group is a unitary 3×3 matrix, U, so that

$$U^\dagger U = U U^\dagger = 1 \,. \tag{6.135}$$

The S in SU(3) stands for 'special', meaning that the determinant of U is unity,

$$\det U = 1 \; . \tag{6.136}$$

If U is represented by $\exp[iM]$, it can be shown that M is a Hermitian ($M = M^\dagger$), traceless matrix.

EXAMPLE: Show this.
If U is a 3×3 matrix, M is also a 3×3 matrix, as is $\exp[iM]$, by extension. From Eq. (6.135), we have

$$\exp[-iM^\dagger]\exp[iM] = 1 \; . \tag{6.137}$$

Multiplying both sides of Eq. (6.137) by $\exp[iM^\dagger]$ gives

$$\begin{aligned} \exp[iM] &= \exp[iM^\dagger] \; , \\ M &= M^\dagger \; . \end{aligned} \tag{6.138}$$

demonstrating the hermiticity of M. To show that M is traceless, we make use of the property of matrix multiplication that say that the determinant of the product of two square ($n \times n$) matrices is equal to the product of the determinants,

$$(\det A)(\det A') = \det(AA') \; . \tag{6.139}$$

Therefore,

$$\begin{aligned} \det U = \det(1 \cdot U) &= \det(CC^{-1}U) = \det(CUC^{-1}) \\ &= \det(C\exp[iM]C^{-1}) \end{aligned} \tag{6.140}$$

where C is any 3×3 matrix and C^{-1} is its inverse. We can write the matrix product $C\exp[iM]C^{-1}$ as

$$C\exp[iM]C^{-1} = C\left(\sum_{n=0}^{\infty}\frac{(iM)^n}{n!}\right)C^{-1} \tag{6.141}$$

but since $CC^{-1} = 1$, we can raise this matrix product to the nth power and multiply it by M directly,

$$C\exp[iM]C^{-1} = \sum_{n=0}^{\infty}\frac{(iCMC^{-1})^n}{n!} = \exp[iCMC^{-1}] \tag{6.142}$$

and

$$\det U = \det(\exp[iM]) = \det(\exp[iCMC^{-1}]) \ . \tag{6.143}$$

It is possible to choose C to diagonalize M so that, for a 3×3 matrix,

$$\begin{aligned} & \det(\exp[iM]) \\ = & \det \begin{pmatrix} \exp[i(CMC^{-1})_{11}] & 0 & 0 \\ 0 & \exp[i(CMC^{-1})_{22}] & 0 \\ 0 & 0 & \exp[i(CMC^{-1})_{33}] \end{pmatrix} \\ = & \exp(i[(CMC^{-1})_{11} + (CMC^{-1})_{22} + (CMC^{-1})_{33}]) \\ = & \exp[i\mathrm{Tr}(CMC^{-1})] = \exp[i\mathrm{Tr}M] \ . \end{aligned} \tag{6.144}$$

The last equality comes about because the order of multiplication inside the trace is cyclic. If $\det U = \det(\exp[iM]) \equiv 1$, then $\mathrm{Tr}M = 0$.

There are eight generators of color SU(3), corresponding to the eight colors of gluons, the gauge fields of QCD. These generators are the Gell-Mann matrices, labeled $\lambda_a/2$ by convention. Thus M can be expanded as a linear combination of the generators so that

$$\begin{aligned} M &= \frac{\lambda_a}{2} M_a \\ U &= \exp\left[i\frac{\lambda_a}{2} M_a\right] \ . \end{aligned} \tag{6.145}$$

The gauge field at point n fixes the rotation of the fermion field in color space at that point so that $\mathbf{A}_\mu(n)$ can be written as

$$\mathbf{A}_\mu(n) = \frac{\lambda_a}{2} \mathbf{A}_\mu^a(n) \ . \tag{6.146}$$

Instead of using the gauge fields as the basic variables on the lattice, link variables, $\mathbf{U}_\mu(n)$, are used since it is simpler to write the Euclidean action on the lattice in terms of the link variables. The link variables, as their name implies, are defined *between* rather than *on* lattice sites, in the form of a path integral of the gauge field between *e.g.* sites n and $n + \mu$,

$$\mathbf{U}_\mu(n) = P \exp\left\{ig \int_n^{n+\mu} \mathbf{A}_\mu(x)\, d\mathbf{x}^\mu\right\} \ . \tag{6.147}$$

The P stand for path ordering. Path-ordered operators are placed left to right with decreasing time so that the operators on the left are later in time than those on the right.

While the link variable for moving from site n to site $n+\mu$ is $\mathbf{U}_\mu(n)$, the link variable for moving back to site n from site $n+\mu$ is $\mathbf{U}_{-\mu}(n+\mu)$. As might be expected,

$$\mathbf{U}_\mu^{-1}(n) = \mathbf{U}_{-\mu}(n+\mu) \tag{6.148}$$

EXAMPLE: Show this using Eq. (6.147).
If we divide the path length from n to $n+\mu$ into N segments of length $\epsilon = \mu/N$ with $d\mathbf{x}^\mu = \epsilon$, writing the integral as a sum, we have

$$\begin{aligned} \mathbf{U}_\mu(n) &= P\exp\{ig\epsilon(\mathbf{A}_\mu(n) + \mathbf{A}_\mu(n+\epsilon) + \mathbf{A}_\mu(n+2\epsilon) + \ldots \\ &\quad + \mathbf{A}_\mu(n+\epsilon(N-1)) + \mathbf{A}_\mu(n+N\epsilon)) \ . \end{aligned} \tag{6.149}$$

Since we are moving from n to $n+\mu$, the path ordering means that the final step is the leftmost, reversing the order in Eq. (6.149) to give

$$\begin{aligned} \mathbf{U}_\mu(n) &= \exp[ig\epsilon\mathbf{A}_\mu(n+N\epsilon)]\exp[ig\epsilon\mathbf{A}_\mu(n+(N-1)\epsilon)] \\ &\quad \times \ldots \exp[ig\epsilon\mathbf{A}_\mu(n)] \ . \end{aligned} \tag{6.150}$$

To return to site n from site $n+\mu$, we define $\mathbf{U}_{-\mu}(n+\mu)$ as

$$\mathbf{U}_{-\mu}(n+\mu) = P\exp\left\{ig\int_{n+\mu}^{n} \mathbf{A}_\mu(x)\, d\mathbf{x}^\mu\right\} \tag{6.151}$$

where now $d\mathbf{x}^\mu = -\epsilon$ and

$$\begin{aligned} \mathbf{U}_{-\mu}(n+\mu) &= P\exp\{-ig\epsilon[\mathbf{A}_\mu(n+N\epsilon) + \mathbf{A}_\mu(n+(N-1)\epsilon)\ldots \\ &\quad + \mathbf{A}_\mu(n+2\epsilon) + \mathbf{A}_\mu(n+\epsilon) + \mathbf{A}_\mu(n)]\} \\ &= \exp[-ig\epsilon\mathbf{A}_\mu(n)]\ldots\exp[-ig\epsilon\mathbf{A}_\mu(n+(N-1)\epsilon)] \\ &\quad \times \exp[-ig\epsilon\mathbf{A}_\mu(n+N\epsilon)] \ . \end{aligned} \tag{6.152}$$

Multiplying Eqs. (6.150) and (6.152), we have

$$\begin{aligned} \mathbf{U}_\mu(n)\mathbf{U}_{-\mu}(n+\mu) &= \exp[ig\epsilon\mathbf{A}_\mu(n+N\epsilon)]\exp[ig\epsilon\mathbf{A}_\mu(n+(N-1)\epsilon)] \\ &\quad \times \ldots\exp[ig\epsilon\mathbf{A}_\mu(n)]\exp[-ig\epsilon\mathbf{A}_\mu(n)]\ldots \\ &\quad \times \exp[-ig\epsilon\mathbf{A}_\mu(n+(N-1)\epsilon)]\exp[-ig\epsilon\mathbf{A}_\mu(n+N\epsilon)] \\ &\equiv 1 \end{aligned} \tag{6.153}$$

since the exponential factors cancel each other term by term. Multiplying both sides of Eq. (6.153) on the left by $\mathbf{U}_{\mu}^{-1}(n)$, we find

$$\mathbf{U}_{-\mu}(n+\mu) = \mathbf{U}_{\mu}^{-1}(n) \ . \tag{6.154}$$

The basic unit of lattice calculations is called the plaquette. In two dimensions, it is a square linking the nearest neighbors, as shown in Fig. 6.4. As we show a bit later, the plaquette is equivalent to $F_{\mu\nu}F^{\mu\nu}$ on the lattice. The plaquette shown in Fig. 6.4 can be evaluated as the

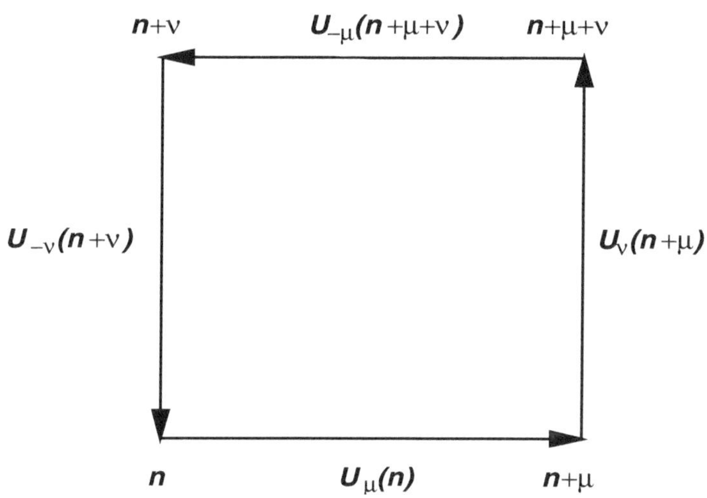

Figure 6.4: A two-dimensional plaquette traced out by moving counterclockwise between sites, starting and ending at $\mathbf{U}_{\mu}(n)$.

path-ordered product of link variables,

$$P_{\Box}(n;\mu,\nu) = \mathbf{U}_{-\nu}(n+\nu)\mathbf{U}_{-\mu}(n+\mu+\nu)\mathbf{U}_{\nu}(n+\mu)\mathbf{U}_{\mu}(n) \ . \tag{6.155}$$

The path around the plaquette is taken in the counter-clockwise direction. A step to the right of n, in the positive 'x' direction, moves the gauge field from n to $n+\mu$ with link variable $\mathbf{U}_{\mu}(n)$. From point $n+\mu$, the gauge field moves in the positive 'y' direction, from $n+\mu$ to $n+\mu+\nu$ by link variable $\mathbf{U}_{\nu}(n+\mu)$. The subscripts μ and ν indicate the 'x' and 'y' direction stepped by the link variable. To move from $n+\mu+\nu$ to $n+\nu$, a step above n in the 'y' direction, we step in the $-\mu$ direction with link variable $\mathbf{U}_{-\mu}(n+\mu+\nu)$. Finally, to return to starting point n from $n+\nu$, we move in the $-\nu$ direction with link variable $\mathbf{U}_{-\nu}(n+\nu)$.

A larger area can be 'tiled' by plaquettes since the links in common between two adjacent plaquettes will integrate to unity, as we later discuss. See Fig. 6.5 for how this works schematically..

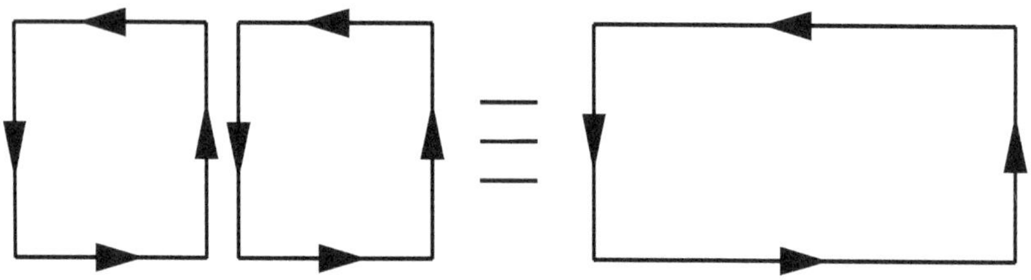

Figure 6.5: An illustration of how a lattice is 'tiled' with two-dimensional plaquettes. The two plaquettes merge into a larger rectangle when the link variables in the center are integrated over, see Eq. (6.319).

Local SU(3) gauge invariance of the lattice action

We now discuss how the various fields transform under rotations in color space to show that the QCD Euclidean action on the lattice [95],

$$\begin{aligned} S_E &= S_E^{\text{fermion}} + S_E^{\text{gauge}} \\ &= \frac{a^3}{2}\sum_{n,\mu}\overline{\psi}(n)\gamma_E^\mu[\mathbf{U}_{-\mu}(n+\mu)\psi(n+\mu) - \mathbf{U}_\mu(n-\mu)\psi(n-\mu)] \\ &\quad + ma^4\sum_n \overline{\psi}(n)\psi(n) \\ &\quad - \frac{1}{2g^2}\sum_{n,\mu,\nu}\mathrm{Tr}\left[\mathbf{U}_{-\nu}(n+\nu)\mathbf{U}_{-\mu}(n+\mu+\nu)\mathbf{U}_\nu(n+\mu)\mathbf{U}_\mu(n)\right] \\ &\quad + \text{hermitian conjugate}\ . \end{aligned} \tag{6.156}$$

is locally gauge invariant. We begin with transformations of the QCD fields, $\psi(n)$, $\overline{\psi}(n)$ and $\mathbf{A}_\mu(n)$ and then turn to the transformations of the link variables. With these transformations in hand, it is straightforward to show the gauge invariance of S_E.

Rotations in color space can be carried out using a unitary 3×3 SU(3) matrix $G(n)$,

$$G(n) = \exp\left[i\frac{\lambda_a}{2}\chi_a(n)\right] \tag{6.157}$$

where $\chi_a(n)$ are components of an eight-dimensional vector in color space, $\vec{\chi}(n)$. Thus the argument of the exponential becomes the scalar product of two vectors,

$$G(n) = \exp\left[i\frac{\vec{\lambda}}{2}\vec{\chi}(n)\right] \tag{6.158}$$

where the eight 3×3 Gell-Mann matrices, $\lambda_a/2$, are multiplied by the rotation angle of the color frame. Thus a vector in color space, V, is transformed as $G(n)V$.

The fermion fields, $\psi(n)$ and $\overline{\psi}(n)$ are vectors in color space, represented by three-component column matrices

$$\psi(n) \quad \rightarrow \quad \psi'(n) = G(n)\psi(n) \ , \tag{6.159}$$
$$\overline{\psi}(n) \quad \rightarrow \quad \overline{\psi}'(n) = \overline{\psi}(n)G^{-1}(n) \ . \tag{6.160}$$

The interaction term in the Euclidean Lagrangian that couples the quarks to the gauge fields (the gluons), $\overline{\psi}\gamma_E^\mu\mathbf{D}_\mu\psi$, contains the covariant derivative, $\mathbf{D}_\mu = \partial_\mu - ig\mathbf{A}_\mu$. We can use the local gauge invariance of the QCD interaction to find out how the gauge field transforms under color rotations.

EXAMPLE: Use the fermion field transformation in Eqs. (6.159) and (6.160) to determine the transformation of the gauge field $\mathbf{A}_\mu(n)$.
The interaction term will be invariant if $\mathbf{D}_\mu(n)\psi(n)$ transforms the same way as $\psi(n)$ or

$$\mathbf{D}_\mu(n)\psi(n) \longrightarrow (\mathbf{D}_\mu(n)\psi(n))' = G(n)\mathbf{D}_\mu(n)\psi(n) \ . \tag{6.161}$$

Then

$$\begin{aligned} G(n)[\partial_\mu - ig\mathbf{A}_\mu(n)]\psi(n) &= [\partial_\mu - ig\mathbf{A}_\mu(n)]'\psi'(n) \\ &= [\partial_\mu - ig\mathbf{A}'_\mu(n)]G(n)\psi(n) \ . \end{aligned} \tag{6.162}$$

We act with the derivative on both $G(n)$ and $\psi(n)$,

$$\partial_\mu[G(n)\psi(n)] = G(n)\partial_\mu\psi(n) + [\partial_\mu G(n)]\psi(n) \ , \tag{6.163}$$

so that

$$\begin{aligned} G(n)[\partial_\mu - ig\mathbf{A}_\mu(n)]\psi(n) &= [G(n)\partial_\mu + \partial_\mu G(n) \\ &\quad - ig\mathbf{A}'_\mu(n)G(n)]\psi(n) \ . \end{aligned} \tag{6.164}$$

The first terms on both the left- and right-hand sides are identical and cancel. Since there are no more derivatives acting on $\psi(n)$, we can drop it, leaving

$$G(n)[-ig\mathbf{A}_\mu(n)] = \partial_\mu G(n) - ig\mathbf{A}'_\mu(n) \ . \tag{6.165}$$

To get $\mathbf{A}'_\mu(n)$ on one side by itself, we subtract $\partial_\mu G(n)$, multiply both sides on the right by $G^{-1}(n)$ and divide by $-ig$, giving

$$\mathbf{A}'_\mu(n) = G(n)\mathbf{A}_\mu(n)G^{-1}(n) - \frac{i}{g}[\partial_\mu G(n)]G^{-1}(n) \ . \tag{6.166}$$

From this it follows that

$$\mathbf{D}'_\mu(n) = G(n)\mathbf{D}_\mu(n)G^{-1}(n) \tag{6.167}$$

since

$$[\mathbf{D}_\mu(n)\psi(n)]' = \mathbf{D}'_\mu(n)\psi'(n) \tag{6.168}$$

$$= \mathbf{D}'_\mu(n)G(n)\psi(n) \tag{6.169}$$

$$= G(n)\mathbf{D}_\mu(n)\psi(n) \ . \tag{6.170}$$

Equation (6.169) comes from Eq. (6.159). Equation (6.170) comes from Eq. (6.161). Dropping the common ψ field between Eqs. (6.169) and (6.170) and multiplying the right sides by $G^{-1}(n)$ gives Eq. (6.167).

The field strength tensor $F_{\mu\nu}(n)$ transforms as

$$F_{\mu\nu}(n) \longrightarrow F'_{\mu\nu}(n) = G(n)F_{\mu\nu}(n)G^{-1}(n) \ . \tag{6.171}$$

This is easy to see using the transformation property of the covariant derivative in Eq. (6.170) and the definition of $F_{\mu\nu}$ as the commutator of covariant derivatives in Eq. (6.74),

$$F_{\mu\nu} = \frac{i}{g}[\mathbf{D}_\mu, \mathbf{D}_\nu] \ . \tag{6.172}$$

Then

$$\begin{aligned} F'_{\mu\nu}(n) &= \frac{i}{g}[\mathbf{D}'_\mu(n)\mathbf{D}'_\nu(n) - \mathbf{D}'_\nu(n)\mathbf{D}'_\mu(n)] \\ &= \frac{i}{g}[G(n)\mathbf{D}_\mu(n)G^{-1}(n)G(n)\mathbf{D}_\nu(n)G^{-1}(n) \\ &\quad - G(n)\mathbf{D}_\nu(n)G^{-1}(n)G(n)\mathbf{D}_\mu(n)G^{-1}(n)] \\ &= G(n)\left[\frac{i}{g}(\mathbf{D}_\mu(n)\mathbf{D}_\nu(n) - \mathbf{D}_\nu(n)\mathbf{D}_\mu(n))\right]G^{-1}(n) \\ &= G(n)F_{\mu\nu}(n)G^{-1}(n) \ . \end{aligned} \tag{6.173}$$

Substituting $\partial_\mu - ig\mathbf{A}_\mu$ for $\mathbf{D}_\mu$, we have

$$\begin{aligned} F_{\mu\nu} &= \frac{i}{g}[\mathbf{D}_\mu\mathbf{D}_\nu - \mathbf{D}_\nu\mathbf{D}_\mu] \\ &= \frac{i}{g}[(\partial_\mu - ig\mathbf{A}_\mu)(\partial_\nu - ig\mathbf{A}_\nu) - (\partial_\nu - ig\mathbf{A}_\nu)(\partial_\mu - ig\mathbf{A}_\mu)] \\ &= \frac{i}{g}[\partial_\mu\partial_\nu - ig\mathbf{A}_\mu\,\partial_\nu - ig\,\partial_\mu\mathbf{A}_\nu - g^2\mathbf{A}_\mu\mathbf{A}_\nu \\ &\quad - \partial_\nu\partial_\mu + ig\mathbf{A}_\nu\,\partial_\mu + ig\,\partial_\nu\mathbf{A}_\mu + g^2\mathbf{A}_\nu\mathbf{A}_\mu] \ . \end{aligned} \tag{6.174}$$

The double derivatives and the $\mathbf{A}_i\partial_j$ terms cancel. Then

$$F_{\mu\nu} = \partial_\mu\mathbf{A}_\nu - \partial_\nu\mathbf{A}_\mu - ig[\mathbf{A}_\nu, \mathbf{A}_\mu] \ . \tag{6.175}$$

Adding the color index a to the field strength tensor gives us

$$F_{\mu\nu} = \frac{\lambda_a}{2}F^a_{\mu\nu} \ . \tag{6.176}$$

Using this, we note that the gauge part of the action can be written equivalently as

$$S_E^{\text{gauge}} = \frac{1}{4}F^a_{\mu\nu}F^{\mu\nu\,a} = \frac{1}{2}\text{Tr}[F_{\mu\nu}F^{\mu\nu}] \ . \tag{6.177}$$

The factor of two arises because

$$
\begin{aligned}
\mathrm{Tr}[F_{\mu\nu}F^{\mu\nu}] &= \mathrm{Tr}\left[\frac{\lambda_a}{2}F^a_{\mu\nu}\frac{\lambda_b}{2}F^{\mu\nu\,b}\right] \\
&= \mathrm{Tr}\left[\frac{\lambda_a}{2}\frac{\lambda_b}{2}\right]F^a_{\mu\nu}F^{\mu\nu\,b} \\
&= \frac{1}{2}F^a_{\mu\nu}F^{\mu\nu\,a} \qquad (6.178)
\end{aligned}
$$

and $\mathrm{Tr}[(\lambda_a/2)(\lambda_b/2)] = \delta_{ab}/2$.

We finally add the color index to the gauge fields as well so that $F_{\mu\nu}$ is

$$
\begin{aligned}
F_{\mu\nu} &= \frac{\lambda_a}{2}F^a_{\mu\nu} \\
&= \frac{\lambda_a}{2}(\partial_\mu \mathbf{A}^a_\nu - \partial_\nu \mathbf{A}^a_\mu) - ig\left[\frac{\lambda_b}{2}\mathbf{A}^b_\mu, \frac{\lambda_c}{2}\mathbf{A}^c_\nu\right] \\
&= \frac{\lambda_a}{2}(\partial_\mu \mathbf{A}^a_\nu - \partial_\nu \mathbf{A}^a_\mu) - ig\left[\frac{\lambda_b}{2}, \frac{\lambda_c}{2}\right]\mathbf{A}^b_\mu \mathbf{A}^c_\nu \,. \qquad (6.179)
\end{aligned}
$$

Using the SU(3) generator commutation relation

$$
\left[\frac{\lambda_b}{2}, \frac{\lambda_c}{2}\right] = i f_{abc}\frac{\lambda_a}{2} \,, \qquad (6.180)
$$

as in Eq. (6.74), factoring out $\lambda_a/2$, we have

$$
\frac{\lambda_a}{2}F^a_{\mu\nu} = \frac{\lambda_a}{2}[\partial_\mu \mathbf{A}^a_\nu - \partial_\nu \mathbf{A}^a_\mu + g f_{abc}\mathbf{A}^b_\mu \mathbf{A}^c_\nu] \,. \qquad (6.181)
$$

The contribution of the gauge field kinetic term to the QCD Lagrangian, Eq. (6.177), transforms as

$$
\begin{aligned}
\mathrm{Tr}[F_{\mu\nu}(n)F^{\mu\nu}(n)] &\rightarrow \mathrm{Tr}[F'_{\mu\nu}F^{\mu\nu\prime}] \\
&= \mathrm{Tr}[G(n)F_{\mu\nu}(n)G^{-1}(n)G(n)F^{\mu\nu}(n)G^{-1}(n)] \\
&= \mathrm{Tr}[G(n)F_{\mu\nu}(n)F^{\mu\nu}(n)G^{-1}(n)] \\
&= \mathrm{Tr}[F_{\mu\nu}(n)F^{\mu\nu}(n)] \,. \qquad (6.182)
\end{aligned}
$$

The last equality in Eq. (6.182) is due to the cyclic property of the trace, $\mathrm{Tr}[abcd] = \mathrm{Tr}[dabc]$.

The final ingredient of the lattice action still needed to be checked for local gauge invariance is the link variable, $\mathbf{U}_\mu(n)$. The link variable transforms as

$$\mathbf{U}_\mu(n) \to G(n+\mu)\mathbf{U}_\mu(n)G^{-1}(n) \ . \tag{6.183}$$

EXAMPLE: Show Eq. (6.183).
To do this proof most effectively, we use the transform of the gauge field and start from the path-ordered expansion of $\mathbf{U}_\mu(n)$ into N equal steps of size ϵ,

$$\begin{aligned}\mathbf{U}_\mu(n) &= \exp[ig\epsilon\mathbf{A}_\mu(n+N\epsilon)]\exp[ig\mathbf{A}_\mu(n+(N-1)\epsilon)] \\ &\quad \times \ldots \exp[ig\epsilon\mathbf{A}_\mu(n)] \ .\end{aligned} \tag{6.184}$$

To simplify notation, if we let $n + k\epsilon \equiv k$ and $\mathbf{A}_\mu(n+k\epsilon) \equiv \mathbf{A}_\mu(k)$ where $0 < k < N$, the path-ordered product in Eq. (6.184) is written as

$$\begin{aligned}\mathbf{U}_\mu(n) &= \exp[ig\epsilon\mathbf{A}_\mu(N)]\exp[ig\mathbf{A}_\mu(N-1)] \\ &\quad \times \ldots \exp[ig\epsilon\mathbf{A}_\mu(1)]\exp[ig\epsilon\mathbf{A}_\mu(0)] \ ,\end{aligned} \tag{6.185}$$

while the transformed link variables is expanded as

$$\begin{aligned}\mathbf{U}'_\mu(n) &= \exp[ig\epsilon\mathbf{A}'_\mu(N)]\exp[ig\mathbf{A}'_\mu(N-1)] \\ &\quad \times \ldots \exp[ig\epsilon\mathbf{A}'_\mu(1)]\exp[ig\epsilon\mathbf{A}'_\mu(0)] \ ,\end{aligned} \tag{6.186}$$

Since ϵ is small,

$$\exp[ig\epsilon\mathbf{A}_\mu(k)] \approx 1 + ig\epsilon\mathbf{A}_\mu(k) \ , \tag{6.187}$$

while, for the transformed gauge field, we have

$$\begin{aligned}\exp[ig\epsilon\mathbf{A}'_\mu(k)] &\approx 1 + ig\epsilon\mathbf{A}'_\mu(k) \\ &= 1 + ig\epsilon\Big[G(k)\mathbf{A}_\mu(k)G^{-1}(k) \\ &\qquad - \frac{i}{g}[\partial_\mu G(k)]G^{-1}(k)\Big] \ .\end{aligned} \tag{6.188}$$

We expand $G(j+1)$ to obtain

$$G(j+1) \approx G(k) + \epsilon\partial_\mu G(k) \ . \tag{6.189}$$

If we act on one component of the gauge field sum making up the link variable with $G(j+1)$ on the left and $G^{-1}(k)$ on the right, we find

$$\begin{aligned} G(j+1)\exp[ig\epsilon\mathbf{A}_\mu(k)]G^{-1}(k) &= [G(k)+\epsilon\,\partial_\mu\,G(k)][1+ig\epsilon\mathbf{A}_\mu(k)]G^{-1}(k) \\ &= G(k)G^{-1}(k)+ig\epsilon G(k)\mathbf{A}_\mu(k)G^{-1}(k) \\ &\quad +\epsilon[\partial_\mu G(k)]G^{-1}(k)+\mathcal{O}(\epsilon^2)\ . \end{aligned} \tag{6.190}$$

Ignoring terms proportional to ϵ^2, we see that

$$\begin{aligned} G(j+1)\exp[ig\epsilon\mathbf{A}_\mu(k)]G^{-1}(k) &= 1+ig\epsilon\Big[G(k)\mathbf{A}_\mu(k)G^{-1}(k) \\ &\quad -\frac{i}{g}[\partial_\mu G(k)]G^{-1}(k)\Big] \\ &= 1+ig\epsilon\mathbf{A}'_\mu(k) \\ &\approx \exp[ig\epsilon\mathbf{A}'_\mu]\ , \end{aligned} \tag{6.191}$$

equivalent to Eq. (6.188), above. Inserting $G(j+1)$ and $G^{-1}(k)$ on either side of each term in Eq. (6.186) gives

$$\begin{aligned} \mathbf{U}'_\mu(n) &= G(N+1)\exp[ig\epsilon\mathbf{A}'_\mu(N)]G^{-1}(N) \\ &\quad \times G(N)\exp[ig\mathbf{A}'_\mu(N-1)]G^{-1}(N-1) \\ &\quad \times \ldots G(2)\exp[ig\epsilon\mathbf{A}'_\mu(1)]G^{-1}(1)G(1)\exp[ig\epsilon\mathbf{A}'_\mu(0)]G^{-1}(0) \\ &= G(N+1)\exp[ig\epsilon\mathbf{A}_\mu(N)]\exp[ig\mathbf{A}_\mu(N-1)] \\ &\quad \times \ldots\exp[ig\epsilon\mathbf{A}_\mu(1)]\exp[ig\epsilon\mathbf{A}_\mu(0)]G^{-1}(0) \\ &\equiv G(n+\mu)\mathbf{U}_\mu(n)G^{-1}(n)\ . \end{aligned} \tag{6.192}$$

The result is similar for link variables with a negative subscript,

$$\mathbf{U}'_{-\mu}(n) = G(n)\mathbf{U}_{-\mu}(n)G^{-1}(n+\mu)\ . \tag{6.193}$$

The plaquette then transforms as

$$\begin{aligned} P'_\Box(n;\mu,\nu) &= G(n)\mathbf{U}_{-\nu}(n+\nu)G^{-1}(n+\nu)G(n+\nu)\mathbf{U}_{-\mu}(n+\mu+\nu) \\ &\quad \times G^{-1}(n+\mu+\nu)G(n+\mu+\nu)\mathbf{U}_\nu(n+\mu)G^{-1}(n+\mu) \\ &\quad \times G(n+\mu)\mathbf{U}_\mu(n)G^{-1}(n) \\ &= G(n)P_\Box(n;\mu,\nu)G^{-1}(n)\ . \end{aligned} \tag{6.194}$$

The cyclic properties of the trace show that

$$\mathrm{Tr}[P'_\Box(n;\mu,\nu)] = \mathrm{Tr}[G(n)P_\Box(n;\mu,\nu)G^{-1}(n)] = \mathrm{Tr}[P_\Box(n;\mu,\nu)] \; . \tag{6.195}$$

Since the trace of the plaquette is the gauge part of the lattice QCD Lagrangian, Eq. (6.195) shows that this part of the Euclidean action is gauge invariant.

Lastly, we show that the fermion part of the action is gauge invariant,

$$\begin{aligned}
&\overline{\psi}(n)\mathbf{U}_{-\mu}(n+\mu)\psi(n+\mu) = \\
&\qquad [\overline{\psi}(n)G^{-1}(n)][G(n)\mathbf{U}_{-\mu}(n+\mu)G^{-1}(n+\mu)] \\
&\qquad \times [G(n+\mu)\psi(n+\mu)] \\
&= \overline{\psi}(n)\mathbf{U}_{-\mu}(n+\mu)\psi(n+\mu) \; , \qquad (6.196) \\
&\overline{\psi}(n)\mathbf{U}_{\mu}(n-\mu)\psi(n-\mu) = \\
&\qquad [\overline{\psi}(n)G^{-1}(n)][G(n)\mathbf{U}_{\mu}(n-\mu)G^{-1}(n-\mu)] \\
&\qquad \times [G(n-\mu)\psi(n-\mu)] \\
&= \overline{\psi}(n)\mathbf{U}_{\mu}(n-\mu)\psi(n+\mu) \; . \qquad (6.197)
\end{aligned}$$

We have finally shown that the lattice action in Euclidean space, Eq. (6.156), is locally gauge invariant under SU(3) color rotations. We next show that, in the continuum limit, we get back the continuum action with the Lagrangian as in Eq. (6.124) [95]. We begin with the gauge part of the action.

Regaining the continuum limit

If the lattice spacing is small, the line integral from n to $n+\mu$ in the path-ordered exponent of the link variable, $\mathbf{U}_\mu(n)$, can be approximated as

$$\int_n^{n+\mu} \mathbf{A}_\mu(\nu)\, d\mathbf{x}^\mu = \frac{a}{2}[\mathbf{A}_\mu(n) + \mathbf{A}_\mu(n+\mu)] \tag{6.198}$$

using the trapezoidal rule. Then the link variable is

$$\begin{aligned}\mathbf{U}_\mu(n) &= P\exp\left(ig\int_n^{n+\mu}\mathbf{A}_\mu(x)\,d\mathbf{x}^\mu\right)\\ &= P\exp\left(ig\frac{a}{2}[\mathbf{A}_\mu(n)+\mathbf{A}_\mu(n+\mu)]\right)\\ &= \exp\left[ig\frac{a}{2}\mathbf{A}_\mu(n+\mu)\right]\exp\left[ig\frac{a}{2}\mathbf{A}_\mu(n)\right]+\cdots\,. \qquad (6.199)\end{aligned}$$

The gauge field at point $n+\mu$ can be expanded around the value of $\mathbf{A}_\mu(n)$, keeping only the first power of a in the expansion,

$$\mathbf{A}_\mu(n+\mu) = \mathbf{A}_\mu(n) + a\,\partial_\mu\,\mathbf{A}_\mu(n) + \cdots\,. \qquad (6.200)$$

Thus the link variable $\mathbf{U}_\mu(n)$ is

$$\begin{aligned}\mathbf{U}_\mu(n) &= \exp\left[ig\frac{a}{2}[\mathbf{A}_\mu(n)+a\,\partial_\mu\mathbf{A}_\mu(n)]\right]\exp\left[ig\frac{a}{2}\mathbf{A}_\mu(n)\right]\\ &= \exp\left[ig\frac{a}{2}[2\mathbf{A}_\mu(n)+a\,\partial_\mu\mathbf{A}_\mu(n)]+\mathcal{O}(a^3)\right]\,. \qquad (6.201)\end{aligned}$$

Note that the final result for the link variable contains terms linear and quadratic in the lattice spacing. Both need to be kept to obtain the correct continuum limit. However, terms proportional to cubic powers of the lattice spacing a may be neglected.

EXAMPLE: Obtain the expanded results for the other three link variables in the plaquette, $\mathbf{U}_\nu(n+\mu)$, $\mathbf{U}_{-\mu}(n+\mu+\nu)$ *and* $\mathbf{U}_{-\nu}(n+\nu)$.

The path-ordered products for the remaining link variables using the trapezoid rule to integrate the gauge fields are:

$$
\begin{aligned}
\mathbf{U}_\nu(n+\mu) &= P\exp\left(ig\int_{n+\mu}^{n+\mu+\nu}\mathbf{A}_\nu(x)\,d\mathbf{x}^\nu\right) &(6.202)\\
&= \exp\left[ig\frac{a}{2}\mathbf{A}_\nu(n+\mu+\nu)\right]\exp\left[ig\frac{a}{2}\mathbf{A}_\nu(n+\mu)\right],\\
\mathbf{U}_{-\mu}(n+\mu+\nu) &= \mathbf{U}_\mu^{-1}(n+\nu) &(6.203)\\
&= P\exp\left(-ig\int_{n+\mu+\nu}^{n+\nu}\mathbf{A}_\mu(x)\,d\mathbf{x}^\mu\right)\\
&= \exp\left[-ig\frac{a}{2}\mathbf{A}_\mu(n+\nu)\right]\exp\left[-ig\frac{a}{2}\mathbf{A}_\mu(n+\mu+\nu)\right],\\
\mathbf{U}_{-\nu}(n+\nu) &= \mathbf{U}_\nu^{-1}(n)\\
&= P\exp\left(-ig\int_{n+\nu}^{n}\mathbf{A}_\nu(x)\,d\mathbf{x}^\nu\right) &(6.204)\\
&= \exp\left[-ig\frac{a}{2}\mathbf{A}_\nu(n)\right]\exp\left[-ig\frac{a}{2}\mathbf{A}_\nu(n+\nu)\right].
\end{aligned}
$$

We also need the expanded gauge fields. In addition to Eq. (6.200), we have

$$
\begin{aligned}
\mathbf{A}_\mu(n+\mu+\nu) &= \mathbf{A}_\mu(n)+a\partial_\mu\mathbf{A}_\mu(n)+a\partial_n u\mathbf{A}_\mu(n)\,, &(6.205)\\
\mathbf{A}_\nu(n+\mu) &= \mathbf{A}_\nu(n)+a\partial_\mu\mathbf{A}_\nu\,. &(6.206)
\end{aligned}
$$

Substituting the expanded gauge fields, Eqs. (6.200), (6.205) and (6.206), into Eqs. (6.202)-(6.204), we have

$$
\begin{aligned}
\mathbf{U}_\nu(n+\mu) &= \exp\left[ig\frac{a}{2}[2\mathbf{A}_\nu(n)+2a\,\partial_\mu\mathbf{A}_\nu(n)+a\,\partial_\nu\mathbf{A}_\nu(n)]\,,\right]\\
\mathbf{U}_{-\mu}(n+\mu+\nu) &= \exp\left[-ig\frac{a}{2}[2\mathbf{A}_\mu(n)+2a\,\partial_\nu\mathbf{A}_\mu(n)+a\,\partial_\mu\mathbf{A}_\mu(n)]\,,\right]\\
\mathbf{U}_{-\nu}(n+\nu) &= \exp\left[-ig\frac{a}{2}[2\mathbf{A}_\nu(n)+a\,\partial_\nu\mathbf{A}_\nu(n)]\right]. &(6.207)
\end{aligned}
$$

To multiply the expanded link variables and obtain the continuum limit of the plaquette, we recall that the multiplication of exponentials of two operators includes not only the sum of the operators but also their commutator:

$$
\exp(x)\exp(y)=\exp\left(x+y+\frac{1}{2}[x,y]\right). \tag{6.208}
$$

The link variables are ordered counterclockwise around the plaquette. Thus, in the path ordering of the plaquette, the first term is the rightmost while the last is the leftmost. We first obtain the product of the first two link variables in the path ordering and then the product of the last two, followed by the product of the pairs. The product of the first two link variables is

$$
\begin{aligned}
&\mathbf{U}_\nu(n+\mu)\mathbf{U}_\mu(n) = \\
&\quad \exp\Bigg[ig\frac{a}{2}\{2\mathbf{A}_\nu(n) + 2a\,\partial_\mu\mathbf{A}_\nu(n) + a\,\partial_\nu\mathbf{A}_\nu(n) + 2\mathbf{A}_\mu(n) + a\,\partial_\mu\mathbf{A}_\mu(n)\} \\
&\quad -\frac{1}{2}\frac{g^2a^2}{4}[2\mathbf{A}_\nu(n) + 2a\,\partial_\mu\mathbf{A}_\nu(n) + a\,\partial_\nu\mathbf{A}_\nu(n), 2\mathbf{A}_\mu(n) + a\,\partial_\mu\mathbf{A}_\mu(n)]\Bigg] \\
&\quad = \exp\Bigg[ig\frac{a}{2}\{2\mathbf{A}_\mu(n) + 2\mathbf{A}_\nu(n) + a\,\partial_\mu\mathbf{A}_\mu(n) + a\,\partial_\nu\mathbf{A}_\nu(n) \\
&\quad + 2a\,\partial_\mu\mathbf{A}_\nu(n)\} - \frac{1}{2}g^2a^2[\mathbf{A}_\nu(n), \mathbf{A}_\mu(n)] + \mathcal{O}(a^3)\Bigg]\,. \qquad (6.209)
\end{aligned}
$$

The derivative terms in the commutator are dropped because they are order a^3 or higher. The product of the last two link variables is

$$
\begin{aligned}
&\mathbf{U}_{-\nu}(n+\nu)\mathbf{U}_{-\mu}(n+\mu+\nu) = \\
&\quad \exp\Bigg[-ig\frac{a}{2}\{2\mathbf{A}_\nu(n) + a\,\partial_\nu\mathbf{A}_\nu(n) + 2\mathbf{A}_\mu(n) + 2a\,\partial_\nu\mathbf{A}_\mu(n) + a\,\partial_\mu\mathbf{A}_\mu(n)\} \\
&\quad -\frac{1}{2}\frac{g^2a^2}{4}[2\mathbf{A}_\nu(n) + a\,\partial_\nu\mathbf{A}_\nu(n), 2\mathbf{A}_\mu(n) + 2a\,\partial_\nu\mathbf{A}_\mu(n) + a\,\partial_\mu\mathbf{A}_\mu(n)]\Bigg] \\
&\quad = \exp\Bigg[-ig\frac{a}{2}\{2\mathbf{A}_\nu(n) + 2\mathbf{A}_\mu(n) + a\,\partial_\nu\mathbf{A}_\nu(n) + a\,\partial_\mu\mathbf{A}_\mu(n) + 2a\,\partial_\nu\mathbf{A}_\mu(n)\} \\
&\quad -\frac{1}{2}g^2a^2[\mathbf{A}_\nu(n), \mathbf{A}_\mu(n)] + \mathcal{O}(a^3)\Bigg]\,. \qquad (6.210)
\end{aligned}
$$

Putting Eqs. (6.209) and (6.210) together gives the result for the plaquette,

$$
\begin{aligned}
P_\Box &= \mathbf{U}_{-\nu}(n+\nu)\mathbf{U}_{-\mu}(n+\mu+\nu)\mathbf{U}_{\nu}(n+\mu)\mathbf{U}_{\mu}(n) \\
&= \exp\Big[ig\frac{a}{2}\{-2\mathbf{A}_\mu(n) - 2\mathbf{A}_\nu(n) - a\,\partial_\mu\mathbf{A}_\mu(n) - a\,\partial_\nu\mathbf{A}_\nu(n) - 2a\,\partial_\nu\mathbf{A}_\mu(n) \\
&+ 2\mathbf{A}_\nu(n) + 2\mathbf{A}_\mu(n) + a\,\partial_\nu\mathbf{A}_\nu(n) + a\,\partial_\mu\mathbf{A}_\mu(n) + 2a\,\partial_\mu\mathbf{A}_\nu(n)\} \\
&+ \frac{1}{2}g^2a^2\frac{1}{4}[2\mathbf{A}_\mu(n) + 2\mathbf{A}_\nu(n) + a\,\partial_\mu\mathbf{A}_\mu(n) + a\,\partial_\nu\mathbf{A}_\mu(n) + 2a\,\partial_\nu\mathbf{A}_\mu(n), \\
&\qquad 2\mathbf{A}_\nu(n) + 2\mathbf{A}_\mu(n) + a\,\partial_\nu\mathbf{A}_\nu(n) + 2a\,\partial_\mu\mathbf{A}_\nu(n)] \\
&- g^2a^2[\mathbf{A}_\mu(n), \mathbf{A}_\nu(n)] + \mathcal{O}(a^3)\Big] \ . \qquad (6.211)
\end{aligned}
$$

The first commutator in Eq. (6.211) reduces to the sum of two commutators of the gauge fields,

$$
\frac{g^2a^2}{2}([\mathbf{A}_\nu(n), \mathbf{A}_\mu(n)] + [\mathbf{A}_\mu(n), \mathbf{A}_\nu(n)]) = 0 \ . \qquad (6.212)
$$

Only the second commutator of Eq. (6.211) is then left. We change the sign of this term by exchanging $\mathbf{A}_\mu$ and $\mathbf{A}_\nu$ in the commutator to obtain

$$
\begin{aligned}
P_\Box &= \exp\left[ia^2g(\partial_\mu\mathbf{A}_\nu - \partial_\nu\mathbf{A}_\mu) + g^2a^2[\mathbf{A}_\mu(n), \mathbf{A}_\nu(n)]\right] \\
&= \exp\left[ia^2g(\partial_\mu\mathbf{A}_\nu - \partial_\nu\mathbf{A}_\mu - ig[\mathbf{A}_\mu, \mathbf{A}_\nu])\right] \\
&= \exp[ia^2gF_{\mu\nu}] \ . \qquad (6.213)
\end{aligned}
$$

The gauge part of the Euclidean action is the trace of the plaquette. To take the trace of the plaquette, we expand the exponential,

$$
\exp[ia^2gF_{\mu\nu}] = 1 + ia^2gF_{\mu\nu} - \frac{a^4g^2}{2}F_{\mu\nu}F^{\mu\nu} \ . \qquad (6.214)
$$

The trace of unity is a constant not important for the action. The trace of $F_{\mu\nu}$ is zero since $F_{\mu\nu} = (\lambda_a/2)F^a_{\mu\nu}$ and the SU(3) generators are traceless,

$$
\mathrm{Tr}(F_{\mu\nu}) = \mathrm{Tr}\left[\frac{\lambda_a}{2}\right]F^a_{\mu\nu} = 0 \ . \qquad (6.215)
$$

Only the last term contributes to the action,

$$\begin{aligned}\mathrm{Tr}\Big[-\frac{a^4g^2}{2}F_{\mu\nu}F^{\mu\nu}\Big] &= \mathrm{Tr}\Big[-\frac{a^4g^2}{2}\frac{\lambda_a}{2}\frac{\lambda_b}{2}F^a_{\mu\nu}F^{\mu\nu\,b}\Big] \\ &= -\frac{a^4g^2}{2}\frac{1}{2}\delta_{ab}F^a_{\mu\nu}F^{\mu\nu\,b} \\ &= -\frac{a^4g^2}{4}F^a_{\mu\nu}F_{\mu\nu\,a}\ . \end{aligned} \tag{6.216}$$

We can finally obtain the gauge part of the Euclidean action,

$$\begin{aligned} S_E^{\mathrm{gauge}} &= -\frac{1}{2g^2}\sum_{n,\mu,\nu}\mathrm{Tr}[P_\Box] + \mathrm{h.\,c.} \\ &= -\frac{1}{2g^2}\cdot 2\cdot\sum_{n,\mu,\nu}\Big[-\frac{a^4g^2}{4}F^a_{\mu\nu}F^{\mu\nu\,a}\Big] \end{aligned} \tag{6.217}$$

where the hermitian conjugate gives the factor 2. The sum over n is converted to a space-time integral,

$$\sum_{n,\mu,\nu}\longrightarrow\int\frac{d\tau\, d^3x}{a^4}\ . \tag{6.218}$$

if all lattice spacings are equal. Thus the final result for the continuum limit of the gauge action is

$$S_E^{\mathrm{gauge}} = \frac{1}{4}\int d^4x\sum_{\mu,\nu}F^a_{\mu\nu}F^{\mu\nu\,a} = \int d^4x\,\mathcal{L}_E^{\mathrm{gauge}}\ . \tag{6.219}$$

We now obtain the continuum limit of the fermion contribution to the Euclidean action,

$$\begin{aligned} S_E^{\mathrm{fermion}} &= S_E^I + S_E^m \\ &= \frac{a^3}{2}\sum_{n,\mu}\overline{\psi}(n)\gamma_E^\mu[\mathbf{U}_{-\mu}(n+\mu)\psi(n+\mu) \\ &\quad - \mathbf{U}_\mu(n-\mu)\psi(n-\mu)] + ma^4\sum_n\overline{\psi}(n)\psi(n)\ . \end{aligned} \tag{6.220}$$

The interaction term employs finite difference techniques to discretize the derivative $\mathbf{D}_\mu$ in terms of the link variables required to connect the fermion field to its conjugate [96]. This finite difference discretization scheme leads to the 'doubling problem' [96], making lattice treatment of fermions difficult.

Since we already have a factor of a^3 in the fermion action, when we expand the link variables, we need only keep terms linear in the lattice spacing a to return to four-dimensional space-time. Then

$$\begin{aligned}
\mathbf{U}_{-\mu}(n+\mu) &\approx \exp\left[-ig\frac{a}{2}[2\mathbf{A}_\mu(n)] + \mathcal{O}(a^2)\right] \approx 1 - iga\mathbf{A}_\mu(n) \, , \\
\mathbf{U}_{\mu}(n-\mu) &\approx \exp\left[ig\frac{a}{2}[2\mathbf{A}_\mu(n)] + \mathcal{O}(a^2)\right] \approx 1 + iga\mathbf{A}_\mu(n) \, .
\end{aligned}$$

The factor of two multiplying the gauge fields comes from integrating the path-ordered exponent using the trapezoidal rule, see Eqs. (6.198)-(6.201).

The interaction term of the fermion contribution to the Euclidean action is then

$$\begin{aligned}
S_E^I &= \frac{a^3}{2}\sum_{n,\mu}\overline{\psi}(n)\gamma_E^\mu[(1 - iga\mathbf{A}_\mu(n))\psi(n+\mu) \\
&\quad - (1 + iga\mathbf{A}_\mu(n))\psi(n-\mu)] \\
&= \frac{a^3}{2}\sum_{n,\mu}\overline{\psi}(n)\gamma_E^\mu[(\psi(n+\mu) - \psi(n-\mu)) \\
&\quad - iga\mathbf{A}_\mu(n)(\psi(n+\mu) + \psi(n-\mu))] \, . \qquad (6.221)
\end{aligned}$$

Expanding $\psi(n+\mu)$ and $\psi(n-\mu)$ about $\psi(n)$ gives

$$\begin{aligned}
\psi(n+\mu) &= \psi(n) + a\partial_\mu\psi(n) \, , \\
\psi(n-\mu) &= \psi(n) - a\partial_\mu\psi(n) \, . \qquad (6.222)
\end{aligned}$$

Thus the sum and difference of the fermion fields in Eq. (6.221) is

$$\begin{aligned}
\psi(n+\mu) - \psi(n-\mu) &= 2a\,\partial_\mu\psi(n) \, , \\
\psi(n+\mu) + \psi(n-\mu) &= 2\psi(n) \, . \qquad (6.223)
\end{aligned}$$

Inserting Eq. (6.223) into Eq. (6.221) gives

$$\begin{aligned} S_E^I &= \frac{a^4}{2}\sum_{n,\mu}\overline{\psi}(n)\gamma_E^\mu[2\,\partial_\mu\psi(n) - 2ig\mathbf{A}_\mu(n)\psi(n)] \\ &= a^4\sum_{n,\mu}\overline{\psi}(n)\gamma_E^\mu\mathbf{D}_\mu(n)\psi(n)\ . \end{aligned} \tag{6.224}$$

The continuum limit of the interaction term is

$$\begin{aligned} S_E^I &= a^4\int\frac{d^4x}{a^4}\overline{\psi}\gamma_E^\mu\mathbf{D}_\mu\psi \\ &= \int d^4x\overline{\psi}\gamma_E^\mu\mathbf{D}_\mu\psi\ . \end{aligned} \tag{6.225}$$

The continuum limit of the mass term is straightforward,

$$S_E^m = ma^4\sum_n\overline{\psi}\psi = ma^4\int\frac{d^4x}{a^4}\overline{\psi}\psi = m\int d^4x\,\overline{\psi}\psi\ . \tag{6.226}$$

Thus the continuum limit of the fermion part of the Euclidean action is

$$\begin{aligned} S_E^{\text{fermion}} &= \int d^4x[\overline{\psi}\gamma_E^\mu\mathbf{D}_\mu\psi + m\overline{\psi}\psi] \\ &= \int d^4x\mathcal{L}_E^{\text{fermion}}\ . \end{aligned} \tag{6.227}$$

Integrating the fermions using Grassmann variables

We will now discuss how the fermionic degrees of freedom are typically dealt with on the lattice. The fermion fields can be integrated out analytically to leave an effective action that depends only on the link variables, $\mathbf{U}_\mu(n)$. We will show how this is done.

The fermion fields are anticommuting classical fields. They anticommute with each other but commute with ordinary numbers. Fields with these properties are called Grassmann variables, see Refs. [92, 97] for more details.

In general, for a single Grassmann variable, θ, we have

$$\{\theta, \theta\} = 0 \tag{6.228}$$

or

$$\theta^2 = 0 \ . \tag{6.229}$$

The differential operator is

$$\left\{ \frac{\partial}{\partial \theta}, \theta \right\} = 1 \tag{6.230}$$

so that

$$\frac{\partial \theta}{\partial \theta} = 1 \qquad \theta \frac{\partial}{\partial \theta} = 0 \ . \tag{6.231}$$

Any function of Grassmann variables has a rather trivial expansion since all higher than linear powers of a Grassmann variable will be zero. Thus any function of variable θ, $f(\theta)$, can be written as

$$f(\theta) = a + b\theta \tag{6.232}$$

where a is an ordinary complex number and b can be either an ordinary number or a Grassmann variable. The rules for derivatives depends on the identity of b. For example, if a and b are both ordinary numbers,

$$\frac{df(\theta)}{d\theta} = b \tag{6.233}$$

since real numbers commute with Grassmann variables. However, if b is a Grassmann variable,

$$\frac{df(\theta)}{d\theta} = -b \tag{6.234}$$

since $\{\theta, b\} = 0$. In either case, whether b is ordinary or Grassmann, we still have

$$\frac{d^2 f(\theta)}{d\theta^2} = 0 \tag{6.235}$$

since

$$\frac{\partial b}{\partial \theta} = 0 \, . \tag{6.236}$$

These last two results in Eqs. (6.235) and (6.236) also imply that

$$\begin{aligned} \left\{ \frac{d}{d\theta}, b \right\} &= 0 \, , \\ \left\{ \frac{d}{d\theta}, \frac{d}{d\theta} \right\} &= 0 \, . \end{aligned} \tag{6.237}$$

There is no real inverse differentiation, *i.e.* the integration of Grassmann variables is not really the inverse of their differentiation. Integration has to be defined formally. The "integration" rules for Grassmann variables are called Berezin integrals and are defined so that

$$\begin{aligned} \int d\theta &= 0 \, , \\ \int d\theta \, \theta &= 1 \, . \end{aligned} \tag{6.238}$$

Thus the integration of Grassmann variables is invariant under translations of the integration variable by a constant.

An arbitrary number of Grassmann variables can be defined, all fulfilling the same criteria, *e.g.* θ_i fields with $i = 1, \ldots n$ where

$$\begin{aligned} \{\theta_i, \theta_j\} &= 0 \, , \\ \left\{ \frac{\partial}{\partial \theta_i}, \theta_j \right\} &= \delta_{ij} \, , \\ \left\{ \frac{\partial}{\partial \theta_i}, \frac{\partial}{\partial \theta_j} \right\} &= 0 \, . \end{aligned} \tag{6.239}$$

The expansion of an arbitrary product of Grassmann variables can be written as

$$\begin{aligned} f(\theta_1, \ldots \theta_n) &= a + b_i\theta_i + c_{ij}\theta_i\theta_j + d_{ijk}\theta_i\theta_j\theta_k + \\ &\quad \cdots + n_{1\cdots n}\theta_1\theta_2\cdots\theta_n \, . \end{aligned} \tag{6.240}$$

Note that no repeated indices are possible for the terms to be nonzero. In general, the Berezin integrals are

$$\begin{aligned}\int d\theta_i &= 0 ,\\ \int d\theta_i\theta_i &= 1 .\end{aligned} \tag{6.241}$$

To integrate over more than one Grassmann variable, it is conventional to nest the integrals so that *e.g.*,

$$\int d\theta_1 d\theta_2\theta_1\theta_2 = -\int d\theta_1\left(\int d\theta_2\theta_2\right)\theta_1 = -\int d\theta_1\theta_1 = -1 \tag{6.242}$$

From here we can begin to see how to integrate the fermion fields on the lattice where we have

$$\int [\mathbf{U}][d\psi][d\overline{\psi}]\exp[-S_E^{\text{fermion}}] . \tag{6.243}$$

Since the fermion fields are Grassmann variables, ψ and $\overline{\psi}$ thus satisfy the following anticommutation relations:

$$\{\psi(n),\psi(m)\} = 0 , \tag{6.244}$$
$$\{\overline{\psi}(n),\overline{\psi}(m)\} = 0 , \tag{6.245}$$
$$\{\psi(n),\overline{\psi}(m)\} = 0 . \tag{6.246}$$

When $n = m$, this implies $\psi^2(n) = \overline{\psi}^2(n) = 0$.

The same integration rules hold as for other Grassmann variables. We can show this in the integral of the mass term with a single value of n,

$$\int d\overline{\psi}(n)d\psi(n)\exp[-m\overline{\psi}(n)\psi(n)] . \tag{6.247}$$

The complete expansion of the exponential gives

$$\begin{aligned}\int d\overline{\psi}(n)d\psi(n)[1-m\overline{\psi}(n)\psi(n)] &= -m\int d\overline{\psi}(n)d\psi(n)\overline{\psi}(n)\psi(n)\\ &= m\int d\overline{\psi}(n)\left(\int d\psi(n)\psi(n)\right)\overline{\psi}(n)\\ &= m .\end{aligned} \tag{6.248}$$

For the more general case where we have a matrix in the exponent, as is the case for the interaction term, we would have

$$\int [d\overline{\psi}][d\psi] \exp[-\sum_{n,n'} \overline{\psi}(n) M_{nn'} \psi(n')] = \det M \tag{6.249}$$

where det M is the determinant of the matrix M.

EXAMPLE: Show that Eq. (6.249) is true for the case of a 2×2 matrix.
Let

$$M = \begin{pmatrix} M_{11} & M_{12} \\ M_{21} & M_{22} \end{pmatrix} \tag{6.250}$$

so that

$$\det M = M_{11} M_{22} - M_{12} M_{21} \; . \tag{6.251}$$

The expansion of $\exp[-\overline{\psi} M \psi]$ is

$$\begin{aligned} \exp[-\overline{\psi} M \psi] &= 1 - \overline{\psi} M \psi + \frac{1}{2}(\overline{\psi} M \psi)(\overline{\psi} M \psi) \\ &= \frac{1}{2}(\overline{\psi}_i M_{ij} \psi_j)(\overline{\psi}_k M_{kl} \psi_l) \end{aligned} \tag{6.252}$$

where $i \neq k$ and $j \neq l$ for Eq. (6.252) to be nonzero. We drop the first two terms since the integral of these over the two fermion fields and the two antifermion fields would be zero. The expansion terminates with the third term since there are no further combinations that would not yield terms proportional to ψ^2 for a 2×2 matrix. The final integrand is then

$$\begin{aligned} \exp[-\overline{\psi} M \psi] &= \frac{1}{2}[\overline{\psi}_1 M_{11} \psi_1 \overline{\psi}_2 M_{22} \psi_2 + \overline{\psi}_1 M_{12} \psi_2 \overline{\psi}_2 M_{21} \psi_1 \\ &\quad + \overline{\psi}_2 M_{21} \psi_1 \overline{\psi}_1 M_{12} \psi_2 + \overline{\psi}_2 M_{22} \psi_2 \overline{\psi}_1 M_{11} \psi_1] \; . \end{aligned} \tag{6.253}$$

Remember that the order of the Grassmann variables is important. The integral of the first term is

$$
\begin{aligned}
&\int d\bar{d}\psi_2 d\psi_2 d\overline{\psi}_1 d\psi_1 [\overline{\psi}_1 M_{11}\psi_1 \overline{\psi}_2 M_{22}\psi_2] \\
&= -M_{11}M_{22}\int d\overline{\psi}_2 d\psi_2 d\overline{\psi}_1 \Big(\int d\psi_1\psi_1\Big)\overline{\psi}_1\overline{\psi}_2\psi_2 \\
&= -M_{11}M_{22}\int d\overline{\psi}_2 d\psi_2 \Big(\int d\overline{\psi}_1\overline{\psi}_1\Big)\overline{\psi}_2\psi_2 \\
&= -M_{11}M_{22}\int d\overline{\psi}_2 d\psi_2 \overline{\psi}_2\psi_2 \\
&= M_{11}M_{22}\int d\overline{\psi}_2 \Big(\int d\overline{\psi}_2\psi_2\Big)\overline{\psi}_2 \\
&= M_{11}M_{22}\int d\overline{\psi}_2\overline{\psi}_2 \\
&= M_{11}M_{22} \ . \qquad (6.254)
\end{aligned}
$$

Note that we can factor out the components of matrix M because they are ordinary numbers and not Grassmann variables. Likewise, the integrals of the other three terms in Eq. (6.253) are

$$\int d\bar{d}\psi_2 d\psi_2 d\overline{\psi}_1 d\psi_1 [\overline{\psi}_2 M_{22}\psi_2 \overline{\psi}_1 M_{11}\psi_1] = M_{11}M_{22} \ , \qquad (6.255)$$

$$\int d\bar{d}\psi_2 d\psi_2 d\overline{\psi}_1 d\psi_1 [\overline{\psi}_1 M_{12}\psi_2 \overline{\psi}_2 M_{21}\psi_1] = -M_{12}M_{21} \ , \qquad (6.256)$$

$$\int d\bar{d}\psi_2 d\psi_2 d\overline{\psi}_1 d\psi_1 [\overline{\psi}_2 M_{21}\psi_1 \overline{\psi}_1 M_{12}\psi_2] = -M_{21}M_{12} \ . \qquad (6.257)$$

Finally, putting all the terms together, we have

$$
\begin{aligned}
\int d\bar{d}\psi_2 d\psi_2 d\overline{\psi}_1 d\psi_1 \exp[-\overline{\psi}M\psi] &= M_{11}M_{22} - M_{12}M_{21} \\
&= \det M \ . \qquad (6.258)
\end{aligned}
$$

Note that while we have chosen a 2×2 matrix for simplicity here, this result holds for an arbitrary number of fermion fields.

We can then write the fermion part of the action suggestively as

$$\sum_{n,n'}^{N} \overline{\psi}(n)[D_{nn'}(\mathbf{U}) + m\delta_{nn'})]\psi(n') \qquad (6.259)$$

where N is the number of independent fermion fields and $D_{nn'}$ includes the contraction of the gamma matrices with the link variables. The fermionic contribution to the partition function is

$$\begin{aligned} Z^{\text{fermion}} &= \int \prod_n^N d\overline{\psi}(n)\, d\psi(n)\, d\mathbf{U}_\mu(n) \exp[-\overline{\psi}(n)(D(\mathbf{U}) + m)\psi] \\ &= \int [d\overline{\psi}][d\psi][d\mathbf{U}] \exp[-\overline{\psi}(n)[D(\mathbf{U}) + m]\psi] \\ &= \int [d\mathbf{U}] \det(D(\mathbf{U}) + m) \\ &= \int [d\mathbf{U}] \exp[\ln(\det[D(\mathbf{U}) + m])] \ . \end{aligned} \tag{6.260}$$

The partition function for the full Euclidean action is

$$\begin{aligned} Z &= \int [d\mathbf{U}] \exp[-S_E^{\text{gauge}} + \ln(\det[D(\mathbf{U}) + m])] \\ &= \int [d\mathbf{U}] \exp[-S_{\text{eff}}(\mathbf{U})] \ . \end{aligned} \tag{6.261}$$

where $S_{\text{eff}}(\mathbf{U}) = S_E^{\text{gauge}} + \ln(\det[D(\mathbf{U}) + m])$ is the effective action.

The effective action is now only dependent on the link variables. Unfortunately, the $N \times N$ determinant generated by the integral over the fermion fields, $\det[D(\mathbf{U}) + m]$, is difficult to deal with, involving an N-fold multiple of the number of lattice sites. The sign of the terms in the determinant changes from positive to negative, depending on the position, making cancellations difficult, particularly in a Monte Carlo calculation where the integral is done by choosing random values over the range of integration (hence Monte Carlo). Many early calculations were done in the 'quenched' approximation, expanding the determinant in powers of $1/m$ and letting $m \to \infty$. Much progress has been made in handling fermions so that calculations with more realistic quark masses are now available. However, the details of these calculations are beyond the scope of an introductory text. See the review in Ref. [98] for some methods.

6.4 Chiral symmetry and spontaneous symmetry breaking

Before showing some general results from lattice QCD at finite temperature, we will introduce the concept of a 'broken' symmetry in nature, chiral symmetry, and discuss its possible restoration at finite temperature.

6.4.1 Chiral symmetry

One symmetry we did not discuss at the beginning of the chapter is chiral (handedness) symmetry, see also Ref. [92]. Chiral symmetry is a flavor symmetry (quark type) rather than a color symmetry. If we use the language of flavor SU(3), the fermions are $\psi(x) = (u(x), d(x), s(x))$ and transform as

$$\psi \to \psi' = \exp[i\vec{\alpha} \cdot (\vec{\lambda}_F/2)]\psi \tag{6.262}$$

so that under SU(3) flavor rotations, a quark field of one flavor could rotate into a quark field of another flavor. We have added the subscript F on the SU(3) flavor generators to better distinguish between flavor and color SU(3). Note that here we work in Minkowski space.

Following the arguments for color SU(3), it is easy to see that the interaction term in the QCD Lagrange density is invariant under flavor SU(3) since

$$\overline{\psi} \to \overline{\psi}' = \overline{\psi} \exp[-i\vec{\alpha} \cdot (\vec{\lambda}_F/2)] \tag{6.263}$$

and the covariant derivative commutes with the flavor generators, $\lambda_F^a/2$. The flavor current, called a vector current because it contains γ^μ, is

$$\mathrm{j}_{Va}^\mu = \overline{\psi}\gamma^\mu \frac{\lambda_F^a}{2}\psi \, . \tag{6.264}$$

The charge associated with this current,

$$Q_{Va} = \int d^3x j_{Va}^0(x) \tag{6.265}$$

satisfies the SU(3) algebra

$$[Q_{Va}, Q_{Vb}] = i f_{abc} Q_{Vc} \ , \tag{6.266}$$

since

$$\left[\frac{\lambda_F^a}{2}, \frac{\lambda_F^b}{2}\right] = i f_{abc} \frac{\lambda_F^c}{2} \ . \tag{6.267}$$

We can also define an 'axial' transformation as

$$\psi \to \psi' = \exp[i(\vec{\alpha} \cdot \vec{\lambda}_F/2)\gamma_5]\psi \tag{6.268}$$

where $\gamma_5 = i\gamma_0\gamma_1\gamma_2\gamma_3$ and, since the gamma matrices anticommute ($\{\gamma^\mu, \gamma^\nu\} = 2g^{\mu\nu}$), γ_5 also anticommutes with the gamma matrices, $\{\gamma_5, \gamma^\mu\} = 0$. The presence of γ_5 means that care must be taken when transforming the adjoint. Thus

$$\begin{aligned} \overline{\psi} \to \overline{\psi}' = (\psi^\dagger \gamma^0)' &= (\exp[i(\vec{\alpha} \cdot \vec{\lambda}_F/2)\gamma_5]\psi)^\dagger \gamma^0 \\ &= \psi^\dagger \exp[-i(\vec{\alpha} \cdot \vec{\lambda}_F/2)\gamma_5]\gamma^0 \\ &= \psi^\dagger \gamma^0 \exp[i(\vec{\alpha} \cdot \vec{\lambda}_F/2)\gamma_5] \\ &= \overline{\psi} \exp[i(\vec{\alpha} \cdot \vec{\lambda}_F/2)\gamma_5] \ . \end{aligned} \tag{6.269}$$

Note that, in this case, the transforms of both ψ and $\overline{\psi}$ have positive signs in the exponent since γ_5 and γ^0 anticommute.

The interaction term of the Lagrange density remains invariant under axial rotations,

$$(\overline{\psi}\gamma^\mu \mathbf{D}_\mu \psi)' = \overline{\psi}'\gamma^\mu \mathbf{D}_\mu \psi' \ , \tag{6.270}$$

because the covariant derivative is invariant under flavor transformations. Then, using Eqs. (6.268) and (6.269),

$$\begin{aligned} (\overline{\psi}\gamma^\mu \mathbf{D}_\mu \psi)' &= \overline{\psi} \exp[i(\vec{\alpha} \cdot \vec{\lambda}_F/2)\gamma_5]\gamma^\mu \mathbf{D}_\mu (\exp[i(\vec{\alpha} \cdot \vec{\lambda}_F/2)\gamma_5]\psi) \\ &= \overline{\psi}\gamma^\mu \exp[-i(\vec{\alpha} \cdot \vec{\lambda}_F/2)\gamma_5] \exp[i(\vec{\alpha} \cdot \vec{\lambda}_F/2)\gamma_5]\mathbf{D}_\mu \psi \\ &= \overline{\psi}\gamma^\mu \mathbf{D}_\mu \psi \ . \end{aligned} \tag{6.271}$$

The last equality comes about both because γ_5 and γ^μ anticommute and because we have not specified any space-time dependence for $\vec{\alpha}$. The corresponding axial current is

$$\mathbf{j}^\mu_{Aa} = \overline{\psi}\gamma^\mu\gamma_5\frac{\lambda^a_F}{2}\psi \tag{6.272}$$

with charge

$$Q_{Aa} = \int d^3x j^0_{Aa}(x) \ . \tag{6.273}$$

The commutation relation for this charge is

$$[Q_{Aa}, Q_{Ab}] = if_{abc}Q_{Vc} \ . \tag{6.274}$$

Since $\gamma_5^2 = 1$, the axial commutator gives back the vector charge. This last result implies that we also have a commutator between the vector and axial-vector charges,

$$[Q_{Va}, Q_{Ab}] = if_{abc}Q_{Ac} \ . \tag{6.275}$$

We can thus linearly combine the vector and axial-vector charges into 'left-handed' and 'right-handed' charges,

$$Q_{La} = \frac{1}{2}(Q_{Va} - Q_{Aa}) \ , \tag{6.276}$$

$$Q_{Ra} = \frac{1}{2}(Q_{Va} + Q_{Aa}) \tag{6.277}$$

which do not mix,

$$[Q_{La}, Q_{Lb}] = if_{abc}Q_{Lc} \ , \tag{6.278}$$

$$[Q_{Ra}, Q_{Rb}] = if_{abc}Q_{Rc} \ , \tag{6.279}$$

$$[Q_{La}, Q_{Rb}] = 0 \ . \tag{6.280}$$

EXAMPLE: Show Eqs. (6.278)-(6.280).

We first look at the commutator of the two left-handed charges. We have

$$\begin{aligned}
[Q_{La}, Q_{Lb}] &= \frac{1}{4}[(Q_{Va} - Q_{Aa}), (Q_{Vb} - Q_{Ab})] \\
&= \frac{1}{4}\Big\{[Q_{Va}, Q_{Vb}] - [Q_{Aa}, Q_{Vb}] - [Q_{Va}, Q_{Ab}] + [Q_{Aa}, Q_{Ab}]\Big\} \\
&= \frac{1}{2}if_{abc}(Q_{Vc} - Q_{Ac}) = if_{abc}Q_{Lc} \ .
\end{aligned} \tag{6.281}$$

We can easily find the similar result for the right-handed charge commutator, Eq. (6.279). However, when we try to mix left- and right-handed charges, we find

$$\begin{aligned}[Q_{La}, Q_{Rb}] &= \frac{1}{4}[(Q_{Va} - Q_{Aa}), (Q_{Vb} + Q_{Ab})] \\ &= \frac{1}{4}\Big\{[Q_{Va}, Q_{Vb}] - [Q_{Aa}, Q_{Vb}] + [Q_{Va}, Q_{Ab}] - [Q_{Aa}, Q_{Ab}]\Big\} \\ &= 0\,. \end{aligned} \tag{6.282}$$

The invariance with respect to left- and right-handed charges is called chiral symmetry. Chiral symmetry is a good symmetry as long as the quark masses are not taken into account.

Looking back at the vector transformation, we see that flavor rotations will not work unless the up, down and strange quark masses are equal. We saw already in Chapter 2 that this is not the case. Instead we have the hierarchy $m_s \gg m_d > m_u$. On the scale of the proton mass, $m_u \sim m_d$, so that, instead of an SU(3) flavor symmetry, we could claim approximate SU(2) flavor symmetry. This SU(2) symmetry is referred to as isospin. In isospin, the proton and neutron form a particle doublet for the 'nucleon'. Their masses would be degenerate if isospin symmetry were exact. In reality, isospin is 'broken' symmetry.

In the axial case, the mass term transforms as

$$\begin{aligned}(m\overline{\psi}\psi)' &\to m\overline{\psi}\exp[i(\vec{\alpha}\cdot\vec{\lambda}_F/2)\gamma_5]\exp[i(\vec{\alpha}\cdot\vec{\lambda}_F/2)\gamma_5]\psi \\ &= m\overline{\psi}\exp[i(\vec{\alpha}\cdot\vec{\lambda}_F)\gamma_5]\psi\,. \end{aligned} \tag{6.283}$$

Here the finite quark mass makes things even worse: the axial transformation does not just require the quark masses to be degenerate for chiral symmetry to be an unbroken symmetry: they must be zero. Thus chiral symmetry is explicitly broken in the real world.

6.4.2 Spontaneous Symmetry Breaking

A nonzero quark mass in the QCD Lagrangian means that chiral symmetry is an explicitly broken symmetry. It is said to be spontaneously broken if it remains broken in the limit of vanishing quark masses. The mass term can be thought of as a perturbation on a Hamiltonian where chiral symmetry is an unbroken symmetry [92]. Thus we can write the true Hamiltonian as the sum of the invariant part (without the mass terms), H_0, and the non-invariant part, H_1,

$$H = H_0 + \lambda H_1 \ , \tag{6.284}$$

where λ is a parameter that quantifies the degree of symmetry breaking. If $\lambda = 0$, the non-invariant part of the Hamiltonian vanishes and the symmetry is good.

Let us now look at a symmetry group U that connects states forming an irreducible representation of the symmetry group,

$$U|A\rangle = |B\rangle \ , \tag{6.285}$$

but leaves H_0 invariant,

$$UH_0U^\dagger = H_0 \ . \tag{6.286}$$

Then $E_A = \langle A|H_0|A\rangle$ and $E_B = \langle B|H_0|B\rangle$ are degenerate energy eigenstates since, using Eq. (6.285),

$$\langle B|H_0|B\rangle = \langle A|U^\dagger H_0 U|A\rangle \ , \tag{6.287}$$

and, using Eq. (6.286), we have

$$\begin{aligned} \langle A|U^\dagger H_0 U|A\rangle &= \langle A|U^\dagger U H_0 U^\dagger U|A\rangle \\ &= \langle A|H_0|A\rangle \ . \end{aligned} \tag{6.288}$$

The degeneracies imply that the ground state is invariant under symmetry transformations. The states $|A\rangle$ and $|B\rangle$ must be related to the ground state by creation operators, ϕ_A and ϕ_B so that

$$|A\rangle = \phi_A|0\rangle \qquad |B\rangle = \phi_B|0\rangle \tag{6.289}$$

with

$$U\phi_A U^\dagger = \phi_B \ . \tag{6.290}$$

It can be shown that Eq. (6.285) is only valid if U leaves the ground state unchanged,

$$U|0\rangle = |0\rangle \ . \tag{6.291}$$

EXAMPLE: Show that Eq. (6.291) is necessary for Eq. (6.285) to hold. Equation (6.285) implies

$$\begin{aligned} U|A\rangle &= |B\rangle \ , \\ U\phi_A|0\rangle &= \phi_B|0\rangle \ . \end{aligned} \tag{6.292}$$

However, Eq. (6.290) says

$$\begin{aligned} \phi_B &= U\phi_A U^\dagger \ , \\ \phi_B U &= U\phi_A \ . \end{aligned} \tag{6.293}$$

Equation (6.293) is obtained when both sides of Eq. (6.290) are operated on by U on the right-hand side. When Eq. (6.293) is inserted on the right-hand side of Eq. (6.292), we have

$$\phi_B U|0\rangle = \phi_B|0\rangle \ . \tag{6.294}$$

For this to be true, the rotation U must leave the ground state unchanged, as in Eq. (6.291).

The symmetry does not hold in degenerate energy levels if

$$U|0\rangle \neq |0\rangle \ . \tag{6.295}$$

This is referred to as spontaneous symmetry breaking. Note, however, that while the symmetry is not manifest in the ground state, the Hamiltonian is still invariant under the transformation. Thus Eq. (6.295) is equivalent to the statement that the vacuum expectation value of an operator is nonzero,

$$\langle 0|\phi_i|0\rangle \neq 0 \ . \tag{6.296}$$

We can use scalar fields as an example and look at the Lagrange density

$$\mathcal{L} = T - V = \frac{1}{2}(\partial_\mu\phi)^2 - V(\phi) \tag{6.297}$$

where

$$V(\phi) = \frac{1}{2}\mu^2\phi^2 + \frac{1}{4}\lambda\phi^4 \tag{6.298}$$

and $\lambda > 0$. We choose $\mathcal{L}$ to be invariant under reflections, $\phi \to -\phi$. If $\mu^2 > 0$, we can identify μ with the mass of a scalar field and the ϕ^4 term with a four-point vertex with coupling strength λ. Thus ϕ is a self-interacting field, like the gluon. The ground state in this case corresponds to $\phi = 0$. We can also have $\mu^2 < 0$ but now the 'mass' term has the wrong sign. We can find the ground state in this case by minimizing the potential with respect to ϕ,

$$\frac{\partial V(\phi)}{\partial \phi} = \phi(\mu^2 + \lambda\phi^2) = 0 \; . \tag{6.299}$$

Aside from a saddle-point solution at $\phi = 0$, we have two lower minima,

$$\phi = \pm v = \pm\sqrt{\frac{-\mu^2}{\lambda}} \; , \tag{6.300}$$

see also Refs. [4, 92]. Figure 6.6 shows the potential in the two cases. When $\mu^2 > 0$, there is only a single minimum at $\phi = 0$. When $\mu^2 < 0$, we have two degenerate ground states but, if we choose one of these, we break reflection symmetry although $V(\phi)$ itself is reflection symmetric.

We expand the solution around one of these minima, *e.g.* $\phi = v$ so that

$$\phi(x) = v + \chi(x) \; . \tag{6.301}$$

where the $\chi(x)$ fields are quantum fluctuations around v. Putting this back into $\mathcal{L}$, we have

$$\begin{aligned} \mathcal{L}' &= \frac{1}{2}(\partial_\mu\chi)^2 - \frac{1}{2}\mu^2(v+\chi)^2 - \frac{1}{4}\lambda(v+\chi)^4 \\ &= \frac{1}{2}(\partial_\mu\chi)^2 - \frac{1}{2}\mu^2(v^2 + 2v\chi + \chi^2) \\ &\quad - \frac{1}{4}\lambda(v^4 + 4v^2\chi + 6v^2\chi^2 + 4v\chi^3 + \chi^4) \; . \end{aligned} \tag{6.302}$$

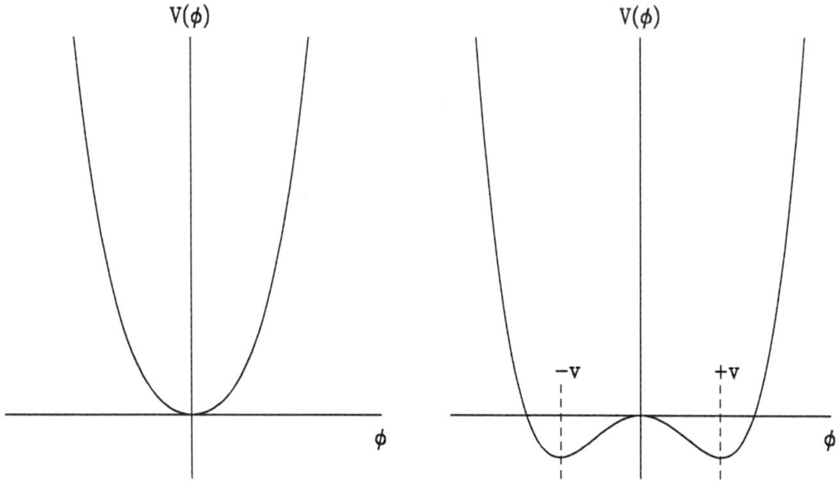

Figure 6.6: The potential of Eq. (6.298) for the symmetric phase, $\mu^2 > 0$, on the left-hand side, and the broken phase, $\mu^2 < 0$, on the right-hand side with solutions at $\pm v$ (Eq. (6.300)).

Substituting Eq. (6.300) for v and canceling common terms, we find

$$\mathcal{L}' = \frac{1}{2}(\partial_\mu \chi)^2 - \lambda v^2 \chi^2 - \lambda v \chi^3 - \frac{1}{4}\lambda \chi^4 + \frac{1}{4}\frac{\mu^4}{\lambda} \,. \tag{6.303}$$

The coefficient of the χ^2 term, which now has the 'right' negative sign, can be identified as a mass with

$$m = \sqrt{2\lambda v^2} = \sqrt{-2\mu^2} \,. \tag{6.304}$$

There is a factor of two because we want the mass term to be of the form

$$-\frac{1}{2}m^2\chi^2 \,. \tag{6.305}$$

The broken symmetry phase, μ^2, then does have a mass but, to see it explicitly, we must expand around a stable minimum.

If we go back and look at the Lagrangian of the complex field at the beginning of the chapter,

$$\mathcal{L} = \frac{1}{2}[(\partial_\mu \phi_1)^2 + (\partial^\mu \phi_2)^2] - \frac{1}{2}\mu^2(\phi_1^2 + \phi_2^2) - \frac{1}{4}\lambda(\phi_1^2 + \phi_2^2)^2 \tag{6.306}$$

choosing $\mu^2 < 0$, we would not have just two minima but a circle of minima at

$$v^2 = \phi_1^2 + \phi_2^2 = -\frac{\mu^2}{\lambda}\,. \tag{6.307}$$

If we, *e.g.* let $\phi_2 = 0$ and expand around $\phi_1 = v$, we can write the complex field as

$$\phi(x) = \frac{1}{\sqrt{2}}[v + \chi(x) + i\zeta(x)] \tag{6.308}$$

where $\chi(x)$ and $\zeta(x)$ are fluctuations in ϕ_1 and ϕ_2 respectively. The shifted Lagrange density,

$$\mathcal{L}' = \frac{1}{2}(\partial_\mu\chi)^2 + \frac{1}{2}(\partial_\mu\zeta)^2 - \lambda v^2\chi^2 + \dots\,, \tag{6.309}$$

thus contains a mass term for the χ field with $m = \sqrt{-2\mu^2}$, as before. There is no mass term for the ζ field. Thus the theory contains a massless scalar field, the Goldstone boson, which can occur when a continuous symmetry is spontaneously broken.

How does this fit into a chapter on lattice gauge theory? Well, if, in the ideal world quarks are massless, we can try to get to this ideal world of the quarks by adding heat to the system. The added heat modifies the potential, perhaps in such a way that it returns to its symmetric state with massless quarks, called chiral symmetry restoration. Such investigations can be carried out on the lattice.

6.5 Selected results from lattice QCD

There are many results from lattice gauge theory. Here we will focus only on the finite temperature aspects, including the $Q\overline{Q}$ potential, deconfinement, chiral symmetry restoration and finite baryon density effects. For more detailed results, beyond the scope of this book, there are a number of good reviews [96, 98].

6.5.1 The quark-antiquark potential

While we have explained that quarks are difficult to deal with on the lattice, the potential between two static, infinitely heavy quarks a distance R apart can be calculated. The expectation value of the trace over a closed loop formed by the product of link variables called the Wilson loop [91, 95],

$$W(c) = \langle \mathrm{Tr}[\prod_c \mathbf{U}_\mu(n)] \rangle \ , \tag{6.310}$$

is evaluated. On the lattice, the expectation value for any operator, $\mathcal{O}$, is defined as

$$\langle \mathcal{O} \rangle = \frac{1}{Z} \int [d\overline{\psi}][d\psi][d\mathbf{U}]\mathcal{O} \exp[-S_E] \ . \tag{6.311}$$

The most trivial way of looking at this potential is to treat the $Q\overline{Q}$ pair as a single system with relative coordinate $x_{\mathrm{rel}} = R$. There is no kinetic energy term since the pair is at rest. The action depends only on $V(R)$, the $Q\overline{Q}$ potential. The action is just

$$\begin{aligned} S_E &= \int_{\tau_A}^{\tau_B} d\tau V(R) \\ &= TV(R) \end{aligned} \tag{6.312}$$

where $V(R)$ is constant in time and $T = \tau_B - \tau_A$. The partition function is then just

$$Z = \exp[-TV(R)] \ . \tag{6.313}$$

Equivalently, one can say that the Q and $\overline{Q}$ generate gauge fields at $x = 0$ and $x = R$. The external current thus created, $\mathbf{j}^\mu(x)$, produces a gauge field that interacts with the $Q\overline{Q}$ pair. The current is independent of time,

$$j^0(x, \tau) = \delta(\vec{x} - \vec{R}) - \delta(\vec{x}) \ . \tag{6.314}$$

The action arising from the interaction of the gauge field and the quark current is

$$\begin{aligned} S_E &= -ig\int d^3x d\tau \mathbf{j}^\mu(x)\mathbf{A}_\mu(x) \\ &= -ig\int d\tau d^3x[\delta(\vec{x}-\vec{R})-\delta(\vec{x})]A_\tau(x) \\ &= -ig\int_0^T d\tau[A_\tau(R)-A_\tau(0)] \\ &= -ig[\int_0^T d\tau A_\tau(R)+\int_T^0 d\tau A_\tau(0)]\ . \end{aligned} \tag{6.315}$$

The effect of this last expression for the action is to form a closed loop in the (x,τ) plane from $(0,0)$ to $(0,R)$ to (T,R) to $(T,0)$ and back to $(0,0)$. We let T be much larger than R. Over the closed path,

$$S_E = -ig\int_c d\mathbf{x}^\mu \mathbf{A}_\mu(x) \tag{6.316}$$

with partition function

$$Z = \lim_{T\to\infty}\exp[-S_E] = \exp[ig\int_c d\mathbf{x}^\mu \mathbf{A}_\mu(x)]\ . \tag{6.317}$$

In the language of the lattice, the closed path can be tiled by N_p plaquettes of area a^2 if a is the lattice spacing in both the space and time directions, for a total area $A = N_p a^2 = RT$. The contour around the loop can be tiled as in Fig. 6.7. When integrating around the contour, the parts of the lattice outside the contour described by a single link variable give zero,

$$\int d\mathbf{U}_\mu(n)[\mathbf{U}_\mu(n)]_{ij} = 0\ , \tag{6.318}$$

where i and j label the components of the $\mathbf{U}$ matrix. Thus points outside the contour do not contribute to it. Those lattice point on the outside edge of the contour or those in the interior have two link variables, the variable and its conjugate, obvious from the directionality of the tiling in Fig. 6.7. For these points,

$$\int d\mathbf{U}_\mu(n)[\mathbf{U}_\mu(n)]_{ij}[\mathbf{U}_\mu(n)]^\dagger_{kl} = \frac{1}{N}\delta_{il}\delta_{jk}\ . \tag{6.319}$$

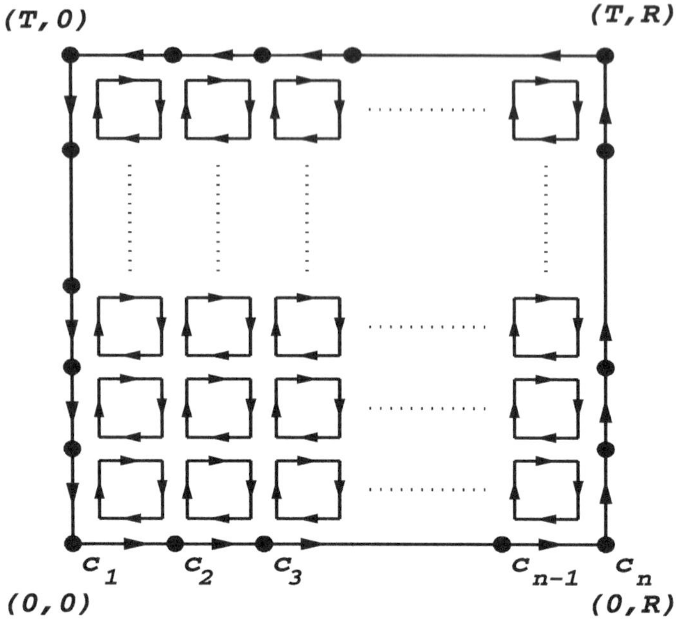

Figure 6.7: The tiling of plaquettes over the Wilson loop of area TR.

Thus the quantity evaluated on the lattice is the average value of the trace of the path-ordered link variables that tile the area,

$$\begin{aligned} W(c) &= \langle \mathrm{Tr}[\prod_{i=1}^{N} \mathbf{U}_{c_i c_{i+1}}(c_i)]\rangle \\ &= \frac{\int \prod_{n,\mu} d\mathbf{U}_\mu(n) \exp[-S_E] \mathrm{Tr}[\prod_{i=1}^{N} \mathbf{U}_{c_i c_{i+1}}(c_i)]}{\int \prod_{n,\mu} d\mathbf{U}_\mu(n) \exp[-S_E]} \,. \end{aligned} \tag{6.320}$$

Integration over the tiled area of the loop, using the integration rules for the link variables, Eqs. (6.318) and (6.319), gives

$$\begin{aligned} W(c) &= \left(\frac{1}{g^2}\right)^{N_p} = \left(\frac{1}{g^2}\right)^{A/a^2} \\ &= \exp[-A \ln g^2/a^2] = \exp\left[-\frac{RT}{a^2}\ln g^2\right] \,. \end{aligned} \tag{6.321}$$

Finally, we relate the trivial result from Eq. (6.313) to that on the lattice, Eq. (6.321), obtaining

$$TV(R) = \frac{RT}{a^2} \ln g^2 \tag{6.322}$$

or

$$V(R) = \sigma R \tag{6.323}$$

if $\sigma = (\ln g^2)/a^2$ where σ is the string tension. The potential is then linearly proportional to the $Q\overline{Q}$ separation. Since $V(R) \to \infty$ as $R \to \infty$, $Z \to 0$, making the Wilson loop a model of confinement.

6.5.2 Deconfinement and Chiral Symmetry Restoration

Recall that for zero chemical potential (zero baryon density), the properties of the phase transition depend on the number of quark flavors. In the high temperature regime, the nature of the transition is controlled by global symmetries which exist only in the limits where $m = 0$ (chiral symmetry) or $m \to \infty$ (pure gauge: the 'quenched' approximation). The order of the phase transition in real QCD, with real physical quark masses is a question that must be addressed numerically. There are ways of handling finite mass fermion (quark) fields on the lattice, see *e.g.* Ref. [98] for a review of some methods.

As we just discussed for static heavy quarks, the $Q\overline{Q}$ potential was studied using the Wilson loop. This calculation, modeling confinement by showing that the potential energy between the two charges grows with distance, Eq. (6.323), was done at zero temperature. To study deconfinement, we want to look at the lattice results for finite temperature to see changes in the linear potential of Eq. (6.323) as a function of temperature. In the imaginary time formalism, changing the temperature can be related to changing the number of time steps.

One method of studying the phase transition is to look at the Wilson line, the trace of the product of link variables that are stationary in space but not in time. At a fixed space point (n_x, n_y, n_z), the Wilson line describing a world line propagating in time is

$$L = \mathrm{Tr}\left[\prod_{n_\tau=1}^{N_\tau} \mathbf{U}_\tau(n_x, n_y, n_z, n_\tau)\right] . \tag{6.324}$$

To relate the line to the partition function, the Wilson line must be periodic in time so that the link variable at $\tau = 0$ must be equal to that at the endpoint, $\tau = 1/T$,

$$\mathbf{U}_\mu(n_x, n_y, n_z, \tau = 0) = \mathbf{U}_\mu(n_x, n_y, n_z, \tau = 1/T) , \tag{6.325}$$

making the Wilson line a gauge invariant closed loop, shown in Fig. 6.8.

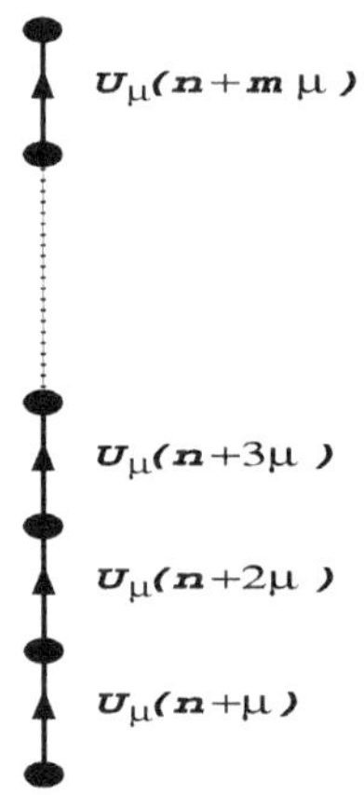

Figure 6.8: A schematic Wilson line L.

We remark that the heavy quark potential can also be extracted from the correlation function of Wilson lines, $\langle L(R)L^\dagger(0)\rangle$, in the T direction and separated by R. This function is the same as the Wilson loop without the sides at times 0 and T. The loop is still closed due to the periodic boundary conditions on the Wilson lines [98].

At finite temperature, the correlator is related to the free energy of the system,

$$\exp\left[-\frac{F_{Q\overline{Q}}(r,T)}{T}\right] = \langle \mathrm{Tr}L\mathrm{Tr}L^\dagger\rangle \to |\langle L\rangle|^2 . \tag{6.326}$$

For large separations, the heavy quark free energy is large and $\langle L \rangle = 0$, corresponding to confinement for $T < T_c$. When $T > T_c$, the free energy is finite, giving a finite expectation value as well, $\langle L \rangle > 0$ in the deconfined phase. A sudden change of $\langle L \rangle$ at $T = T_c$ is then an indicator of a first-order phase transition.

A quantity that makes a sudden change across a phase boundary is known as an order parameter. Thus the Wilson loop and Wilson line are good order parameters for deconfinement.

The sudden change in $\langle L \rangle$ is most apparent in the heavy quark mass limit, the pure gauge case. However, when light quarks are also present, this correlator is not an indicator of the partonic degrees of freedom since no sharp change in the free energy occurs across T_c for finite light quark masses [96].

One might expect that at high temperatures chiral symmetry may be restored. This possibility can also be calculated on the lattice if dynamical (light) quarks are included. It may be expected that a good order parameter for chiral symmetry restoration is the expectation value of $\overline{\psi}\psi$ since this multiplies the mass in the fermion action. In the broken phase, the zero temperature world, $\langle \overline{\psi}\psi \rangle$ has a finite value. Chiral restoration at high temperature would give $\langle \overline{\psi}\psi \rangle = 0$. On the lattice,

$$\langle \overline{\psi}\psi \rangle = \frac{1}{N_x^3 N_\tau} \frac{\partial \ln Z}{\partial m_q} \; . \tag{6.327}$$

The order of the phase transition is expected to depend on the number of light quark degrees of freedom. It is possible to write down an effective Lagrangian with the same global symmetry as the QCD Lagrangian [99]

$$\begin{aligned} \mathcal{L}_{\rm eff} &= -\frac{1}{2}{\rm Tr}(\partial_\mu \phi^\dagger \partial^\mu \phi) - \frac{1}{2} m^2 {\rm Tr}(\phi^\dagger \phi) \\ &\quad + \frac{\pi^2}{3}[g_1({\rm Tr}(\Phi^\dagger \Phi))^2 + g_2 {\rm Tr}((\Phi^\dagger \Phi)^2)] \\ &\quad + c(\det \Phi + \det \Phi^\dagger) \; . \end{aligned} \tag{6.328}$$

where $\Phi \equiv \Phi_{ij}$ and i and j run over the number of flavors. Studies of the flavor dependence of Eq. (6.328) show a first-order transition for three or more flavors while it is second order for two flavors. This is the pattern observed in lattice calculations.

We have now discussed both chiral symmetry restoration and deconfinement. It is not clear whether these are two distinct transitions or whether they occur at the same temperature. General arguments suggest that the chiral transition temperature should be higher than the deconfinement temperature. However, over a wide range of quark masses, no singular behavior is observed, only a rapid crossover. Thus it is difficult to distinguish which properties have to do with chiral symmetry restoration and which with deconfinement.

Related quantities that might help sort out whether or not the two transitions occur together are the susceptibilities: the Polyakov loop susceptibility, χ_L, related to the Wilson line, $\langle L \rangle$, and the chiral susceptibility, χ_m, related to $\langle \overline{\psi}\psi \rangle$,

$$\begin{aligned} \chi_L &= N_x^3(\langle L^2 \rangle - \langle L \rangle^2) \, , \\ \chi_m &= \frac{\partial \langle \overline{\psi}\psi \rangle}{\partial m_q} \, . \end{aligned} \tag{6.329}$$

Instead of a sudden change from a zero to a nonzero value such as in $\langle L \rangle$ and $\langle \overline{\psi}\psi \rangle$, the susceptibility passes through a maximum at the transition point. The deconfinement and chiral symmetry order parameters, as well as the corresponding susceptibilities, are shown together in Fig. 6.9 as functions of T/T_c. Even if the transition shown is not a discontinuity, it still takes place rapidly over a small interval of temperature.

6.5.3 Bulk Thermodynamic Properties

Bulk thermodynamic properties of the system should reflect the number of degrees of freedom in the high temperature limit. For example, the pressure should approach the ideal gas value as $T \to \infty$. Recall from Chapter 4 that the pressure at zero chemical potential is

$$\begin{aligned} \frac{P_\pi}{T^4} &= \frac{3\pi^2}{90} \qquad T \to 0 \, , \\ \frac{P_{\rm qgp}}{T^4} &= (16 + \frac{21}{2} n_f) \frac{\pi^2}{90} \qquad T \to \infty \, . \end{aligned} \tag{6.330}$$

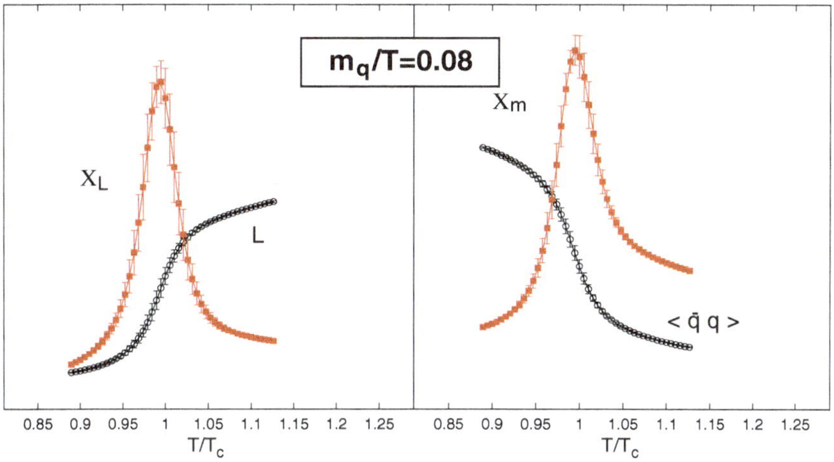

Figure 6.9: An example of deconfinement and chiral symmetry restoration in a calculation with two quark flavors. The left-hand side shows the order parameter for deconfinement in the quenched approximation, $\langle L \rangle$, and the corresponding Polyakov loop susceptibility, χ_L. The right-hand side shows the chiral symmetry order parameter, $\langle \overline{\psi}\psi \rangle \equiv \overline{q}q$, and the chiral susceptibility, χ_m [100].

We take the pion degrees of freedom for $T \to 0$ since they are the dominant contribution to produced particles in low energy collisions. The pressure is related to the free energy density by

$$P = -\frac{T}{V} \ln Z = f \ . \tag{6.331}$$

This can also be related to the free energy density through

$$\frac{\epsilon - 3P}{T^4} = T \frac{d}{dT} \left(\frac{P}{T^4} \right) \tag{6.332}$$

which, in the ideal gas limit above, gives

$$\epsilon = 3P = -3f \ . \tag{6.333}$$

Classically, the free energy density is obtained using differences at two temperatures. Here we let $x = T$ and

$$\frac{\partial(T \ln Z/T^4)}{\partial T} = \frac{\partial(x^{-3} \ln Z)}{\partial x} \tag{6.334}$$

so that

$$\left.\frac{f}{T^4}\right|_{T_0}^{T} = -\frac{1}{V} \int_{T_0}^{T} dx \frac{\partial(x^{-3} \ln Z)}{\partial x} . \tag{6.335}$$

To calculate this quantity on the lattice, instead of taking the derivative with respect to the temperature, the derivative with respect to the gauge coupling, g, is used. Letting $\beta = 6/g^2$, the derivative above is replaced by finite differences of the expectation values of the Euclidean action,

$$\left.\frac{f}{T^4}\right|_{\beta_0}^{\beta} = N_\tau^4 \int_{\beta_0}^{\beta} d\beta' [\langle \tilde{S}_E \rangle - \langle \tilde{S}_E(T = 0) \rangle] , \tag{6.336}$$

where

$$\langle \tilde{S}_E \rangle = -\frac{1}{N_x^3 N_\tau} \frac{\partial \ln Z}{\partial \beta} . \tag{6.337}$$

The expectation value of the action is calculated on a lattice of three spatial dimensions with N_x lattice points in each dimension and N_τ points in imaginary time. The zero temperature result, $\langle \tilde{S}_E(T = 0) \rangle$, is often approximated by a symmetric lattice with $N_x = N_\tau$. The lower integration limit, β_0, is chosen at low temperature so that f/T_0^4 is negligible [96].

The resulting pressure is shown in Fig. 6.10. Note that while no sharp change is generally seen, the transition occurs for different temperatures, depending on the number of quark degrees of freedom. In the high temperature regime, the asymptotic value of P/T^4 reflects the number of quark flavors.

The exact value of T_c depends both on whether or not quarks are included as well as the number of light quark flavors. In pure gauge

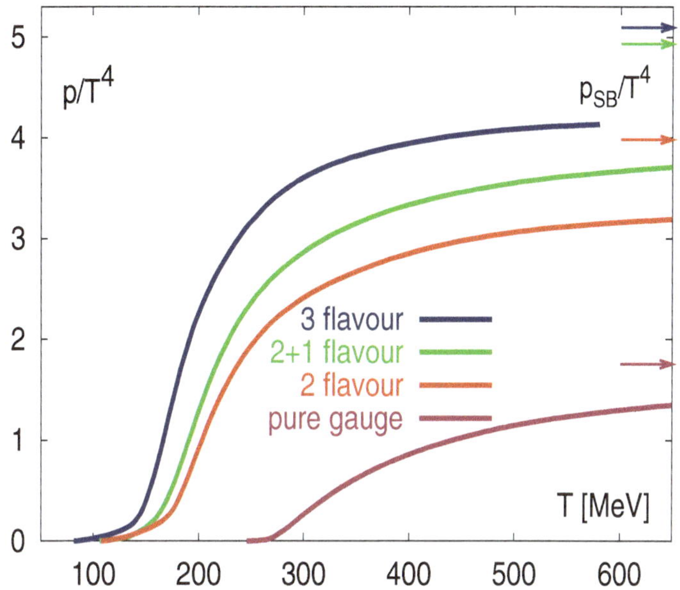

Figure 6.10: The QCD pressure with different number of quark degrees of freedom. In the $2+1$ flavor case, the calculation is done for two light quarks and a strange quark with a mass four times larger than the light quark mass. Reprinted from Ref. [101] with permission from Elsevier.

calculations, the $m \to \infty$ limit, T_c is generally higher than in the case where quarks are included [101],

$$T_c(\text{gauge}) = 271 \pm 2\,\text{MeV} . \tag{6.338}$$

Several quark cases have been explored: two and three light flavors of equal mass as well as the case where two of the quarks are light and one is heavier, closer to the real world case. The two and three flavor results are [96]

$$T_c(2\,\text{flavors}) = 173 \pm 4\,\text{MeV} , \tag{6.339}$$

$$T_c(3\,\text{flavors}) = 154 \pm 8\,\text{MeV} . \tag{6.340}$$

The $2+1$ results are similar to the two flavor results.

The fact that T_c depends on the number of quark flavors is not surprising. We found the same result in Chapter 4. The greater the number of degrees of freedom included in the calculation, the larger the thermodynamic properties on the high temperature side and the smaller T_c, also seen in Fig. 6.10. A QCD phase diagram for various quark masses is shown in Fig. 6.11.

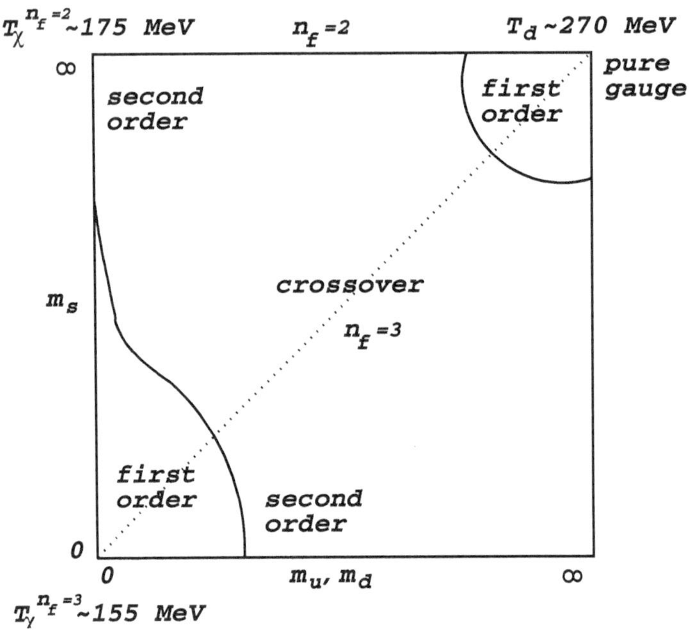

Figure 6.11: The QCD phase diagram with three quarks of various masses. Following [102] with permission from Elsevier.

6.5.4 Finite baryon density effects

So far, everything we have discussed about the lattice, whether quark degrees of freedom are included or not, are on the finite temperature, zero baryon density side of the phase diagram. When non-zero chemical potential is introduced on the lattice by $\mu\overline{\psi}\gamma^0\psi$ (since μ/T corresponds to fields in the time direction only), this new term is complex. The argument of the fermion determinant becomes complex and the fermion matrix is no longer Hermitian. If the fermion determinant, used as a weight factor, has complex components, the oscillating real parts cancel in observables, known as the 'sign' problem. Thus traditional Monte Carlo methods no longer apply. It is only recently that ways of overcoming this difficulty have been developed, including overlap improving multi-parameter reweighting [103], Taylor expansion [104] and analytic continuation from imaginary to real chemical potential [105]. We briefly discuss some of these results to end this chapter.

The overlap improving multi-parameter reweighting method calculates an ensemble of lattice configurations at $\mu = 0$ and finite temperature. Then reweighting factors are calculated at finite μ and a lower value of T. Thus

$$\begin{aligned} Z(\mu, T) &= \int [d\mathbf{U}] \exp[-S_E^{\mathrm{gauge}}(T)] \det M(\mu, m) \qquad (6.341) \\ &= \int [d\mathbf{U}] \exp[-S_E^{\mathrm{gauge}}(T_0)] \det M(\mu = 0, m) \\ &\quad \times \left[\exp[-S_E^{\mathrm{gauge}}(T) + S_E^{\mathrm{gauge}}(T_0)] \frac{\det M(\mu, m)}{\det M(\mu = 0, m)} \right] . \end{aligned}$$

At $\mu \neq 0$, $\det M$ is complex. Thus the integral is rewritten so that the fermion determinant at $\mu = 0$ is used as an integration measure since Monte Carlo techniques work at $\mu = 0$. The contribution to the partition function in square brackets is a weight factor, w. The reweighting is done along best weight lines in the (μ, T) plane determined by minimizing the spread of $\ln w$. The paths of integration for $T < T_c$, $T = T_c$ and $T > T_c$ are shown in Fig. 6.12.

In the calculation, μ and T are changed together. While this technique works well for moderate values of μ, it is still based on a $\mu = 0$ calculation. Thus it cannot describe the interesting phases conjectured at large μ [59]. Using this technique, Fodor and Katz found the endpoint of first-order transitions to be at $T = 160 \pm 3.5$ MeV and $\mu_B = 725 \pm 35$ MeV in $2+1$ flavor QCD. At $\mu_B = 0$, they obtained, $T_c = 172 \pm 3$ MeV, consistent with other $\mu = 0$ calculations. The result is shown in Fig. 6.13. Recall that $\mu_B = 3\mu_q$. Since these are early results, the final value of μ_B at the endpoint may well change.

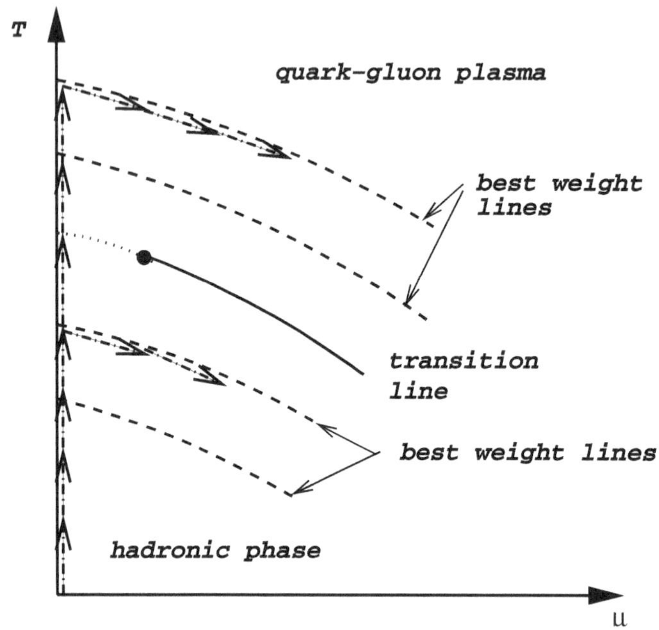

Figure 6.12: The best weight lines in the (μ, T) plane where the reweighting is performed. The transition line is in the middle. The dotted line is the crossover region and the blob shows the critical end-point where first-order transitions begin. Above the transition line, the system is a quark-gluon plasma while below it is a region of hadron gas. The integration paths are indicated by arrows along the $\mu = 0$ axis and the best weight lines. Modified from Ref. [106] with permission from Elsevier.

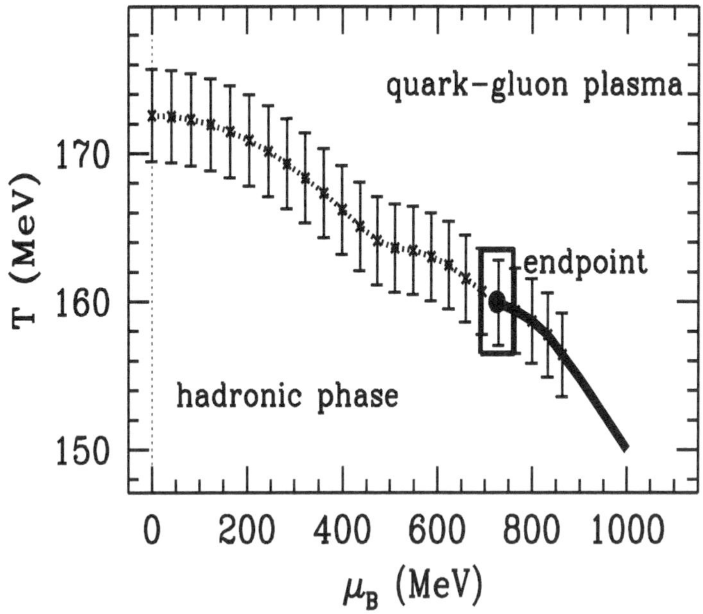

Figure 6.13: The (μ, T) diagram showing the results of Fodor and Katz along the crossover region. The box shows the quoted uncertainties in the endpoint of first-order transitions. Reprinted from [103] with permission from Elsevier.

The Taylor expansion technique expands the partition function in Eq. (6.341), estimating the derivatives of $\det M(\mu = 0)$ with respect to chemical potential, μ,

$$\ln\left(\frac{\det M(\mu)}{\det M(0)}\right) = \sum_{n=1}^{\infty} \frac{\mu^n}{n!} \frac{\partial^n \ln \det M(0)}{\partial^n \mu} \ . \tag{6.342}$$

Results using this method agree with those of Fodor and Katz in the region where the two calculations overlap. The uncertainties increase with μ.

Exercises

1. Show that $j^\mu = -(\partial^\mu \phi_1)\phi_2 + (\partial^\mu \phi_2)\phi_1$ for the real fields in $\mathcal{L} = (\partial_\mu \phi)(\partial^\mu \phi^*) - \mu^2(\phi^* \phi) - \lambda(\phi^* \phi)^2$.

2. Show that $-(1/4)F_{\mu\nu}F^{\mu\nu}$ is locally gauge invariant.

3. Using the characteristics of the Grassmann variables, check Eqs. (6.255)-(6.257).

4. Derive the vector and axial vector flavor currents, Eqs. (6.264) and (6.272), from the Euler-Lagrange equation of motion.

Part II

Probes

Chapter 7

Thermal dileptons

7.1 Introduction

Ideal probes of the temperature of the system created in a heavy-ion collision should have small interaction cross sections and consequently long mean-free paths in matter (recall that the mean-free path, λ, is proportional to the inverse interaction cross section of the probe, $1/\sigma$). This criteria is difficult to satisfy for hadrons because they interact strongly with cross sections on the order of several to tens of millibarns. Thus, ideal probes are those that do not interact via the strong force, such as leptons and photons which interact electromagnetically. We focus on the production of high mass thermal lepton pairs here since it is a useful example calculation.

The structure of the calculation of thermal dilepton production is very similar to that of the Drell-Yan dilepton calculation. The main difference in the plasma phase is that the initial quark and antiquark distributions that interact to produce the lepton pair are not the parton distributions of a cold nucleus, such as those shown in Chapter 2 with parton distribution functions modified by cold nuclear matter effects such as shadowing, but those of a thermal system. As for Drell-Yan dilepton production, thermal dileptons are not emitted at one fixed mass but over a range of masses. The slope of the thermal distribution, the inverse temperature, changes with time as the temperature decreases due to the expansion of the system. As the temperature decreases, the contribution to the lepton pair mass distribution at that temperature should become steeper, resulting in negligible contributions near freezeout. Thus the quark-gluon plasma phase, where the temperature is highest, should provide the largest mass contributions to thermal dilepton production.

We focus on pairs with masses greater than 1 GeV so that we can compare the results directly to the Drell-Yan rate and also avoid the low mass resonance region where contributions from ρ, ω and ϕ decays are important.

To measure the dilepton signal ($l\;l$ pairs, known as 'unlike-sign' pairs), the background must be subtracted. Experimentally, the dilepton invariant mass is constructed by combining two leptons, either of the same sign ($l\;l$ pairs, known as 'like-sign' pairs) or of opposite sign, and constructing their invariant mass. The source of each pair is either 'correlated', arising from as single source, such as in the Drell-Yan process, or 'uncorrelated', with each lepton of the pair arising from different sources. The uncorrelated lepton pairs, which come from unrelated combinations of hadronic decays *e.g.* $\pi\;\pi$, $\pi\;\pi$, $K\;K$, $K\;K$, $\pi\;K$ and $\pi\;K$ decays are both opposite sign and like sign. Taking advantage of this, a 'like-sign subtraction' can be performed, leaving only the correlated signal behind.

7.2 High mass thermal dilepton rate

In this section, we make our way through a simple calculation of the lepton pair invariant mass distribution for lepton pairs of mass M produced from a system in thermal equilibrium [65, 107, 108, 109]. As an aside on nomenclature, we note that lepton pair and dilepton are used interchangeably to denote an $l\;l$ pair while 'dilepton pair' is not good usage since it would refer to four leptons.

To make the initial derivation of the rate transparent, we first focus on a system at rest. The number of particles produced per time and volume is

$$\frac{N}{tV} = \rho_1 \rho_2 v \quad \sigma(1 + 2 \to X) \tag{7.1}$$

where ρ_1 and ρ_2 are the densities of the interacting particles, v is their relative velocity and the cross section for the production of final state X is denoted $\sigma(1 + 2 \to X)$.

The thermal dileptons, while most plentiful in the plasma phase, can also arise from the mixed phase and the hadron gas phase, if the phase transition is of first order. In the quark-gluon plasma phase, the reaction is the same as in the Drell-Yan process, $q\overline{q} \to \gamma \quad \to l \quad l$. The cross section in Eq. (7.1) is the same as the Drell-Yan cross section,

$$\sigma \quad (M^2) = \frac{4\pi\alpha^2}{9M^2} \quad e^2 \, . \tag{7.2}$$

Note that here the cross section depends inversely on M^2 while the formula in Chapter 2 is differential with respect to M, hence there is a factor of $2M$ difference between the right-hand side of the cross section here and that in Chapter 2. In Chapter 2, the differential form of the cross section was given, along with the delta function, $\delta(M^2 \quad s)$. We have not included the delta function here since it is introduced below. In the hadron gas phase, the largest contributions will come from low mass pairs since the temperature is low, between the critical temperature and the freezeout temperature. Hadronic resonances that decay to lepton pairs typically have masses much greater than the temperature of the system (the ρ, ω and ϕ at the expected initial temperatures reached in heavy-ion collisions at the LHC being an exception). The most likely channel is $\pi \quad \pi \quad \to \rho^0 \to l \quad l$. Contributions from such resonance decays will have cross sections proportional to a mass-dependent form factor, $G(M^2)$, so that

$$\sigma \quad (M^2) = \frac{4\pi\alpha^2}{9M^2} G(M^2) \, . \tag{7.3}$$

In the case of $\pi \quad \pi$ annihilation, the form factor will be the Breit-Wigner resonance distribution shown in Fig. 2.9 of Chapter 2. Finally, in the mixed phase, the lepton pair production cross section will have contributions from both the quark-gluon plasma and from the hadron gas. The relative contributions will change with proper time during the lifetime of the mixed phase. If $f(\tau)$ is the fraction of the system in a quark-gluon plasma as a function of time, then we can represent the lepton pair production in the mixed phase as

$$f(\tau)\sigma \quad (M^2) + (1 \quad f(\tau))\sigma \quad (M^2) \; . \tag{7.4}$$

The densities of the annihilation partners (quarks and antiquarks in the quark-gluon plasma and π , π in the hadron gas) are proportional to the number per volume. We write the volume elements in momentum space, $d^3p/(2\pi)^3$, to better describe the reaction. The average number of fermions/bosons in a thermal system is the Fermi-Dirac/Bose-Einstein distribution, as shown in Chapter 4. The densities also include the degeneracy factors, g. The degeneracy factor is six in the quark-gluon plasma: three colors and two active flavors. In the hadron gas the factor is unity: the pions must be of equal and opposite charge. Spin is not included in the degeneracy factor of ρ_1 and ρ_2 because the cross section includes an average over spin. Then replacing $dt\,dV = dt\,d^3x$ by d^4x, we have

$$\begin{aligned}\frac{dN}{d^4x} &= g\int \frac{d^3p_1}{(2\pi)^3}\frac{1}{\exp(E_1/T) \quad 1} \quad g\int \frac{d^3p_2}{(2\pi)^3}\frac{1}{\exp(E_2/T) \quad 1} \\ &\quad v \quad \sigma(l \quad l \quad) \; .\end{aligned} \tag{7.5}$$

We have denoted the thermal dilepton production cross section generically as $\sigma(l \quad l \quad)$. The relative velocity is

$$v = \frac{\sqrt{M^2(M^2 \quad 4m^2)}}{2E_1E_2} \qquad \frac{M^2}{2E_1E_2} \tag{7.6}$$

when the lepton mass, m, is ignored. Here $s = M^2 = (E_1 + E_2)^2$ $(\vec{p}_1 + \vec{p}_2)^2$. We use the lower case s for the Mandelstam invariant since in the quark-gluon plasma phase the lepton pairs are produced by partons. Note, however, that we cannot straightforwardly related the parton four-momentum to the initial four-momentum of the colliding hadrons since equilibration of the thermal system has disrupted the hadron wavefunction and wiped out the memory of the initial hadron momentum. The rate can be calculated analytically, as we will do here, if $M \quad T$ so that the Fermi-Dirac/Bose-Einstein distributions can be replaced by Maxwell-Boltzmann distributions. Then

$$\frac{dN}{d^4x} = g^2 \int \frac{d^3p_1}{(2\pi)^3} \exp(\quad E_1/T) \int \frac{d^3p_2}{(2\pi)^3} \exp(\quad E_2/T) \\ \frac{M^2}{2E_1E_2} \sigma(l \quad l \quad) \ . \tag{7.7}$$

To determine the rate for a given dilepton mass and momentum, we multiply the rate by unity by including two delta functions, one for the squared invariant mass and one for the lepton pair three momentum,

$$1 = \int dM^2 d^3p \delta(M^2 \quad s) \delta^3(\vec{p} \quad \vec{p}_1 \quad \vec{p}_2) \tag{7.8}$$

where $\vec{p}_1$ and $\vec{p}_2$ are the three momenta of the two leptons. The full differential rate is then

$$\frac{dN}{dM^2 d^3p d^4x} = g^2 \int \frac{d^3p_1}{(2\pi)^3} \frac{d^3p_2}{(2\pi)^3} \exp[\quad (E_1 + E_2)/T] \\ \frac{M^2}{2E_1E_2} \sigma(l \quad l \quad) \delta(M^2 \quad s) \delta^3(\vec{p} \quad \vec{p}_1 \quad \vec{p}_2) \ . \tag{7.9}$$

It is trivial to integrate over d^3p_1 using the delta function for the lepton pair three-momentum. We obtain

$$\frac{dN}{dM^2 d^3p d^4x} = g^2 \int \frac{d^3p_2}{(2\pi)} \exp[\quad (E_1 + E_2)/T] \\ \frac{M^2}{2E_1E_2} \sigma(l \quad l \quad) \delta(M^2 \quad s) \ . \tag{7.10}$$

Note that we still have the factor of E_1 in the integrand. Given that the initial-state quarks are assumed to be massless, $E_1 \quad p_1 = p \quad p_2$. As in Chapter 2, we write

$$\frac{1}{2E_1} = \int dE_1 \delta(E_1^2 \quad p_1^2) \theta(E_1) \ . \tag{7.11}$$

Then

$$\frac{dN}{dM^2 d^3p d^4x} = g^2 \int \frac{d^3p_2 dE_1}{(2\pi)} \exp[\ \ (E_1 + E_2)/T] \frac{M^2}{E_2} \sigma(l\ \ l\ \ ) \\ \delta(M^2\ \ \ (E_1+E_2)^2\ \ \ p^2)\delta(E_1^2\ \ \ p_1^2)\ . \quad (7.12)$$

We now integrate over the delta function containing M^2. Let the argument of the delta function be $f(E_1) = M^2\ \ \ (E_1+E_2)^2\ \ \ p^2$. Solving for E_1 when $f(E_1) = 0$ gives $E_{1_0} = \sqrt{M^2+p^2}\ \ \ E_2 = M_T\ \ \ E_2$. The absolute value of the derivative with respect to E_1 is $2(E_1+E_2)$. Then

$$\delta(M^2\ \ \ (E_1+E_2)^2\ \ \ p^2) = \frac{\delta(E_1\ \ \ E_{1_0})}{2(E_1+E_2)}\ . \quad (7.13)$$

The rate then becomes

$$\frac{dN}{dM^2 d^3p d^4x} = g^2 \int \frac{d^3p_2}{(2\pi)} \frac{dE_1}{2(E_1+E_2)} \delta(E_1\ \ \ E_{1_0}) \\ \exp[\ \ (E_1+E_2)/T] \frac{M^2}{E_2} \sigma(l\ \ l\ \ )\delta(E_1^2\ \ \ p_1^2) \\ = g^2 \int \frac{d^3p_2}{(2\pi)} \exp[\ \ (E_{1_0}+E_2)/T] \frac{M^2\sigma(l\ \ l\ \ )}{2(E_{1_0}+E_2)E_2} \\ \delta((M_T\ \ \ E_2)^2\ \ \ (\vec{p}\ \ \ \vec{p}_2)^2)\ . \quad (7.14)$$

We next replace d^3p_2 by $2\pi d\cos\theta p_2^2 dp_2$. We use the remaining delta function to integrate over $d\cos\theta$. Let the argument of the delta function be

$$f(\cos\theta) = (M_T\ \ \ E_2)^2\ \ \ (\vec{p}\ \ \ \vec{p}_2)^2 = M^2 + 2pp_2\cos\theta\ \ \ 2M_T p_2 \quad (7.15)$$

Setting $f(\cos\theta) = 0$ gives

$$\cos\theta_0 = 2M_T p_2\ \ \ M^2/(2pp_2)\ , \\ \frac{\partial f(\cos\theta)}{\partial\cos\theta|_{c\ \ \theta_0}} = 2pp_2 \quad (7.16)$$

so that

$$\delta((M_T\ \ \ E_2)^2\ \ \ (\vec{p}\ \ \ \vec{p}_2)^2) = \frac{\delta(\cos\theta\ \ \ \cos\theta_0)}{2pp_2}\ . \quad (7.17)$$

We then have

$$\begin{aligned}
\frac{dN}{dM^2 d^3p d^4x} &= g^2 \int \frac{dp_2 p_2^2}{(2\pi)} \frac{d\cos\theta}{2pp_2} \delta(\cos\theta \quad \cos\theta_0) \\
&\qquad \exp[\;\; (E_{1_0} + E_2)/T] \frac{M^2 \sigma(l \;\; l \;\;)}{2(E_{1_0} + E_2)E_2} \\
&= g^2 \int \frac{dp_2 p_2^2}{(2\pi)} \exp[\;\; (E_{1_0} + E_2)/T] \\
&\qquad \frac{M^2 \sigma(l \;\; l \;\;)}{4pp_2(E_{1_0} + E_2)E_2} \sigma(l \;\; l \;\;) \, .
\end{aligned} \tag{7.18}$$

Finally we can do the integral over p_2 with the limits of integration determined from the limits on $\cos\theta$,

$$1 \quad \frac{2M_T p_2 \quad M^2}{2pp_2} \quad 1 \, . \tag{7.19}$$

To satisfy the upper limit of Eq. (7.19), we need

$$\begin{aligned}
M^2 &= 2p_2^{\max}(M_T \quad p) \, , \\
p_2^{\max} &= 0.5(M_T + p) \, .
\end{aligned} \tag{7.20}$$

Likewise, for the lower limit,

$$\begin{aligned}
M^2 &= 2p_2^{\mathrm{m}} \;\; (M_T + p) \, , \\
p_2^{\mathrm{m}} &= 0.5(M_T \quad p) \, .
\end{aligned} \tag{7.21}$$

Since the $p_2 E_2 = p_2^2$ in the denominator of Eq. (7.18) cancels the factor of p_2^2 in the numerator, the integral over dp_2 is

$$\int_{{}_{2}^{\mathrm{m}}}^{{}_{2}^{\mathrm{ma}}} dp_2 = p_2^{\max} \quad p_2^{\mathrm{m}} \;\; = p \, . \tag{7.22}$$

This factor of p cancels the one in the denominator of Eq. (7.18), leaving the factor $E_{1_0} + E_2 \quad E$, the energy of the lepton pair. We also make this replacement in the Maxwell-Boltzmann distribution. Bringing the factor of E in the denominator to the left-hand side, we form the invariant rate for lepton pair production

$$E \frac{dN}{dM^2 d^3p d^4x} = \frac{g^2 M^2}{4(2\pi)} \exp(\;\; E/T) \sigma(l \;\; l \;\;) \, . \tag{7.23}$$

To obtain an observable dilepton production rate that is comparable to the Drell-Yan rate, it is necessary to integrate over the space-time history of the collision. Note that this form of the rate is a bit different than the Drell-Yan cross section since here the annihilating particles (either quarks or hadrons) need not have zero transverse momentum before the collision. Consequently the pair has finite p_T already at leading order. By integrating over the p_T of the pair as well, the thermal contribution to lepton pair production as a function of lepton pair mass and rapidity can be obtained.

Recall that we have calculated the dilepton rate in the rest frame of the system. To obtain the rate in an expanding system, we must replace the E in the Maxwell-Boltzmann factor by $\mathbf{p}\ \ \mathbf{u}$ where $\mathbf{u}$ is the fluid velocity four-vector and

$$\mathbf{p} = (M_T \cosh y, p_T, M_T \sinh y) \ . \tag{7.24}$$

For a longitudinally expanding medium,

$$\mathbf{u} = (\cosh \eta, 0, \sinh \eta) \tag{7.25}$$

where the fluid velocity unit vector is given in (t, x_T, z) coordinates instead of the cylindrical coordinates used in Chapter 5 to be compatible with the momentum four-vector. Then

$$\mathbf{p}\ \ \mathbf{u} = M_T(\cosh y \cosh \eta \quad \sinh y \sinh \eta) = M_T \cosh(y \quad \eta) \tag{7.26}$$

and

$$E\frac{dN}{dM^2 d^3p d^4x} = \frac{g^2 M^2}{(2\pi)} \exp[\ \ M_T \cosh(y \quad \eta)/T]\sigma(l\ \ l\ \ ) \ . \tag{7.27}$$

Assuming the transverse size is fixed,

$$d^4x = \tau d\tau d\eta d^2x = \pi R^2 \tau d\tau d\eta \tag{7.28}$$

where R is the radius of the system, usually approximated as the nuclear radius. Recalling that $d^3p/E = 2\pi p_T dp_T dy$, we can write the dilepton rate as a function of M^2 and y,

$$\begin{aligned} \frac{dN}{dM^2 dy} &= \frac{g^2 M^2 R^2}{32\pi^3} \sigma(l\ \ l\ \ ) \int_{\ Y}^{Y} d\eta \int^{\ f} d\tau\tau \int_0 dp_T p_T \\ &\quad \exp[\ \ M_T \cosh(y \quad \eta)/T] \ . \end{aligned} \tag{7.29}$$

where Y is the extent of the system in rapidity. Given the appearance of M_T in the exponential of Eq. (7.29), it is much more convenient to change variables from $p_T dp_T$ to $M_T dM_T$. The lower limit of the M_T integral is M at $p_T = 0$ so that

$$\frac{dN}{dM^2dy} = \frac{g^2M^2R^2}{32\pi^3}\sigma(l \quad l \quad) \int_{\quad Y}^{Y} d\eta \int^{\quad f} d\tau\tau \int \quad dM_T M_T \\ \exp[\quad M_T \cosh(y \quad \eta)/T] \,. \quad (7.30)$$

The integrals have been arranged so that the last integral, over the space-time rapidity, is the one that must be done numerically.

The M_T integration is done first. If we define $a = \cosh(y \quad \eta)/T$, the M_T integral alone is

$$\begin{aligned} &\int \quad dM_T M_T \exp(\quad M_T a) \\ &= \quad \frac{M_T}{a}\exp(\quad M_T a) \quad + \frac{1}{a}\int \quad dM_T \exp(\quad M_T a) \\ &= \frac{M}{a}\exp(\quad Ma) \quad \frac{1}{a^2}\exp(\quad M_T a) \\ &= \left(\frac{M}{a} + \frac{1}{a^2}\right)\exp(\quad Ma) \,. \end{aligned} \quad (7.31)$$

Then

$$\frac{dN}{dM^2dy} = \frac{g^2M^2R^2}{32\pi^3}\sigma(l \quad l \quad) \int_{\quad Y}^{Y} d\eta \int^{\quad f} d\tau\tau \left(\frac{MT}{\cosh(y \quad \eta)} \right. \\ \left. + \frac{T^2}{\cosh^2(y \quad \eta)}\right) \exp[\quad M\cosh(y \quad \eta)/T] \,. \quad (7.32)$$

Before going further, we replace the generic factors of the squared degeneracy times the cross section,

$$g^2\sigma(l \quad l \quad) \quad \frac{\alpha^2}{8\pi^4}F \quad (7.33)$$

where

$$F = \begin{cases} F \quad = \sum \quad e^2 & \text{plasma} \\ F \quad = \sum \quad e^2 f(\tau) + G(M^2)(1 \quad f(\tau)) & \text{mixed phase} \\ F \quad = G(M^2) & \text{hadron gas} \end{cases} \quad (7.34)$$

The next step is to integrate over the time evolution of the system. The integral over τ must be done in stages since, in the quark-gluon plasma and hadron gas phases the temperature changes with time while in the mixed phase the temperature is equal to T over the lifetime of the mixed phase. The proper time integration may then be broken up into three intervals: $\tau < \tau < \tau$ in the quark-gluon phase; $\tau < \tau < \tau$ in the mixed phase and $\tau < \tau < \tau$ in the hadron gas. The quark-gluon plasma and hadron gas phases are handled identically modulo the lepton pair production cross section. We discuss these integrations first, in general terms, and return to the mixed phase shortly.

We have two different components in Eq. (7.32), one proportional to T and one to T^2. Since the temperature depends on proper time, we change integration variables from τ to T. Recall that for a system of massless particles,

$$\tau_1 T^3(\tau_1) = \tau_2 T^3(\tau_2) \tag{7.35}$$

where the subscripts 1 and 2 label different times. If

$$\tau = \tau_1 \left(\frac{T(\tau_1)}{T(\tau)} \right)^3 , \tag{7.36}$$

then

$$d\tau = \quad 3\tau_1 \frac{T^3(\tau_1)}{T^4(\tau)} dT \tag{7.37}$$

and

$$\tau d\tau = \quad 3(\tau_1 T^3(\tau_1))^2 \frac{dT}{T \;(\tau)} . \tag{7.38}$$

We now let $x = M\cosh(y \quad \eta)/T$ and $dx = \quad MdT\cosh(y \quad \eta)/T^2$ so that $T = M\cosh(y \quad \eta)/x$ and $dT = \quad Mdx\cosh(y \quad \eta)/x^2$, giving

$$\tau d\tau = 3[\tau_1 T^3(\tau_1)]^2 \frac{dx x}{M \quad \cosh\;(y \quad \eta)} . \tag{7.39}$$

Then the proper time integration turns into an integration over x where the limits are changed from $\tau < \tau < \tau$ to $x < x < x$ where $x =$

$M\cosh(y\quad\eta)/T(\tau\ )$ and $x\quad = M\cosh(y\quad\eta)/T(\tau\ )$. The τ integral then takes the form

$$\frac{3[\tau\ T^2(\tau\ )]^2}{M^4\cosh\ (y\quad\eta)}\int_{x}^{x_f} dx(x^4+x^3)\exp(\quad x)\ . \tag{7.40}$$

Recall that this form of the time dependence only holds for the quark-gluon plasma phase and the hadron gas phase and not for the mixed phase in the case of a first-order transition. Thus in the plasma phase, $T(\tau\ ) = T(\tau_0)$ and $T(\tau\ ) = T(\tau\ )$, while in the hadron gas phase, $T(\tau\ ) = T(\tau\ )$ and $T(\tau\ ) = T(\tau\ )$. The integral over x can be done by parts. The x^4 integration is

$$\begin{aligned}&\int_{x}^{x_f} dx x^4\exp(\quad x)\\ =\quad & x^4\exp(\quad x)\Big|_{x}^{x_f} + 4\int_{x}^{x_f} dx x^3\exp(\quad x)\end{aligned} \tag{7.41}$$

so that

$$\begin{aligned}&\int_{x}^{x_f} dx(x^4+x^3)\exp(\quad x)\\ =\quad & x^4\exp(\quad x)\Big|_{x}^{x_f} + 5\int_{x}^{x_f} dx x^3\exp(\quad x)\ .\end{aligned} \tag{7.42}$$

The x^3 integration gives

$$\int_{x}^{x_f} dx x^3\exp(\quad x) = \quad (x^3+3x^2+6x+6)\exp(\quad x)\Big|_{x}^{x_f}\ . \tag{7.43}$$

The full τ integration in the quark-gluon plasma and hadron gas phases is then

$$\frac{3[\tau\ T^2(\tau\ )]^2}{M^4\cosh\ (y\quad\eta)}(x^4+5x^3+15x^2+30x+30)\exp(\quad x)\Big|_{x}^{x_f}. \tag{7.44}$$

Note that the negative sign does not mean that the rate is negative since, when the values of x and x are inserted, the rates will become positive because $\exp(\quad x\ ) > \exp(\quad x\ )$.

We now turn to the mixed phase. From Eq. (7.34), we see that the proper time integrals are $d\tau\tau f(\tau)$ and $d\tau\tau(1 \quad f(\tau))$ for the quark-gluon plasma and hadron gas contributions to the mixed phase respectively. The entropy density in the mixed phase includes both contributions, $s \quad = f(\tau)s \quad + (1 \quad f(\tau))s$. During the lifetime of the mixed phase, $\tau \quad < \tau < \tau$, entropy current conservation relates the entropy current in the mixed phase to the initial condition,

$$\tau_0 s(\tau_0) = \tau[f(\tau)s \quad + (1 \quad f(\tau))s \quad] \tag{7.45}$$

where $s \quad = 4g \quad \pi^2 T^3/90$ and $s \quad = 4g \; \pi^2 T^3/90$, see Chapter 5. Solving for the quark-gluon plasma contribution, $f(\tau)$, using Eq. (7.45) gives

$$f(\tau) \quad = \quad \frac{\tau_0}{\tau}\frac{s(\tau_0)}{s \qquad s} \qquad \frac{s}{s \qquad s} \; . \tag{7.46}$$

From this result, we also obtain hadron contribution to the mixed phase,

$$1 \quad f(\tau) \quad = \quad \frac{s}{s \qquad s} \qquad \frac{\tau_0}{\tau}\frac{s(\tau_0)}{s \qquad s} \; . \tag{7.47}$$

At the beginning of the mixed phase, $\tau_0 = \tau$ and, if the initial temperature is high enough for a quark-gluon plasma to be formed, $s_0 = s$. It is left as an exercise to show that

$$\int_m^H d\tau\tau f(\tau) \quad = \quad \frac{\tau^2}{2} f_0^2(r \quad 1) \; , \tag{7.48}$$

$$\int_m^H d\tau\tau(1 \quad f(\tau)) \quad = \quad \frac{\tau^2}{2} f_0(r \quad 1)[f_0(r \quad 2) + 2] \tag{7.49}$$

where $r = s \; /s$ and $f_0 \quad f(\tau_0) = (s(\tau_0) \quad s \;)/(s \qquad s \;)$ using Eq. (7.46). We can write the final results in Eqs. (7.48) and (7.49) in terms of τ alone since, at the end of the mixed phase the entropy current is $\tau \; s$ and $\tau \quad = \tau \; (s_0/s \;)$.

The integral over the fluid rapidity, $Y \qquad \eta \qquad Y$, is left to do numerically. The expression for the thermal dilepton rate at a given mass and rapidity is

$$\frac{dN}{dM^2\,dy} =$$

$$\frac{\alpha^2 R^2}{4\pi^2} \; \Theta(s \quad s\;) \quad e^2 \frac{3(\tau\, T^3(\tau\,))^2}{M^4} \int_{\;Y}^{Y} \frac{d\eta}{\cosh\;(y \quad \eta)}$$

$$(\;\; x^4 \quad 5x^3 \quad 15x^2 \quad 30x \quad 30)\exp(\;\; x)\Big|_{x_0}^{x_m}$$

$$+\,\Theta(s \quad s\;)\frac{\tau_{\mathrm{m}}^2}{2} f_0(r \quad 1)[\quad e^2 f_0 + ((r \quad 2) f_0 + 2) G(M^2)]$$

$$\int_{\;Y}^{Y} d\eta \quad \frac{MT}{\cosh(y \quad \eta)} + \frac{T^2}{\cosh^2(y \quad \eta)}$$

$$\exp[\;\; M\cosh(y \quad \eta)/T\,]$$

$$+\,\Theta(s \quad s\;)G(M^2)\frac{3(\tau\;\; T^3(\tau\;\;))^2}{M^4} \int_{\;Y}^{Y} \frac{d\eta}{\cosh\;(\eta \quad y)}$$

$$(\;\; x^4 \quad 5x^3 \quad 15x^2 \quad 30x \quad 30)\exp(\;\; x)\Big|_{x_H}^{x_{\mathrm{fo}}} \quad . \qquad (7.50)$$

The limits of the x integration are defined as $x_0 = M\cosh(y \quad \eta)/T(\tau_0)$ $x \;\; = x \;\;\; = M\cosh(y \quad \eta)/T$ and $x \;\;\; = M\cosh(y \quad \eta)/T(\tau\;\;)$.

7.3 Initial conditions

We use Bjorken scaling, assuming longitudinal expansion, to determine the initial conditions. Recall from Chapter 4 that the final-state multiplicity is related to the initial number density and hence the initial entropy density. The entropy and number density are related since the entropy per particle is a constant, labeled ξ, so that $s/n = \xi$ where $\xi = 3.92$ for a quark-gluon plasma with two massless flavors and 3.6 for a massless pion gas. Entropy current conservation relates the initial and final entropy densities by $s(\tau_0)\tau_0 = s(\tau)\tau$. The number density at freezeout is related to the multiplicity by $n = N/V$ where

$V = \pi R^2 \tau \Delta y$. Thus, in a given rapidity interval,

$$\frac{dN}{dy} = n(\tau)\tau\pi R^2 \tag{7.51}$$

since we let $N/\Delta y \quad dN/dy$. If $s = aT^3$, then $n = s/\xi = (a/\xi)T^3$. If we assume $\tau_0 = 1$ fm, regardless of which state the system finds itself in after equilibration (quark-gluon plasma, mixed phase, or hadron gas), then there are several densities which can be compared to the multiplicity for zero chemical potential:

$$n \quad = \quad a \quad T^3 = \frac{4}{\xi}\frac{\pi^2 g}{90} T^3 \ , \tag{7.52}$$

$$n \quad = \quad a \ T^3 = \frac{4}{\xi}\frac{\pi^2 g}{90} T^3 \ , \tag{7.53}$$

$$n \quad = \quad a \ T^3 = \frac{4}{\xi}\frac{\pi^2 g}{90} T^3 \ . \tag{7.54}$$

These densities are, respectively, the beginning of the mixed phase where the degrees of freedom are only quarks, antiquarks and gluons, Eq. (7.52), the end of the mixed phase when only hadronic degrees of freedom remain Eq. (7.53), both at $T = T$, and the density at freezeout when $T = T$. Thus, if the final-state multiplicity is such that

$$\frac{dN}{dy} \quad \pi R^2 n \quad , \tag{7.55}$$

the system begins in the quark-gluon plasma phase while, if

$$\pi R^2 n \quad < \frac{dN}{dy} \quad \pi R^2 n \quad , \tag{7.56}$$

the system begins in the mixed phase with $T = T$, and finally, if

$$\pi R^2 n \quad > \frac{dN}{dy} \ , \tag{7.57}$$

the system begins in the hadron gas. Whatever the state of matter after thermalization, for fixed τ_0, we can write

$$s = aT^3 = \xi n = \frac{dN}{dy}\frac{\xi}{\pi R^2 \tau_0} \quad , \tag{7.58}$$

giving us an initial temperature of

$$T(\tau_0) = \left(\frac{dN}{dy} \frac{\xi}{a\pi R^2 \tau_0} \right)^{1/3} . \tag{7.59}$$

An alternative method, which allows simultaneous determination of the initial proper time and the initial temperature is using the uncertainty principle as a bound, $\tau_0 E_0 \approx 3\tau_0 T(\tau_0) \gtrsim 1$. The lower limit, when $3\tau_0 T(\tau_0) = 1$, gives a lower bound on τ_0 and an upper bound on $T(\tau_0)$. In this case [110, 111], we can write Eq. (7.58) as

$$\tau_0 T^3(\tau_0) = \frac{dN}{dy} \frac{\xi}{a\pi R^2} . \tag{7.60}$$

We replace $\tau_0 T(\tau_0)$ by 1/3, leaving only the square of the temperature on the left-hand side, giving an initial temperature of

$$T(\tau_0) = \left(\frac{dN}{dy} \frac{3\xi}{a\pi R^2} \right)^{1/2} . \tag{7.61}$$

The initial proper time then follows from the uncertainty relation.

The initial temperature is quite sensitive to the multiplicity, dN/dy. Although the assumption of longitudinal scaling assumes a plateau in rapidity, the most likely ansatz is that of a Gaussian-shaped multiplicity distribution,

$$\frac{dN}{d\eta} = \left(\frac{dN}{d\eta} \right) |_0 \exp(-\eta^2/2\sigma^2) \tag{7.62}$$

where $(dN/d\eta)|_0$ is the multiplicity at $\eta = 0$ and σ is the width of the Gaussian. We write η here for the rapidity of the produced fluid and identify $(dN/d\eta)|_0$ with dN/dy at midrapidity. In Ref. [111], before RHIC data were available, $(dN/d\eta)|_0 = 2000$ and 5000 were assumed for RHIC and the LHC respectively. Experience from RHIC has shown that these multiplicities are likely somewhat too high but we will use these estimates here for illustration. The width of the Gaussian is estimated by integrating over all phase space and using $\langle m_T \rangle$ to obtain the total energy. The values used in Ref. [111] were $\sigma = 3$ for RHIC and 5 for the LHC. Note also that using the Gaussian form for $dN/d\eta$

makes it possible to fix the initial temperature over all rapidity space, not just at $y = 0$.

Of the two scenarios for estimating the initial temperature, the one with τ_0 1 fm gives the lowest initial temperatures and thus the lowest thermal dilepton yield, $T(\tau_0)$ 255 MeV at RHIC and 350 MeV at the LHC [111] with our values of $(dN/d\eta)|_0$. On the other hand, the uncertainty bound gives initial temperatures as high as $T(\tau_0)$ 515 MeV at RHIC and 810 MeV at the LHC with corresponding initial times of $\tau_0 = 0.2$ and 0.08 fm respectively. These initial times are similar to those assumed for minijet production, as discussed in Chapter 5. At such high initial temperatures, some thermal charm production may also be possible [112]. Because charm is produced by strong interactions and the strong coupling constant is much great than the electromagnetic coupling constant thermal charm production and decay to lepton pairs might be competitive with thermal dilepton production [113].

Once the initial time and temperature are fixed by dN/dy, the subsequent values of τ , τ and τ follow since $\tau \ = \tau_0(T(\tau_0)/T\ )^3$, $\tau \ = \tau\ (g \quad /g\ )$ and $\tau \ = \tau\ (T\ /T(\tau\ ))^3$ for systems initially in the quark-gluon plasma phase. If the initial temperature is not high enough for a quark-gluon plasma to be formed, then the system evolves through either both the mixed phase and hadron gas or the hadron gas alone, depending on the temperature, with $\tau_0 = 1$ fm. If the system is formed in a mixed phase, $\tau_0 = \tau$ and the initial entropy density contains both quark-gluon plasma and hadron gas contributions, according to Eq. (7.45). If the system is formed in a hadron gas, $\tau_0 = \tau$ and the system evolves from the hadron gas to freezeout. In all cases, the thermal dilepton rate can be calculated from Eq. (7.50).

7.4 Numerical results

Numerical calculations of dN/dM in a given rapidity interval or dN/dy over a given mass range using Eq. (7.50) are straightforward. Results for a quark-gluon plasma with two light flavors ($g \quad = 37$), $g \ = 3$, $T \ = 170$ MeV and $T(\tau\ ) = 100$ MeV are shown in Figs. 7.1 and 7.2 for $(dN/d\eta)|_0 = 2000$ and $\sigma = 3$ at RHIC and $(dN/d\eta)|_0 = 5000$ and

$\sigma = 5$ at the LHC. The initial temperature has been determined both with fixed $\tau_0 = 1$ fm and with the upper bound from the uncertainty principle. The upper bound gives the largest thermal dilepton rate, especially for large masses. The mass scale on the x axis has been somewhat truncated since we want to emphasize the region over which discussion of thermal dilepton production is most meaningful.

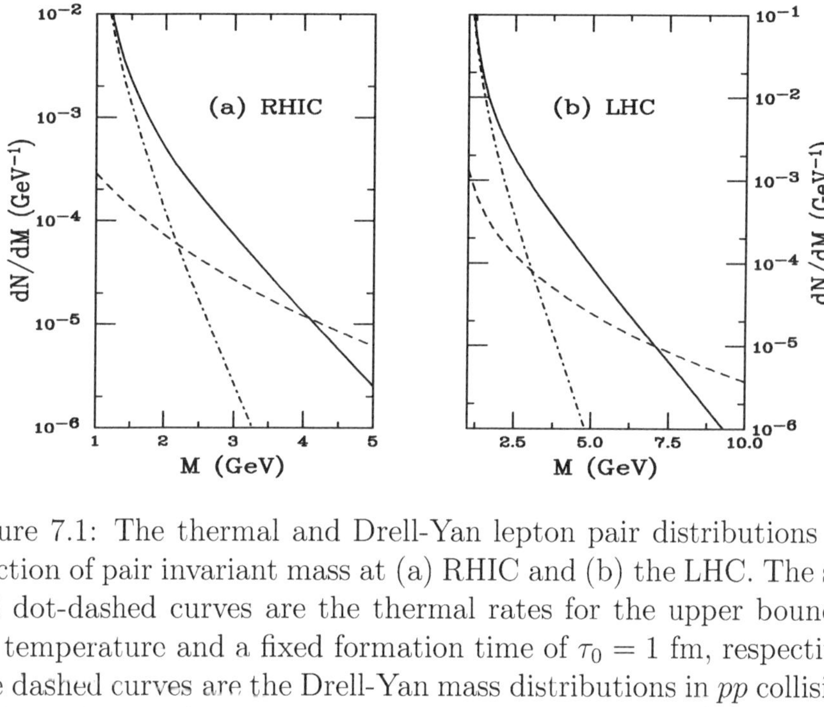

Figure 7.1: The thermal and Drell-Yan lepton pair distributions as a function of pair invariant mass at (a) RHIC and (b) the LHC. The solid and dot-dashed curves are the thermal rates for the upper bound on the temperature and a fixed formation time of $\tau_0 = 1$ fm, respectively. The dashed curves are the Drell-Yan mass distributions in pp collisions. Both results are shown in the pair rapidity interval of $|y|$ 1.

To discuss the observability of thermal dilepton production, it is important to consider other sources of dilepton production. Drell-Yan pair production is an obvious background to thermal dilepton production. The number of Drell-Yan pairs at a given pair mass and rapidity in an AB collision at zero impact parameter is

$$\frac{dN}{dMdy} = T \quad (0) \frac{d\sigma}{dMdy} \ . \tag{7.63}$$

The Drell-Yan rate in pp collisions at $\sqrt{S}$ = 200 GeV and 5.5 TeV is also shown in Figs. 7.1 and 7.2.

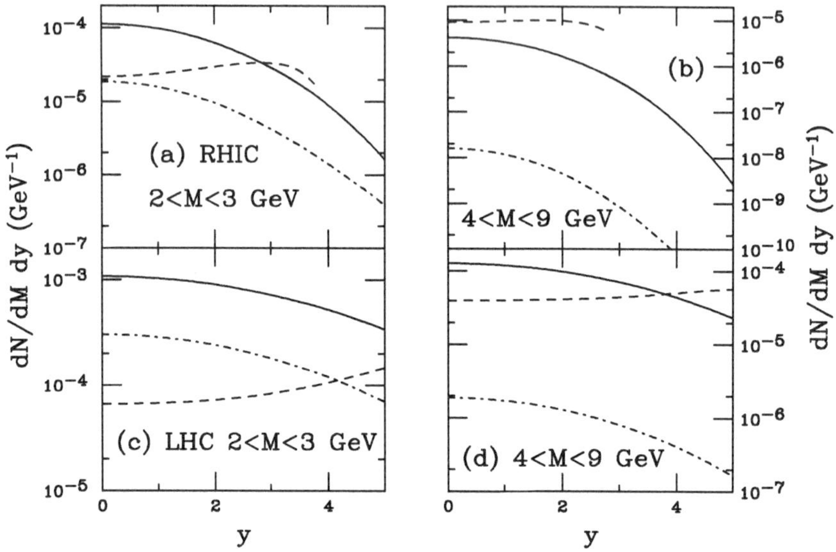

Figure 7.2: The thermal and Drell-Yan lepton pair distributions as a function of pair rapidity at RHIC (upper plots) and the LHC (lower plots). On the left-hand side, the distributions are given in the mass range $2 < M < 3$ GeV while on the right-hand side, $4 < M < 9$ GeV is shown. The solid and dot-dashed curves are the thermal rates for the upper bound on the temperature and a fixed formation time of $\tau_0 = 1$ fm, respectively. The dashed curves are the Drell-Yan rapidity distributions in pp collisions.

Figure 7.1 shows the pair mass distributions in the lepton pair rapidity interval of $|y|$ 1. The thermal dilepton and Drell-Yan rates are both calculated down to $M = 1$ GeV although the perturbative Drell-Yan cross section is expected to be less accurate at such low masses. The rates are shown for masses where thermal dilepton production is larger than the Drell-Yan rate, $M < 4$ GeV at RHIC and $M < 7.5$ GeV at the LHC with the upper bound on the temperature. The thermal rates drop more steeply with pair mass than does the Drell-Yan rate since the thermal distribution drops exponentially with mass while the

Drell-Yan cross section goes as $1/M$ in this region. The higher temperatures obtained using the uncertainty principle bound (solid curves) give a larger contribution relative to Drell-Yan production over a broader mass range than does the rate for the fixed formation time (dot-dashed curves).

The relative proportions of dileptons from the quark-gluon plasma, mixed phase and hadron gas are shown for the fixed formation time case in Figs. 7.3-7.5. The mass ratios, $R(M)$, in Fig. 7.3, show that the quark-gluon plasma phase is not dominant at the lowest masses considered but constitutes the entire contribution by $M > 4$ GeV at RHIC and $M > 3$ GeV at the LHC. This is not surprising because high temperatures are needed to produce high mass lepton pairs. The quark-gluon plasma contribution takes over at lower masses at the LHC because the initial temperature is higher. The hadron gas, latest in time and lowest in temperature, provides about half the thermal dileptons at $M = 1$ GeV but contributes negligibly to the total by $M = 2$ GeV. The rather long-lived mixed phase gives the largest contribution between 1 and 2 GeV. However, at the LHC where the temperature is higher due to the increased multiplicity, the mixed phase never contributes more than 50% of the total.

The ratios for the upper bound on the temperature are similar except that, in this case, due to the higher initial temperature the quark-gluon plasma phase dominates at still lower masses. Interestingly enough, these results suggest that in the region below the J/ψ peak in the dilepton mass spectrum at M 3.1 GeV, where thermal dilepton production may be visible above the Drell-Yan background, these thermal pairs may not dominantly arise from the quark-gluon plasma itself.

Figure 7.2 compares the thermal dilepton and Drell-Yan rates as a function of rapidity. Lepton pair invariant mass bins of $2 < M < 3$ GeV and $4 < M < 9$ GeV are shown. These bins are chosen since they are below ($2 < M < 3$ GeV) and above ($4 < M < 9$ GeV) the J/ψ mass region. Since the decay, $J/\psi \to l \; l$, produces a peak near the J/ψ mass, no additional sources of lepton pair production are easily separable.

There are two obvious differences in the rapidity distributions from

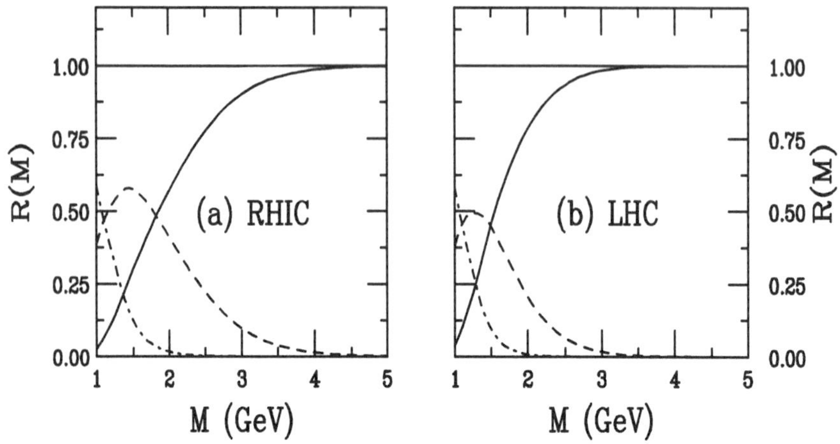

Figure 7.3: The fractional contribution of the different phases of system evolution are shown as a function of pair mass for (a) RHIC and (b) the LHC. The quark-gluon plasma contribution is given by the solid curve, the mixed phase by the dashed curve and the hadron gas by the dot-dashed curve. The results are given for $\tau_0 = 1$ fm.

the two dilepton sources. The first, particularly visible at RHIC, is the shorter range of the Drell-Yan rapidity distribution. This is rather trivially a consequence of the maximum possible rapidity on the nucleon-nucleon center-of-mass energy, $y_{\rm max} = \ln(\sqrt{S}/M)$. The upper limit of the mass range provides the cutoff shown in Fig. 7.2. There is no energy constraint on the rapidity except that imposed by the value of σ in the Gaussian parameterization of the multiplicity distribution. This places a limit on the extent of the produced system in rapidity space but does not place any additional constraint on the rapidity of a lepton pair produced from that system. The second obvious difference between the lepton sources is the shape: the thermal dilepton rapidity distribution is concave as a function of rapidity while the Drell-Yan distribution is convex. The increase of the Drell-Yan distribution with rapidity is due to the growing importance of the valence quark distributions in the $q\overline{q}$ convolution, as discussed in Chapter 2, as the kinematic limit is approached and $x \rightarrow 1$ in the projectile nucleus (moving in the direction of positive rapidity). On the other hand, the thermal dilepton rapidity distribution is shaped by the parameterization of the multiplicity.

The distributions in Fig. 7.2 reflect the mass distributions shown in Fig. 7.1. At RHIC, the high temperature scenario gives a larger rate than the Drell-Yan process in the lowest mass bin, $2 < M < 3$ GeV, while when the initial temperature is determined from $\tau_0 = 1$ fm, the thermal rate is similar only at $y = 0$ and then drops below the Drell-Yan rate. On Fig. 7.1(a), the thermal dilepton rate with $\tau_0 = 1$ fm becomes less than the Drell-Yan at M 2.2 GeV for $|y|$ 1. Thus, in the mass averaged bin, the Drell-Yan rate is larger. In the larger mass bin, above the J/ψ mass region, the Drell-Yan rate is larger over all rapidity. At the higher energy of the LHC, the thermal dilepton rate is larger than the Drell-Yan rate for $y < 3$ in both scenarios of the initial temperature. Even when $4 < M < 9$ GeV, the midrapidity thermal dilepton rate is higher than the Drell-Yan rate when the initial temperature is determined from the uncertainty bound. Note also that while the two thermal dilepton scenarios are nearly parallel as a function of rapidity, the curves become further apart as the mass increases.

The contributions to the thermal dilepton rates as a function of rapidity shown in Figs. 7.4 and 7.5 for the case $\tau_0 = 1$ fm reflect the shape of the multiplicity distribution. While the hadron gas contribution is always negligible and, in fact, almost invisible in Figs. 7.4 and 7.5, the contribution from the mixed phase becomes larger as the pair rapidity increases. In the lower mass bin at RHIC, Fig. 7.4(a), the crossover of the two contributions occurs at $y = 3$, corresponding to the width of the Gaussian multiplicity distribution. This must occur because the multiplicity drops with increasing rapidity until the multiplicity becomes too low for the temperature to be high enough for quark-gluon plasma production. Only the lowest mass bin has been shown for the LHC in Fig. 7.5 since here quark-gluon plasma dominance is more complete at lower masses, as also reflected in Fig. 7.3.

So far, all the results have been shown for two light flavors (g $=$ 37), g $= 3$, T $= 170$ MeV and $T(\tau$ $) = 100$ MeV and $(dN/d\eta)|_0 =$ 2000, $\sigma = 3$ at RHIC and $(dN/d\eta)|_0 = 5000$, $\sigma = 5$ at the LHC. We now discuss how the results change if one of these parameters is

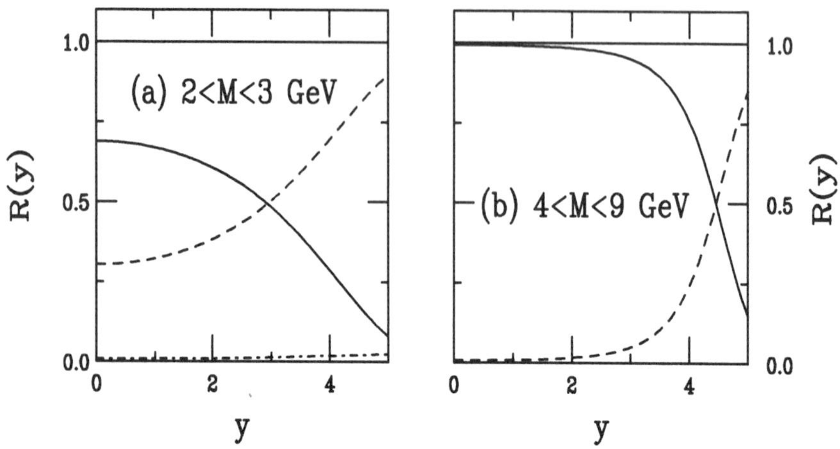

Figure 7.4: The fractional contribution of the different phases of system evolution are shown as a function of pair rapidity for RHIC with mass ranges $2 < M < 3$ GeV (a) and $4 < M < 9$ GeV (b). The quark-gluon plasma contribution is given by the solid curve, the mixed phase by the dashed curve and the hadron gas by the dot-dashed curve (usually not visible). The results are given for $\tau_0 = 1$ fm.

changed at a time.

If we have three light flavors instead of two, g becomes 47.5. Such a change, without a corresponding increase in final-state multiplicity, would have the effect of decreasing the initial temperature since g is in the denominator of Eqs. (7.59) and (7.61), thus reducing the dilepton yield. This would also have the effect of lengthening the mixed phase since $\tau \; = \tau \; (g \; / g \;)$. The effect on the thermal dilepton rate is thus an overall decrease in the dilepton yields but since the mixed phase also gets longer, contributing more lepton pairs, the reduction is smaller at low masses.

If the number of hadronic degrees of freedom is increased from $g \; =$ 3 to an average of 6.6, the thermal rate is also decreased. The change only affects the low masses since the initial temperature is unaffected. Increasing the hadronic degrees of freedom reduces the time spent in the mixed phase, only of relevance at low masses.

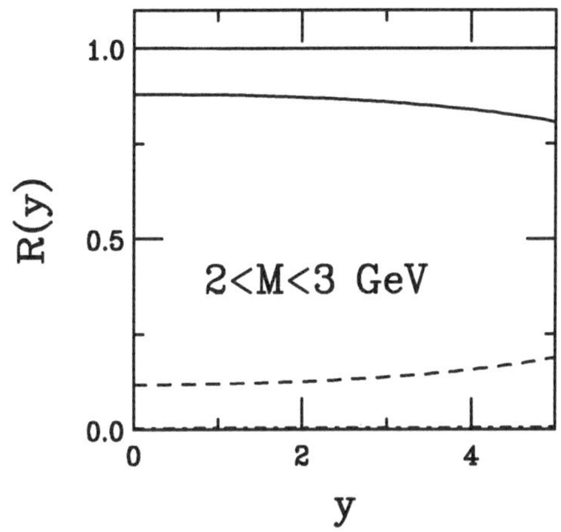

Figure 7.5: The fractional contribution of the different phases of system evolution are shown as a function of pair rapidity for $2 < M < 3$ GeV at the LHC for $\tau_0 = 1$ fm. The quark-gluon plasma contribution is given by the solid curve, the mixed phase by the dashed curve and the hadron gas by the dot-dashed curve (along the lower x axis).

Increasing T from 170 MeV to 200 MeV increases the low mass rate since it enhances the effect of the mixed phase. At higher rapidities, the larger T would increase the contribution of the mixed phase further at low mass since the higher T makes it more difficult to produce a quark-gluon plasma initially with lower multiplicities.

Changing the freezeout temperature from 100 MeV to 140 MeV would decrease the low mass rate since it would reduce the already small effect of the hadron gas on the total rate. Note that neither changing T or $T(\tau\ )$ has an effect on the initial temperature and would thus have little effect on the quark-gluon plasma contribution.

The thermal rate is most sensitive to $(dN/d\eta)|_0$. If $(dN/d\eta)|_0$ is increased by a factor of two, the initial temperature is increased over all rapidity. The increase in rate is only a factor of several at the lowest masses but at high masses the rate can be increased by orders of magnitude. On the other hand, decreasing $(dN/d\eta)|_0$ by a factor of two would reduce the initial temperature, decreasing the chance of the system to initially be in a quark-gluon plasma. The high mass rate would decrease by orders of magnitude in this case.

If the width of the multiplicity distribution is changed, the greatest effect is on the thermal dilepton rapidity distributions. There is no effect at midrapidity if $(dN/d\eta)|_0$ is unchanged but, if the width is decreased, the rapidity distributions will decrease more rapidly with increasing y. The contribution from the mixed phase at higher rapidity increases since the initial temperature decreases more quickly with rapidity.

7.5 Other dilepton sources

We end this chapter with some discussion of competing dilepton sources which affect the observation of thermal dilepton production. While comparison to the Drell-Yan rate seems to indicate that the thermal dilepton yield might rise above the Drell-Yan background at low pair masses, the Drell-Yan process is not the only significant background. Indeed, there are more lepton pairs from decays of charm and bottom hadrons than the thermal yields at collider energies. The heavy flavor hadron, H, yields are large at collider energies and could even be large enough for multiple $H\overline{H}$ pairs to be produced, especially at the LHC, leading to both correlated and uncorrelated contributions to the dilepton yield.

If the average number of $H\overline{H}$ pairs,

$$N - = T \quad (0)\sigma - , \tag{7.64}$$

produced in central collisions is small, $N - \quad 1$, the lepton pairs will be correlated with

$$N^{\rm c} \quad = N - B^2(H/\overline{H} \to l \;\; X) \tag{7.65}$$

where B is the branching ratio for decays to leptons. However, if $N - > 1$, opposite sign lepton pairs from uncorrelated $H\overline{H}$ pair decays need to be taken into account. When $N - \quad 1$, the average number of uncorrelated lepton pairs is

$$N \;\; ^{\rm c} \quad = N - (N - \quad 1)B^2(H\overline{H} \to l \;\; X) \; . \tag{7.66}$$

If $N -$ 1, a distribution in $N -$ must be considered to calculate the uncorrelated pairs [57, 113]. Significant uncertainties still remain in the heavy flavor yields [114], making measurements imperative to determine whether the heavy flavor decays can be subtracted out, allowing a thermal signal to be observed. While a charm signal has been observed in pp collisions at $\sqrt{S} = 200$ GeV at RHIC by reconstructing hadronic decays such as $D^0 \rightarrow K$ π , the uncertainties are still large [115].

Exercises

1. Calculate the plasma fraction and hadronic fraction in the mixed phase as a function of τ , f_0, s and s .

2. For systems produced in a quark-gluon plasma with $dN/dy =$ 200, 1000 and 3000, calculate the initial temperature of the system using Eq. (7.59) with $\tau_0 = 1$ fm and Eq. (7.61). When obtaining the initial temperature with Eq. (7.61), also determine τ_0. Use $R = R \quad = r_0 A^{1/3}$. Do not forget to include the appropriate factors of $\hbar c$ to get the proper units. Once the initial conditions have been determined, use them to find τ , τ and τ assuming $T = 170$ MeV and $T(\tau\ ) = 100$ MeV.

3. Let $T \quad (0) = 29.3/\mathrm{mb}$ and $B(D \rightarrow lX) \quad 10\%$. Calculate N^{c} and $N \quad ^{\mathrm{c}}$ for $\sigma_{D\overline{D}} = 4\ \mu$b, 300 μb and 5 mb.

Chapter 8

Quarkonium

8.1 Introduction to quarkonium in heavy-ion collisions

As we just showed in the previous chapter, lepton pairs are an advantageous probe since they only interact weakly with matter. However, they are difficult to detect since they are produced with a continuum of masses rather than a fixed value. Quarkonium states (bound states of heavy quark-antiquark pairs) decay electromagnetically to lepton pairs of definite mass: the pair has the mass of the decaying bound state. The peak in the dilepton continuum at the quarkonium mass is relatively easy to see after subtraction of the like-sign pairs. A disadvantage of the quarkonium states is the fact that they are hadrons. They will interact strongly with other hadrons in the medium. Their interaction cross sections are typically lower than those of normal mesons due to their smaller radii. This last statement is particularly true for the most tightly bound quarkonium states, the J/ψ (a $c\bar{c}$ bound state with a mass of 3.097 GeV) and the Υ (a $b\bar{b}$ bound state with a mass of 9.46 GeV).

The discovery of the J/ψ was announced virtually simultaneously at Brookhaven National Laboratory [116] (in the reaction $p\mathrm{Be} \rightarrow e^+e^-$) and the Stanford Linear Accelerator Center [117] (in the process $e^+e^- \rightarrow \mu^+\mu^-$) in 1974. The existence of the J/ψ was not itself controversial since the charm quark had been predicted to explain the absence of certain neutral kaon decay channels [118]. However, the priority for its discovery was disputed so that it was given two names, J at Brookhaven and ψ by SLAC: thus the J/ψ. The higher-lying $c\bar{c}$ bound states, including the $\psi(2S) \equiv \psi'$ and $\psi(1P) \equiv \chi_c$ states, were discovered later at SLAC. There are other, more massive, charmonium bound states above the $D\overline{D}$ threshold. However, they decay rather rarely to lepton pairs and thus do not produce peaks in the dilepton continuum. We discuss why this happens a bit later in the chapter. The $1S$ state will be referred to here as either J/ψ or ψ. A few years later, the even more massive bottom quarks and the Υ family ($b\bar{b}$ or bottomonium states) were also discovered [119]. In the case of the Υ states, in addition to the $\Upsilon(1S)$, $\Upsilon(2S)$ and $\chi_b(1P)$, there are more $b\bar{b}$ states below the $B\overline{B}$ threshold, particularly the $\Upsilon(3S)$ and $\chi_b(2P)$.

While the discovery of the J/ψ confirmed the basic validity of the quark model, its absence was suggested as a means of confirming quark-gluon plasma production. In 1986, Matsui and Satz [120] predicted that quarkonium production would be suppressed by color screening in a quark-gluon plasma. Suppression does not imply that the $Q\overline{Q}$ will not be produced, but rather that the observed quarkonium yield will be depleted relative to the expected yield either because the $Q\overline{Q}$ pair fails to form a quarkonium state or the state itself is destroyed through its subsequent interactions.

Since the J/ψ-nucleon cross section extracted from photoproduction measurements was small, $\approx$ 1-3 mb [121], hadronic mechanisms of J/ψ suppression were assumed to be negligible. However, this small cross section is still large enough to make the A dependence nonlinear in hadron-nucleus interactions, see *e.g.* Refs. [122, 123, 124, 125].

The J/ψ has been studied in nuclear collisions since the first measurements at the CERN SPS in the late 1980s [126]. These results are now supplemented by RHIC measurements [125] with more data to come from the LHC on both J/ψ and Υ production [127].

We first discuss the quarkonium levels at zero temperature and describe quarkonium production by perturbative QCD in elementary collisions. We then discuss the prediction of quarkonium suppression in nuclear collisions and its implications for measurements in nucleus-

nucleus collisions. We will also discuss some competing quarkonium suppression mechanisms in cold nuclear matter and production mechanisms in hot matter.

8.2 Quarkonium levels at $T = 0$

At $T = 0$, as in pp and pA interactions, the quarkonium bound state spectra can be described by the nonrelativistic potential,

$$V(r, T = 0) = \sigma r - \frac{\alpha_c}{r} , \tag{8.1}$$

where r is the separation between the Q and $\overline{Q}$. This naive potential does not account for spin-orbit or spin-spin couplings needed to separate the three χ_c states or to separate the J/ψ from the η_c respectively but is sufficient for our purposes.

The $1/r$ term is Coulomb-like and governs the short distance behavior of the potential. It arises from the exchange of a gluon between the Q and $\overline{Q}$. The coupling in the Coulomb-like term, α_c, can be related to the strong coupling constant with a color factor to account for summation over color indices, $\alpha_c = (4/3)\alpha_s$. The shorter the distance scale, corresponding to increasing momentum scales, the weaker the coupling. While the charmonium and bottomonium states have been fit to a single value of α_c in Ref. [128], in principle the bottomonium coupling could be smaller.

The linear term corresponds to the confining potential obtained in the calculation of the Wilson loop in Chapter 6. The strength of the confining term is determined by the string tension, σ. The total angular momentum, J, of hadrons with a certain internal symmetry is proportional to the square of energy of the state, E, such that

$$J \sim \alpha E^2 . \tag{8.2}$$

The slope α is approximately 1 GeV^{-2} [129]. If two massless quarks are connected by a taut string of radius r_0 rotating with the speed of light, the velocity of the string at a point $r < r_0$ is $v = r/r_0$ in $c = 1$

units. The total energy in the string is the integral over the potential, $V \sim \sigma\gamma r$ from the center of the string to the endpoint r_0,

$$E = 2\int_0^{r_0} \frac{dr\sigma}{\sqrt{1-v^2}} \ . \tag{8.3}$$

The factor of two is needed to account for both ends of the string and $\gamma = 1/\sqrt{1-v^2}$. Replacing v with $r/r_0 = \sin\theta$, the upper limit of the integral becomes $\pi/2$. Then the integral is

$$E = 2\sigma\int_0^{\pi/2} r_0 \frac{\cos\theta d\theta}{\sqrt{1-\sin^2\theta}} = \pi\sigma r_0 \ . \tag{8.4}$$

Likewise, the angular momentum is

$$J = 2\int_0^{r_0} \frac{dr\sigma v r}{\sqrt{1-v^2}} \ . \tag{8.5}$$

Making the same substitutions as in Eq. (8.4), we have

$$\begin{aligned} J &= 2\int_0^{\pi/2} \sigma r_0^2 \sin^2\theta d\theta \\ &= 2\sigma r_0^2 \frac{1}{2}(\theta - \sin\theta\cos\theta)\Big|_0^{\pi/2} = \frac{\pi}{2}\sigma r_0^2 \ . \end{aligned} \tag{8.6}$$

Replacing Eqs. (8.4) and (8.6) for E and J in Eq. (8.2), we find

$$\sigma = \frac{1}{2\pi\alpha} \sim 0.2\,\mathrm{GeV}^2 \tag{8.7}$$

for $\alpha = 1$ GeV^{-2}.

The quarkonium levels can be calculated by solving the Schrodinger equation with the potential in Eq. (8.1). The binding energy, E_b, is related to the quarkonium mass, M, by

$$M = 2m_Q + E_b \tag{8.8}$$

where m_Q is the heavy quark mass. Because the potential has both positive and negative components, E_b can be either positive or negative depending on the value of m_Q relative to the mass of the quarkonium state, M_L. If $m_Q \geq M_L/2$, $E_b < 0$ while if $m_Q < M_L/2$, $E_b > 0$. The parameters that best fit the general features of the spectrum are [128]

$$\sigma = 0.192\,\mathrm{GeV}^2\ , \tag{8.9}$$
$$\alpha_c = 0.471\ , \tag{8.10}$$
$$m_c = 1.32\ ,\mathrm{GeV} \tag{8.11}$$
$$m_b = 4.75\,\mathrm{GeV}\ . \tag{8.12}$$

Note that $m_c < m_\psi/2$ and $m_b > m_\Upsilon/2$. Thus the binding energy for the J/ψ is positive while that of the Υ is negative.

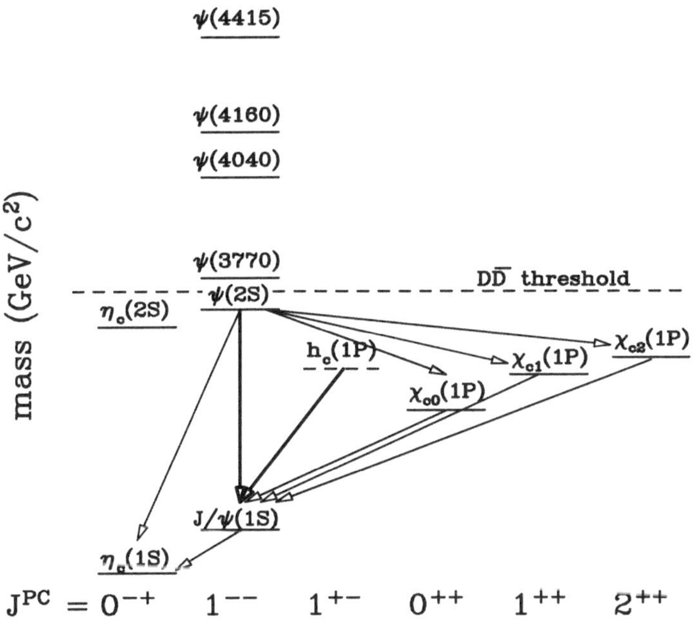

Figure 8.1: The charmonium family with quantum numbers. The thick transition lines indicate hadronic feed-down decays while the thinner lines indicate radiative decays. Unconfirmed states are shown as dashed lines. The $D\overline{D}$ threshold is also shown.

The charmonium and bottomonium families are shown in Figs. 8.1 and 8.2 respectively, along with common feed-down channels. The mass of a pair of open heavy flavor mesons is also indicated. Below this $H\overline{H}$ threshold, the quarkonium states mainly decay to other, lower mass, quarkonium states or to light hadrons. The transitions involving hadrons such as $\psi' \rightarrow J/\psi\pi\pi$ are indicated by thick arrows while radiative transitions such as $\chi_c \rightarrow J/\psi\gamma$ are denoted by thinner lines.

The ψ' contributes to indirect J/ψ production in two ways: through direct decays to hadrons, as already mentioned, and radiative decays to the χ_c states ($\psi' \rightarrow \chi_c\gamma$) which then decay radiatively to J/ψ ($\chi_c \rightarrow J/\psi\gamma$). The χ_c, $1P$ charmonium states contribute about 30% of inclusive J/ψ production. (Inclusive means all observed final-state J/ψs, regardless of the source.) The χ_c states are identified in experiments by first detecting the J/ψ and subsequently finding the accompanying photon with momentum $m_{\chi_c} - m_{J/\psi}$, approximately 300 - 430 MeV, depending on the identity of the decaying χ_c. The J/ψ and ψ' are readily identified through their electromagnetic decays, $J/\psi(\psi') \rightarrow l^+l^-$, where l is a lepton. Besides the electromagnetic channel, the J/ψ decays to light hadrons. The J/ψ has a very narrow decay width, indicating a long lifetime. When first measured at SLAC, the real J/ψ decay width is 87 keV [3] while the width of the ψ' is 277 keV.

The quarkonium levels are reminiscent of the positronium (e^+e^- bound states) spectrum. The η_c and J/ψ correspond to parapositronium ($^1S_0$ spin singlet state) and orthopositronium ($^3S_1$ spin triplet state) respectively. In the positronium system, the "ortho" state typically de-excites by photon emission to parapositronium which then decays to two photons. Likewise, the J/ψ decays 1.3% of the time to $\eta_c\gamma$. There is no hadronic $J/\psi \rightarrow \eta_c X$ decay channel since the mass difference, $m_{J/\psi} - m_{\eta_c} = 118$ MeV, is smaller than the mass of any hadron. The η_c can decay to $\gamma\gamma$ but with a small branching ratio of 0.03%. The lifetimes are also different: in the positronium system, the ortho state is longer lived, 1.4×10^{-7} s relative to 1.25×10^{-10} s for the parapositronium state. In the charmonium system, the J/ψ, with its small width, has a lifetime of 2.27×10^3 fm or 7.56×10^{-21} s. On the other hand, the η_c width is 13.2 MeV, giving a lifetime of 15.2 fm or 5.06×10^{-23} s. The large differences in the positronium and quarkonium lifetimes may be attributed to the difference in time scales for electromagnetic and strong interaction processes.

Above the $D\overline{D}$ threshold, the spin one S and D states, sharing the same J^{PC} assignments as the J/ψ and ψ', the $1S$ and $2S$ charmonium states, are the only ones known since they have small branching ratios to lepton pairs. If any P states exist above the $D\overline{D}$ threshold, they are unknown. These higher mass resonances are all broader (shorter lived) than the states below the threshold. For example, the $\psi(3770)$ has a width of 25 MeV. These resonances dominantly decay to open charm hadrons. Decays in other channels are unknown.

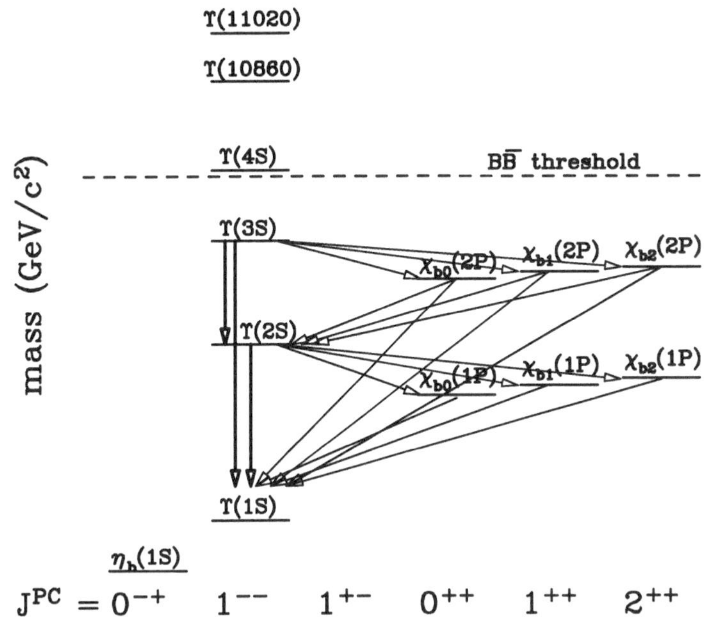

Figure 8.2: The bottomonium family with quantum numbers. The thick transition lines indicate hadronic feed-down decays while the thinner lines indicate radiative decays. Unconfirmed states are shown as dashed lines. Speculative but unmeasured states are not shown. The $B\overline{B}$ threshold is also shown.

The bottomonium levels are more complicated since there are 3 S states and two sets of χ_b states below the $B\overline{B}$ threshold. The higher mass $\Upsilon(2S)$ and $\Upsilon(3S)$ states can feed down to the lower mass $\Upsilon(1S)$ and χ_b states. The η_b states associated with the $\Upsilon(2S)$ and $\Upsilon(3S)$ states and the h_b states are postulated to exist but are not shown due to the lack of experimental confirmation. While studies of charmonium production in $\bar{p}p$ collisions at the Fermilab Tevatron with center-of-mass energy $\sqrt{S} = 1.8$ TeV have been able to extract the direct J/ψ production from the ψ' and χ_c contributions at high p_T [130], extraction of direct $\Upsilon(1S)$ production is more difficult even though the χ_b states have also been measured at the Tevatron [131].

Since the $\Upsilon(1S)$ is composed of heavier quarks than the J/ψ, it is more tightly bound. It is also a narrower resonance: the widths of the $\Upsilon(1S)$, $\Upsilon(2S)$ and $\Upsilon(3S)$ are 52, 44 and 26 keV respectively, significantly less than those of the J/ψ and ψ'. As is the case for the charmonium family, more bottomonium states lie above the $B\overline{B}$ threshold which decay primarily to pairs of bottom mesons.

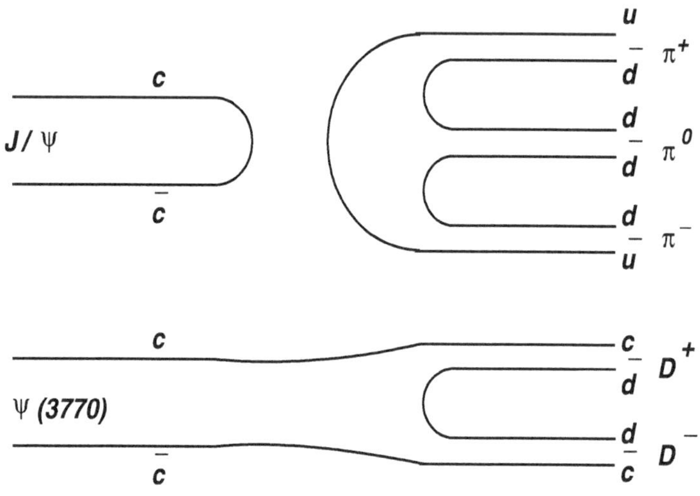

Figure 8.3: An illustration of the OZI rule in charmonium decays. Decays of J/ψ to hadrons are OZI suppressed since the quark lines are disconnected. Charmonium states below the $D\overline{D}$ threshold are forced to annihilate to gluons to decay to lighter hadrons (upper figure). Decays to charm mesons with connected quark lines (lower figure), possible for the $\psi(3770)$, is OZI favored.

The inability of the J/ψ and Υ to decay directly to charm and bottom hadrons was compared to $\phi \rightarrow K\overline{K}$ decays in Chapter 2.

The width of the ϕ is also broader, 42 MeV. The differences in the decay channels can be empirically described by the Okuba-Zweig-Iiluka (OZI) rule [132], illustrated in Fig. 8.3 for the J/ψ and the $\psi(3770)$. Since the J/ψ lies below the $D\overline{D}$ threshold, the c and $\overline{c}$ quarks must first annihilate into either a photon (the leptonic decay channels, l^+l^-) or gluons (light hadron decay channels such as $\pi^+\pi^0\pi^-$ in Fig. 8.3). The quark lines on the left and right halves of the upper figure are not connected. Thus this decay is OZI suppressed, giving the state a longer lifetime. Hence, also, the plethora of J/ψ decay channels. On the other hand, the $\psi(3770)$ has a mass above $2m_D$ so that the c and $\overline{c}$ carry through into the final state in the decay $\psi(3770) \rightarrow D^+D^-$. Thus the quark lines in the initial and final states are connected. This is an OZI favored decay and the $\psi(3770)$ decays almost exclusively to $D\overline{D}$. Since the ϕ mass is already above $2m_K$, its decays are primarily to K^+K^- and $K^0\overline{K}^0$ with a small branching to l^+l^-, making them OZI favored.

This OZI suppression for quarkonium states with masses below the $H\overline{H}$ threshold leads to much larger branching ratios to lepton pairs than states above the open heavy flavor thresholds. Thus these states are readily visible as peaks in the l^+l^- continuum while states above threshold are not.

8.3 Quarkonium production

Quarkonium production is somewhat more difficult to calculate in perturbative QCD than other hard processes. It is easy enough to calculate the production of heavy quarks, the tricky part is obtaining the bound quarkonium state from the $Q\overline{Q}$ pair, particularly if the bound state is to be produced with the right angular momentum and spin quantum numbers. In addition, most $Q\overline{Q}$ pairs are not produced as color singlets, as required for bound states.

Several model approaches have been employed for quarkonium production including the color singlet model (CSM) [133], nonrelativistic QCD (NRQCD) [134], the color evaporation model (CEM) [135] and the comover enhancement scenario (CES) [136] which remains on a rather qualitative level. Perhaps the most fundamentally appealing is the nonrelativistic QCD approach. It works best for high p_T quarkonium production. However, the color evaporation model is better able to predict both the total yields and the rapidity distributions. Below we provide a little more detail about the CEM and NRQCD approaches. For more quantitative comparisons and detailed discussions, see Refs. [127, 137].

8.3.1 Color evaporation model

In the color evaporation model, quarkonium production is treated identically to open heavy quark production with the exception that, in the case of quarkonium, the invariant mass of the heavy quark pair is restricted to be less than twice the mass of the lowest mass meson that can be formed with the heavy quark. For charmonium, the upper limit on the $c\bar{c}$ pair mass is then $2m_D$.

The color evaporation model assumes that the quarkonium dynamics are identical to low invariant mass $Q\overline{Q}$ pairs although the $Q\overline{Q}$ pairs are typically produced in different color, angular momentum, and spin states than the final charmonium states. The hadronization of the quarkonium states from the $Q\overline{Q}$ pairs is nonperturbative, usually involving the emission of one or more soft gluons. The produced $Q\overline{Q}$ pair then neutralizes its color by interaction with the collision-induced color field—"color evaporation". The Q and the $\overline{Q}$ either combine with light quarks to produce heavy flavored hadrons or bind with each other to form quarkonium. The additional energy needed to produce heavy flavored hadrons when the partonic center-of-mass energy, $\sqrt{s}$, is less than $2m_H$, the heavy hadron threshold, is obtained nonperturbatively from the color field in the interaction region. Thus the yield of all quarkonium states may be only a small fraction of the total $Q\overline{Q}$ cross section below $2m_H$.

At leading order, the production cross section of quarkonium state

C in an AB collision is

$$\sigma_C^{\rm CEM} = F_C \sum_{i,j} \int_{4m_Q^2}^{4m_H^2} ds \int dx_1 dx_2 \; f_{i/A}(x_1,\mu^2) \; f_{j/B}(x_2,\mu^2) \times \sigma_{ij}(s) \; \delta(s - x_1 x_2 S) \; , \quad (8.13)$$

where A and B can be any hadron or nucleus, $ij = q\overline{q}$ or gg, $\sigma_{ij}(s)$ is the $ij \rightarrow Q\overline{Q}$ subprocess cross section and $f_{i/A}(x,\mu^2)$ is the parton density in the hadron or nucleus. The total $Q\overline{Q}$ cross section has $s \rightarrow S$.

Depending on the quantum numbers of the initial $Q\overline{Q}$ pair and the final-state quarkonium, a different nonperturbative matrix element is needed for the production of the quarkonium state. The average of these nonperturbative matrix elements are combined into the universal factor F_C which depends on m_Q, the scale of α_s and the parton densities. Once F_C has been fixed for a given state, *e.g.* J/ψ, ψ', or χ_c, the model successfully predicts the energy and momentum dependencies [138]. Note that *e.g.* F_ψ includes both direct J/ψ production and indirect production through radiative decays of the χ_c states and hadronic ψ' decays.

Since F_C must be a constant for the model to have any predictive power, the relative differential and integrated quarkonium production rates should be independent of projectile, target, and energy. While this appears to be the case for current quarkonium measurements [138], better data over a range of energies would either help make a stronger case or disprove the model predictions.

The leading order calculation of the quarkonium cross section is now outlined. At leading order, heavy quark hadroproduction is the sum of contributions from $q\overline{q}$ annihilation and gg fusion. If y is the $Q\overline{Q}$ rapidity in the AB center-of-mass frame, the cross section for free $Q\overline{Q}$ pairs of invariant mass m is [139]

$$\begin{aligned} \frac{d\sigma}{dydm^2} &= \frac{1}{S}\int_0^1 dx_1 \int_0^1 dx_2 \delta(x_1 x_2 S - m^2) \\ &\quad \times \delta\left(y - \frac{1}{2}\ln\left(\frac{x_1}{x_2}\right)\right) H_{AB}(x_1, x_2; \mu^2) \\ &= \frac{1}{S} H_{AB}(x_{01}, x_{02}; \mu^2) \; , \end{aligned} \quad (8.14)$$

where $x_{01,02} = (m/\sqrt{S})\exp(\pm y)$ and

$$H_{AB}(x_1, x_2; \mu^2) = f_g^A(x_1, \mu^2) f_g^B(x_2, \mu^2) \sigma_{gg}(m^2) + \sum_{q=u,d,s} (f_q^A(x_1, \mu^2) f_{\overline{q}}^B(x_2, \mu^2) + f_{\overline{q}}^A(x_1, \mu^2) f_q^B(x_2, \mu^2)) \sigma_{q\overline{q}}(m^2) \; . \tag{8.15}$$

The leading order partonic cross sections are [139]

$$\sigma_{gg}(m^2) = \frac{\pi\alpha_s^2}{3m^2} \left\{ \left(1 + \frac{4m_Q^2}{m^2} + \frac{m_Q^4}{m^4}\right) \ln\left(\frac{1+\lambda}{1-\lambda}\right) - \frac{1}{4}\left(7 + \frac{31m_Q^2}{m^2}\right)\lambda \right\} \; , \tag{8.16}$$

$$\sigma_{q\overline{q}}(m^2) = \frac{8\pi\alpha_s^2}{27m^2}\left(1 + \frac{2m_Q^2}{m^2}\right)\lambda \; , \tag{8.17}$$

where $\lambda = \sqrt{1 - 4m_Q^2/m^2}$. The leading order strong coupling constant is

$$\alpha_s(\mu^2) = \frac{12\pi}{(33 - 2n_f)\ln(\mu^2/\Lambda^2)} \tag{8.18}$$

where Λ is determined by the chosen set of parton densities and n_f is the number of light flavors. The cross section can also be expressed as a function of the fraction of the total longitudinal momentum carried by the $Q\overline{Q}$ pair, $x_F = x_1 - x_2$,

$$\frac{d\sigma}{dm dx_F} = \frac{2m}{\sqrt{x_F^2 s + 4m^2}} H_{AB}(x_{01}, x_{02}; x_{01}x_{02}s) \; , \tag{8.19}$$

where now $x_{01,02} = \frac{1}{2}(\pm x_F + \sqrt{x_F^2 + 4m^2/S})$.

The leading order quarkonium cross section x_F distribution is obtained by integrating Eq. (8.19) over mass, as in Eq. (8.13). Then

$$\frac{d\sigma_C}{dx_F} = 2F_C \int_{2m_Q}^{2m_H} m\, dm \frac{H_{AB}(x_{01}, x_{02}; x_{01}x_{02}s)}{\sqrt{x_F^2 S + 4m^2}} \; . \tag{8.20}$$

The leading order calculation in Eq. (8.13) is insufficient to describe high p_T quarkonium production since the $Q\overline{Q}$ pair p_T is zero at leading order. Therefore, the CEM was taken to next-to-leading order in the total cross section [138, 140] using the exclusive $Q\overline{Q}$ hadroproduction code of Ref. [141] with the same upper limit on the $Q\overline{Q}$ pair mass. When the next-to-leading order contribution is included, the p_T dependence of J/ψ and Υ production at high energy hadron colliders has also been shown to agree with the model calculations [140]. The agreement is improved if some transverse momentum broadening of the parton distributions in the initial state is included [137]. The mass and scale parameters used to fix F_C are the same as those used to empirically fit the $Q\overline{Q}$ total cross sections [127].

The extraction of F_C for the individual quarkonium states is now described. The J/ψ has been measured in pp and pA interactions up to $\sqrt{S} = 63$ GeV. The data are of two types: the forward cross section, $\sigma(x_F > 0)$, and the cross section at zero rapidity, $d\sigma/dy|_{y=0}$. All the cross sections are inclusive with feed down from χ_c and ψ' decays. The value of $F_{J/\psi}$ for inclusive J/ψ production is obtained by fitting the normalization of Eq. (8.13) using the $c\bar{c}$ parameters in Ref. [127]. The ratios of the direct production cross sections to the inclusive J/ψ cross section can be determined from data on inclusive cross section ratios and branching fractions. These direct ratios, R_C, given in Table 8.1, are multiplied by the inclusive fitted $F_{J/\psi}$ to obtain the direct production fractions, $F_C^{\rm dir} = F_{J/\psi} R_C$.

Table 8.1: Direct quarkonium production ratios, $R_C = \sigma_C^{\rm dir}/\sigma_{C'}^{\rm inc}$ where $C' = J/\psi$ and Υ. From Ref. [142].

	J/ψ	ψ'	χ_{c1}	χ_{c2}	Υ	Υ'	Υ''	$\chi_b(1P)$	$\chi_b(2P)$
R_C	0.62	0.14	0.60	0.99	0.52	0.33	0.20	1.08	0.84

The same procedure, albeit somewhat more complicated due to the larger number of bottomonium states below the $B\overline{B}$ threshold, is followed for the bottomonium. For most data below $\sqrt{S} = 100$ GeV, the three bottomonium S states were either not separated or their sum was reported. No x_F-integrated cross sections were available so that the CEM Υ cross section was fit to the effective lepton pair cross section at $y = 0$ for the three $\Upsilon(S)$ states. The extracted fit fraction is labeled $F_{\sum\Upsilon}$. Using the individual branching ratios of the $\Upsilon(1S)$, $\Upsilon(2S)$ and $\Upsilon(3S)$ to lepton pairs and the total cross sections reported in $\overline{p}p$ collisions at $\sqrt{S} = 1.8$ TeV by the CDF collaboration [143], it is possible to extract the inclusive $\Upsilon(1S)$ fit fraction, F_Υ. The resulting direct to inclusive $\Upsilon(1S)$ ratios, R_C, are also given in Table 8.1. The sub-threshold $b\overline{b}$ cross section is then multiplied by $F_C^{\rm dir} = F_\Upsilon R_C$ to obtain the direct bottomonium cross sections.

As an example calculation, the charmonium x_F distributions are shown in Fig. 8.4 for fixed-target interactions with $p_{\rm lab} = 920$ GeV. The cross sections are identical save for the scale factor, F_C.

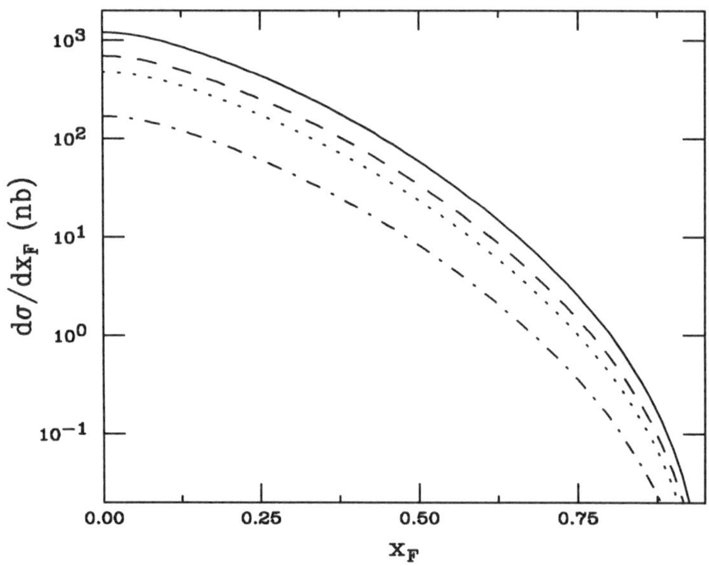

Figure 8.4: Charmonium x_F distributions in pp interactions at 920 GeV in the CEM. The total J/ψ (solid), direct J/ψ (dashed), ψ' (dot-dashed) and χ_c (dotted) distributions are given. The distributions are symmetric around $x_F = 0$.

8.3.2 Nonrelativistic QCD

In both heavy-quarkonium decays and hard-scattering production, large energy-momentum scales appear. The heavy quark mass, m_Q, is much larger than Λ_{QCD}, and, in the case of production, the transverse momentum p_T can be much larger than Λ_{QCD} as well. Thus the associated values of α_s are much less than one, $\alpha_s(m_c) \approx 0.25$ and $\alpha_s(m_b) \approx 0.18$. Therefore, one might hope that it would be possible to calculate the rates for heavy quarkonium production and decay accurately in perturbation theory. However, there are clearly low momentum, nonperturbative effects associated with the dynamics of the quarkonium bound state that make the direct application of perturbation theory questionable.

In order to make use of perturbative methods, one must first factorize the short distance/high momentum, perturbative effects from the long distance/low momentum, nonperturbative effects. One convenient way to carry out this separation is through the use of the effective field theory Nonrelativistic QCD (NRQCD) [134, 144, 145]. NRQCD reproduces full QCD accurately at momentum scales of order $m_Q v$ and smaller, where v is heavy quark velocity in the bound state in the center-of-mass frame, $v^2 \approx 0.3$ for charmonium, and $v^2 \approx 0.1$ for bottomonium. Virtual processes involving momentum scales of order m_Q and larger can affect the lower momentum processes. Their effects are taken into account through the short distance coefficients of the operators that appear in the NRQCD action.

In this approach, the partonic quarkonium production cross section can be written as a sum of the products of NRQCD matrix elements and short-distance coefficients:

$$\sigma(C) = \sum_n \frac{F_n(\Lambda)}{m_Q^{d_n-4}} \langle 0|\mathcal{O}_n^C(\Lambda)|0\rangle \ . \tag{8.21}$$

Here C is the quarkonium state, Λ is the ultraviolet cutoff of the effective theory, the F_n are short distance, perturbative coefficients, and the $\mathcal{O}_n^H$ are nonperturbative operators with mass dimensions d_n.

The short distance coefficients $F_n(\Lambda)$ are essentially the process-dependent partonic cross sections to make a $Q\overline{Q}$ pair. The $Q\overline{Q}$ pair can be produced in a color singlet state or in a color octet state. The short distance coefficients are determined by matching the square of the production amplitude in NRQCD to full QCD. Because the $Q\overline{Q}$

production scale is of order m_Q or greater, this matching can be carried out in perturbation theory.

The operators in Eq. (8.21) create a $Q\overline{Q}$ pair, project it onto an intermediate state that consists of a heavy quarkonium plus anything, and then annihilate the $Q\overline{Q}$ pair. The vacuum matrix element of such an operator is the probability for a $Q\overline{Q}$ pair to form a quarkonium plus anything. These matrix elements are somewhat analogous to fragmentation functions, discussed in the next chapter, that contain all of the nonperturbative physics associated with evolution of the $Q\overline{Q}$ pair into a quarkonium state.

Both color singlet and color octet operators appear in Eq. (8.21). They correspond, respectively, to the evolution of a $Q\overline{Q}$ pair in a relative color singlet or a color octet state into color singlet quarkonium. Note that if all the color octet contributions in Eq. (8.21) are excluded, we have the color singlet model [146].

The x_F distribution of charmonium state C in NRQCD is

$$\begin{aligned} \frac{d\sigma_C}{dx_F} &= \sum_{i,j} \sum_n \int_0^1 dx_1 dx_2 \delta(x_F - x_1 + x_2) \\ &\quad \times f_i^p(x_1, \mu^2) f_j^A(x_2, \mu^2) \frac{F_n^C(\Lambda)}{m_Q^{d_n - 1}} \langle \mathcal{O}_n^C \rangle \,. \end{aligned} \tag{8.22}$$

The parameters determined for fixed-target hadroproduction in Ref. [147] are used. The expansion coefficients, $\langle \mathcal{O}_n^C \rangle$, are fit to data. The octet contributions calculated for fixed-target total cross sections [147] are quite different than the octet contributions found by fitting the high p_T Tevatron data [127]. That is why we remarked at the beginning of the section that the CEM can best reproduce the total yields. So far, for hadroproduction, the CEM gives both the total yields and the p_T distributions with the same parameters while the NRQCD approach does not. However, that could change.

Direct J/ψ production has only contributions from gg fusion and $q\overline{q}$ annihilation, as in the CEM. The $q\overline{q}$ contribution is all octet while the gg component is a combination of octet and singlet production. On the other hand, a color singlet χ_c can be formed from two gluons so that χ_c production is predominantly color singlet. In addition, the

χ_{c1} has a singlet contribution from gq scattering at $\mathcal{O}(\alpha_s^3)$. Only the $q\overline{q}$ channel contributes to color octet χ_c production. Thus the largest singlet contribution to total J/ψ production is from χ_{cJ} decays. The total J/ψ x_F distribution then includes radiative decays of the three χ_{cJ} states and hadronic decays of the ψ',

$$\begin{aligned}\frac{d\sigma_{J/\psi}}{dx_F} &= \frac{d\sigma^{\rm dir}_{J/\psi}}{dx_F} + \sum_{J=0}^{2} B(\chi_{cJ} \to J/\psi X)\frac{d\sigma_{\chi_{cJ}}}{dx_F} \\ &\quad + B(\psi' \to J/\psi X)\frac{d\sigma_{\psi'}}{dx_F} \ . \end{aligned} \tag{8.23}$$

In Fig. 8.5 we show an example of the relative singlet and octet contributions to total J/ψ, direct J/ψ, ψ' and the sum of the three χ_c contributions to J/ψ production at $p_{\rm lab} = 920$ GeV. For predictions of the NRQCD p_T distributions at nuclear collider energies, see Ref. [127].

8.4 Quarkonium suppression by a quark-gluon plasma

In this section, J/ψ break up in a quark-gluon plasma is discussed. The mechanism of plasma screening is first introduced, followed by the characteristics of the suppression as a function of quarkonium p_T and plasma energy density, ϵ, for initial conditions similar to those expected at the CERN SPS.

8.4.1 Plasma screening

The $T = 0$ potential in Eq. (8.1) is modified at finite temperature due to color screening. Color screening can be quantified in the potential either through the Debye screening radius, $r_D(T)$, or screening mass, $\mu(T) \sim 1/r_D(T)$. In a quark-gluon plasma, when the screening radius of the quark color charge is less than the hadronic radius, r_C, the valence quarks of the quarkonium state can no longer 'feel' each other and are unable to form a bound state.

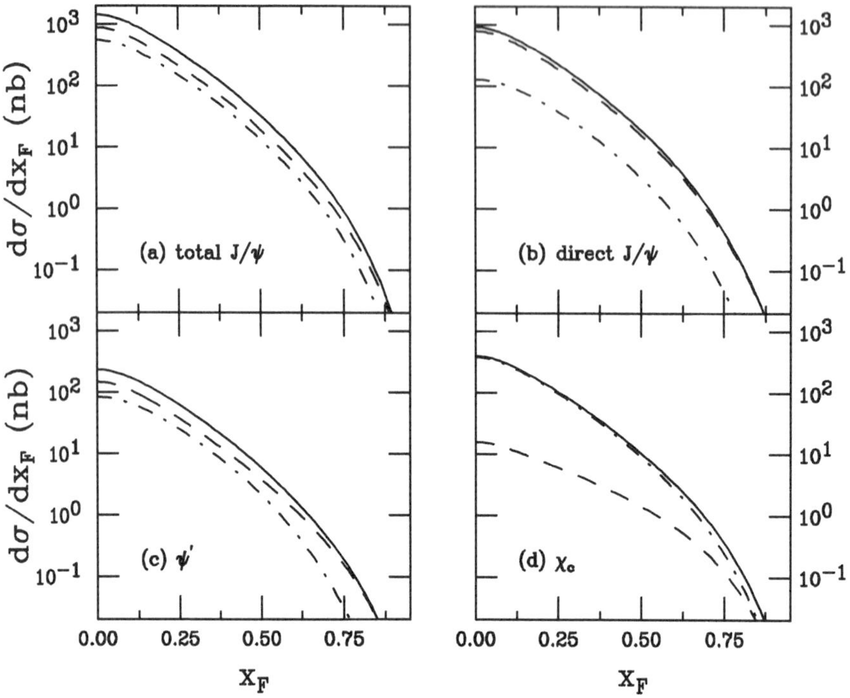

Figure 8.5: The NRQCD charmonium x_F distributions in pp interactions at 920 GeV. The contributions from octet (dashed) and singlet (dot-dashed) production are given along with the total (solid). The total J/ψ (a), direct J/ψ (b), ψ' (c) and the three χ_c states (d) x_F distributions are shown. The distributions are all symmetric around $x_F = 0$.

The Debye screening radius is determined in lattice gauge theory by calculating the correlation function between a static $Q\overline{Q}$ pair in a gluon heat bath. There are no dynamical quarks in the system for the pair to interact with to directly screen the color charge, similar to the calculation of the Wilson loop in Chapter 6. However, one-gluon exchanges between the quarks will be affected by the heat bath since the exchanged gluons acquire a mass at finite temperature. The correlation function, $\Gamma(r, T)$, is expressed as

$$\Gamma(r, T) \sim \exp[-r/r_D(T)] \tag{8.24}$$

where r is the separation between the Q and $\overline{Q}$, T is the temperature of the heat bath and $r_D(T)$ is the screening radius as a function of temperature. When $T \leq T_c$, $r_D \to \infty$, implying no screening. When $T > T_c$, $r_D(T)$ decreases rapidly. At $T/T_c \approx 1.5$, $r_D \sim 0.2 - 0.3$ fm. We can use the idea of screening to naively modify the quark-antiquark potential. The Coulomb-like part of Eq. (8.1) then becomes

$$-\frac{\alpha_c}{r} \exp[-r/r_D(T)] \ . \tag{8.25}$$

The confining term is likewise modified so that

$$\sigma r_D(T)(1 - \exp[-r/r_D(T)]) \ . \tag{8.26}$$

The full finite-temperature potential is the sum of the two terms,

$$\begin{aligned} V(r,T) &= \sigma r_D(T)(1 - \exp[-r/r_D(T)]) \\ &\quad - \frac{\alpha_c}{r} \exp[-r/r_D(T)] \ . \end{aligned} \tag{8.27}$$

EXAMPLE: Show that when $r_D \to \infty$, the $T = 0$ result is obtained. When $r_D \to \infty$, $r/r_D \to 0$. The confining term can be expanded, keeping only the first two terms,

$$\sigma r_D(T)(1 - \exp[-r/r_D(T)]) \to \tag{8.28}$$

$$\sigma r_D(T)[1 - (1 - r/r_D + \ldots)] = \sigma r \ . \tag{8.29}$$

The Coulomb-like term is

$$-\frac{\alpha_c}{r} \exp[-r/r_D(T)] \to -\frac{\alpha_c}{r} \ . \tag{8.30}$$

Thus Eq. (8.1) is regained.

It is generally more convenient to write the potential in terms of the screening mass since it is linearly proportional to the temperature. In this case [128]

$$V(r,T) = \frac{\sigma}{\mu(T)}(1 - e^{-\mu(T)r}) - \frac{\alpha_c}{r} e^{-\mu(T)r} \ . \tag{8.31}$$

For increasing values of $\mu(T)$, the confining potential is less effective. Figure 8.6 shows the potential in Eq. (8.31) as a function of distance at both $T = 0$ (solid curve) and for increasing values of $\mu(T)$. As the temperature increases, the potential becomes increasingly more Coulomb-like with a lower barrier at large r. The finite-temperature potential has also been calculated more quantitatively on the lattice [148].

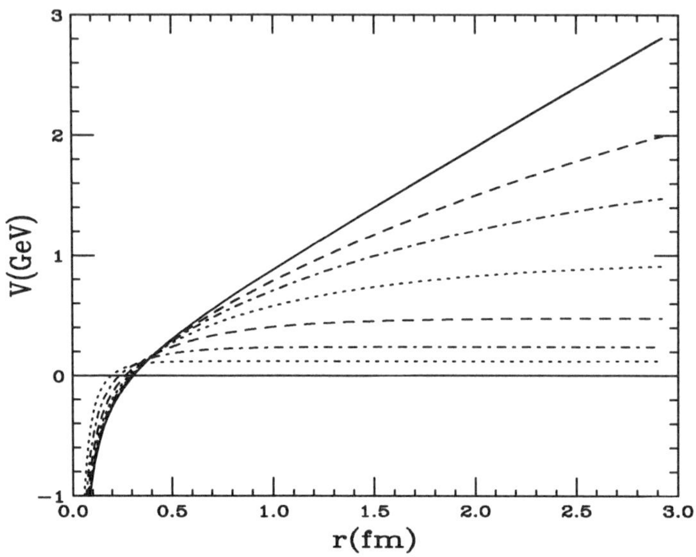

Figure 8.6: The potential from Eq. (8.31) for $T = 0$ (solid curve) and for finite temperature with $\mu = 0.05$, 0.1, 0.2, 0.4, 0.8 and 1.6 GeV respectively.

The temperature at which the bound state radius, r_C, is equal to the screening length r_D is the dissociation temperature, T_D. The value of T_D is very sensitive to the functional form of $\mu(T)$, as we will discuss in more detail soon. When $r_C > r_D(T)$ at $T > T_c$, the bound state will not be formed. Since the quarkonium radii differ, some states will break up at lower temperatures than others. The lowest mass S states are more tightly bound. Hence the J/ψ and $\Upsilon(1S)$, with the smallest values of r_C at $T = 0$, have higher values of T_D. One can envision a hierarchy of suppression with the higher mass states which feed down to the J/ψ and $\Upsilon(1S)$ being broken up first. Their contribution to the inclusive

J/ψ and $\Upsilon(1S)$ rates would thus be lost before the temperature would be high enough to break up these lower mass states.

The time scale is also an important part of quarkonium binding and dissociation. The time needed to produce the bound state, the formation time, τ_F, is much longer than the production time scale, τ_P. The production time depends on the quark mass,

$$\tau_P \sim \frac{1}{2m_Q} . \tag{8.32}$$

For charm quarks, this time is $\approx 0.055 - 0.08$ with $1.2 < m_c < 1.8$ GeV and, for bottom quarks it is ≈ 0.021 with $4.5 < m_b < 5$ GeV. On the other hand, the formation time is sensitive to the identity of the final quarkonium state. It depends on the bound-state radius and the relative velocity of the $Q\overline{Q}$ pair,

$$\tau_F = r_C \frac{E_{Q\overline{Q}}}{p_{Q\overline{Q}}} . \tag{8.33}$$

The likelihood of color screening of quarkonium production also depends on the formation time of the quark-gluon plasma. If the pre-quarkonium state does not interact with its environment until after τ_F, it could escape the system without interaction if the quark-gluon plasma formation time is less than τ_F.

The binding energy for a given state is

$$E(r,T) = 2m_Q + \frac{c}{m_Q r^2} + V(r,T) . \tag{8.34}$$

The constant c in the kinetic energy is of order 1, $\langle p^2 \rangle \langle r^2 \rangle = c \approx \mathcal{O}(1)$ [128]. Minimizing $E(r,T)$ gives the radius of the bound state at each T. For $\mu(T)$ above the critical value, μ_D, there is no longer a minimum and the screening has become strong enough to prevent the formation of the resonance at temperature T_D where $\mu(T_D) = \mu_D$. The values of μ_D for the quarkonium states are given in Table 8.2 along with the mass, radius and formation time of the state. (Note that this radius may actually be more like the diameter of the state. Later calculations of the radii give about half these values [149].

Table 8.2: Properties of the quarkonium states both at $T = 0$ and $T = T_D$, taken from Ref. [128]. The masses, radii and formation times at $T = 0$ are obtained from the solution to the Schrodinger equation with the potential of Eq. (8.1). The screening mass μ_D is independent of the actual value of T_D determined by the functional form of $\mu(T)$.

	J/ψ	ψ'	$\chi_c(1P)$	$\Upsilon(1S)$	$\Upsilon(2S)$	$\chi_b(1P)$
M (GeV)	3.07	3.698	3.5	9.445	10.004	9.897
r (fm)	0.453	0.875	0.696	0.226	0.509	0.408
τ_F (fm)	0.89	1.5	2.0	0.76	1.9	2.6
μ_D (GeV)	0.699	0.357	0.342	1.565	0.671	0.558

Perturbative estimates of screening suggest that $\mu(T) \propto gT$ [150],

$$\frac{\mu(T)}{T_c} = \sqrt{1 + \frac{n_f}{6}} g \left(\frac{T}{T_c}\right) \frac{T}{T_c} \tag{8.35}$$

where the temperature-dependent running coupling constant is

$$g^2 \left(\frac{T}{T_c}\right) = \frac{48\pi^2}{(33 - 2n_f) \ln F^2} , \tag{8.36}$$

with $F = K(T/T_c)(T_c/\Lambda_{\overline{MS}})$ and K is also in principle temperature dependent [151]. (Note that $\alpha_s = g^2/4\pi$.) In SU(3) gauge theory, $T_c = 260$ MeV [152] and $T_c/\Lambda_{\overline{\rm MS}} = 1.03 \pm 0.19$ [153]. A fit to the heavy quark potential in the high temperature limit, $T \gg T_c$, yields the constant $K \approx 33.8$ [151]. Depending on the form of $K(T/T_c)$ and the number of light quark flavors, only some or all of the charmonium states will break up near $T_c \sim 170$ MeV. For $n_f = 3$, the χ_c and ψ' will break sequentially while the J/ψ will not.

In their prediction of J/ψ suppression, Matsui and Satz [120] used a parameterization based on SU(N) lattice simulations [154]

$$\frac{\mu(T)}{T_c} \simeq 4\frac{T}{T_c} . \tag{8.37}$$

The two evaluations of $\mu(T)$ are compared in Fig. 8.7. The largest difference in the two results at $T = T_c$ is due to the higher T_c in the

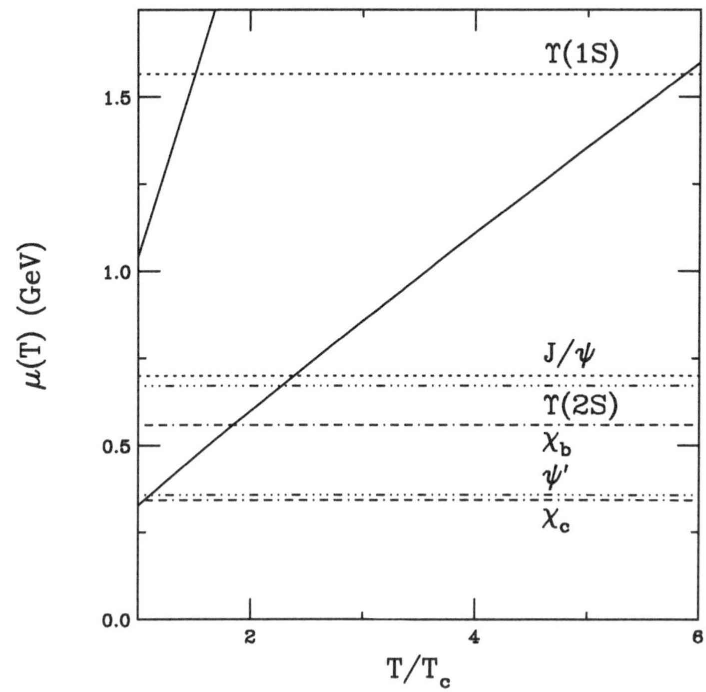

Figure 8.7: The lower solid curve is calculated with Eq. (8.35), using $n_f = 3$, $T_c = 170$ MeV and $K = 33.8$ while the upper solid curve shows the result for Eq. (8.37) with $T_c = 260$ MeV. The values of μ_D for the quarkonium states are indicated by the dotted lines for $1S$ states, dot-dot-dot-dashed lines for $2S$ states and dot-dash-dashed lines for the $1P$ states.

pure gauge case. If $T_c = 170$ MeV rather than 260 MeV, the J/ψ would not be suppressed by color screening at $T = T_c$, as is the case for the pure gauge T_c. Since the sequential screening case seems to be the most compatible with the CERN SPS data, we take $n_f = 3$, $T_c = 170$ MeV and $\mu(T) \propto gT$ for further illustration. The actual values of T_D for these two schematic cases are given in Table 8.3.

Note that the results from Ref. [128] are representative of a wide range of calculations, some more sophisticated than others. Newer results along the same lines, also including the $\chi_b(2P)$ and $\Upsilon(3S)$ states,

Table 8.3: The values of T_D for the two choices of $\mu(T)$, Eq. (8.35) with $n_f = 3$ and $K = 33.8$ and the pure gluon SU(N) case with no light quarks, Eq. (8.37).

State	$n_f = 3$	Eq. (8.37)
J/ψ	406	260
ψ'	189	260
χ_c	178	260
$\Upsilon(1S)$	994	391
$\Upsilon(2S)$	386	260
χ_b	314	260

can be found in Ref. [149]. The trend remains relatively the same: The χ_c and ψ' states, as well as the $\chi_b(2P)$ and $\Upsilon(3S)$, break up relatively close to T_c. The $\chi_b(1P)$ and $\Upsilon(2S)$ break up due to color screening at $T < 2T_c$. The $1S$ states remain bound to higher temperatures: the J/ψ until $T > 2T_c$ and the $\Upsilon(1S)$ until $T > 4T_c$.

Of course, static calculations alone are not the whole story. The reduction of the potential barrier shown in Fig. 8.6 will lead to a lower binding energy and larger radii. These more weakly bound states will be more susceptible to the surrounding medium and may not survive their trip through the medium. Instead they may break up in interactions with thermalized quarks and gluons.

8.4.2 Transverse momentum and energy density dependence

Total J/ψ suppression [120] would only occur if the quark-gluon plasma were of infinite temporal and spatial extent because the $Q\overline{Q}$ pairs would not form resonances as long as $T > T_D$. The suppression would then be essentially independent of the quark-gluon plasma energy density, ϵ, and the transverse momentum of the bound state if $T > T_D$. Of course, both the lifetime and the volume of the quark-gluon plasma are finite. In the simple model illustrated here, the formation time of the bound state, τ_F, is particularly important. The competition between τ_F and the lifetime of the quark-gluon plasma affect the p_T and ϵ dependence of the quarkonium states.

The finite lifetime of the plasma sets an upper limit on the p_T at which quarkonium states are suppressed. In the quark-gluon plasma rest frame, a transversely moving $Q\overline{Q}$ will form a bound state at

$$t_F = \gamma\tau_F = \tau_F\sqrt{1+(p_T/M)^2} \ . \tag{8.38}$$

When the quark-gluon plasma has cooled below T_D, bound state formation is again possible. The time at which this occurs can be estimated from the Bjorken model [74] since

$$s_D t_D = s_0 t_0 \ . \tag{8.39}$$

Because the entropy density, s, is proportional to T^3, the time at which the temperature drops below T_D is

$$t_D = t_0 \left(\frac{T(\tau_0)}{T_D}\right)^3 \ . \tag{8.40}$$

Thus color screening will be effective for the time interval $t_D - t_0$. If this time is longer than the quarkonium formation time, $t_D/t_F > 1$, the bound states will not form and quarkonium production will be suppressed. The maximum p_T at which suppression is possible can be determined from the inversion of Eq. (8.38),

$$p_{T\mathrm{m}} = M\sqrt{(t_D/\tau_F)^2 - 1} \ . \tag{8.41}$$

EXAMPLE: Calculate t_D and $p_{T\mathrm{m}}$ for the charmonium states at the CERN SPS assuming that the formation time of the quark-gluon plasma is $t_0 = 1$ fm and the initial temperature is 300 MeV.
According to Table 8.3, for $n_f = 3$, the J/ψ, ψ' and χ_c break up at $T_D = 406$, 189 and 178 MeV respectively. If the initial temperature is $T(\tau_0) = 300$ MeV, the J/ψ will remain bound. The ψ' and χ_c will, however, break up by color screening. The ψ' can be suppressed until $t_D = (300/189)^3 = 4$ fm while the lower T_D for the χ_c suggests suppression until $t_D = 4.8$ fm.

In both cases, $t_D > \tau_F$. The shorter formation time of the ψ' as well as its higher mass means that it will experience plasma suppression at higher

p_T than will the χ_c. Here $p_{T_{\rm m}} = 3.686\sqrt{(4/1.5)^2 - 1} = 9.14$ GeV for the ψ' while $p_{T_{\rm m}} = 7.61$ GeV for the χ_c.

In finite systems, such as those produced in heavy-ion collisions, the suppression also depends on the size of the system. Schematic calculations of the p_T dependence of the suppression following Ref. [155] are presented here. (See also Refs. [156, 157].) The $Q\overline{Q}$ pair is assumed to move only in the transverse plane. The initial entropy density and breakup time depend on the plasma radius,

$$\frac{s_0(r)}{s_0} = \frac{t_D(r)}{t_D(0)} = \left(1 - \left(\frac{r}{R}\right)^2\right)^{1/4} , \tag{8.42}$$

where $t_D(0)$ is the value of t_D calculated in Eq. (8.40) for resonances produced at the center of the system and R is the transverse radius of the plasma, assumed to be no larger than the radius of the nucleus, $R = R_A \simeq 1.2A^{1/3}$ fm. At any time t during the evolution of the quark-gluon plasma, the spatial boundary of the screening region is located at the transverse radius r_S given by $t_D(r_S) = t$,

$$r_S = R\left[1 - \left(\frac{t}{t_D(0)}\right)^{1/\beta}\right]^{1/2} . \tag{8.43}$$

In general, if the $Q\overline{Q}$ pair is created with transverse position $\mathbf{x}^\mu = (0, \vec{r}, 0)$ and momentum $\mathbf{p}^\mu = (\sqrt{M^2 + p_T^2}, \vec{p}_T, 0)$, at $t = \tau_F$ the pair will form a bound state at $\mathbf{x}^\mu = (\tau_F\sqrt{1 + (p_T/M)^2}, \vec{r} + \tau_F\vec{p}_T/M, 0)$. If $|\vec{r} + \tau_F\vec{p}_T/M| \geq r_S$, the $Q\overline{Q}$ pair escapes and forms the bound state. The probability that the $Q\overline{Q}$ pair forms a bound state, $S(p_T)$, is the ratio of the bound states produced by the escaped pairs relative to the maximum number of bound states that would be formed in the absence of a plasma,

$$S(p_T) = \frac{\int_0^R dr r\rho(r)\theta(r, p_T)}{\pi\int_0^R dr r\rho(r)} , \tag{8.44}$$

where the plasma density profile is $\rho(r) = (1 - (r/R)^2)^{1/2}$ and

$$\theta(r, p_T) = \begin{cases} \pi & z \leq -1 \\ \cos^{-1} z & |z| < 1 \\ 0 & z \geq 1 \end{cases} , \tag{8.45}$$

with

$$z = \frac{r_S^2 - r^2 - (\tau_F p_T/M)^2}{2\tau_F r(p_T/M)} \ . \tag{8.46}$$

When $z \geq 1$ the pair cannot escape whereas when $z \leq -1$ the pair always escapes.

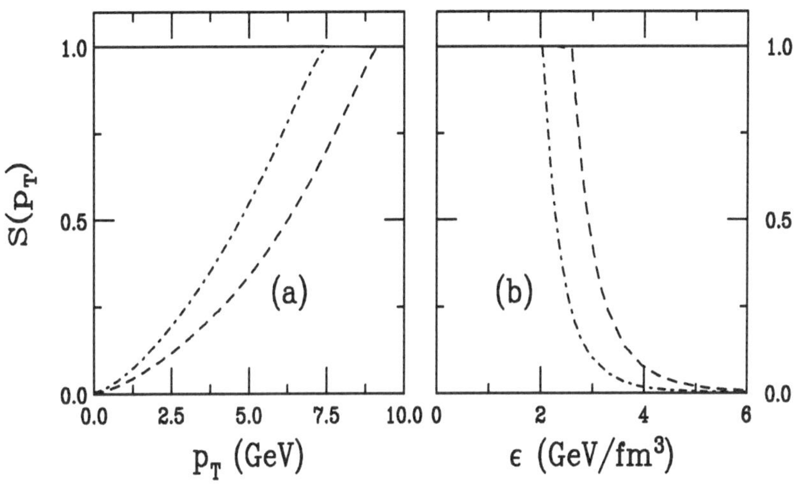

Figure 8.8: The survival probability as a function of (a) p_T and (b) ϵ for $T(\tau_0) = 300$ MeV, $t_0 = 1$ fm and $R = R_S$. The solid curves show the result for the J/ψ, the dashed for the ψ' and the dot-dashed the χ_c. When $S(p_T) = 1$ there is no suppression.

The survival probability, $S(p_T)$, for the charmonium resonances assuming $T_0 = 300$ MeV, $t_0 = 1$ fm and a quark-gluon plasma with transverse radius equal to the radius of a sulfur nucleus ($A = 32$) is shown in Fig. 8.8(a). The suppression, when it occurs, is nearly total at $p_T \approx 0$. No details of the collision geometry are included which would tend to reduce the overall suppression near $p_T = 0$. The J/ψ is not suppressed because the initial temperature is not high enough.

The survival probability as a function of the plasma energy density, ϵ, can be calculated analogously to Eq. (8.44). Since the entropy is proportional to the entropy density, $s \propto \epsilon/T$ for an ideal gas of massless gluons and/or quarks, $t/t_D(0)$ can be replaced in Eq. (8.43) by $\epsilon_D(0)/\epsilon$ where ϵ_D is the energy density at $T = T_D$. In an ideal gas, the energy density is proportional to the number of degrees of freedom in the plasma, $\gamma_{\rm qgp}$,

$$\epsilon = \frac{g_{\rm qgp}\pi^2}{30}\left(\frac{T}{\hbar c}\right)^3 T \, . \tag{8.47}$$

Suppression will occur when the energy density is above that needed to breakup the bound state, $\epsilon > \epsilon_D$, and also above the critical energy density for plasma production, $\epsilon > \epsilon_c$. When $T_c = 170$ MeV, the critical energy density is $1.3 - 2.1$ GeV/fm^3, depending on n_f. In Fig. 8.8(b), $S(\epsilon)$ is shown for the same initial conditions as in Fig. 8.8 with $p_T \approx 0$ to illustrate the maximum suppression. The χ_c is suppressed first, at $\epsilon_D^{\chi_c} \approx 2$ GeV/fm^3, followed by the ψ' at $\epsilon_D^{\psi'} \approx 2.6$ GeV/fm^3. Recall that although ϵ is not a direct observable, in the scaling expansion it is proportional to the global transverse energy of the system, E_T. Realistic assumptions of the initial energy density would therefore result in less than total suppression at $p_T \approx 0$. Then, for $n_f = 3$ and $\epsilon = 3$ GeV/fm^3, $\approx 60\%$ of the ψ' states are suppressed while $\approx 90\%$ of the χ_c states are suppressed.

8.4.3 Quarkonium regeneration

At the CERN SPS, quarkonium production is truly rare. For $\sqrt{S_{NN}} = 20$ GeV, the total $c\bar{c}$ cross section is on the order of a few microbarns while the J/ψ production cross section is about a factor of 50 smaller. A $c\bar{c}$ pair will be produced in one out of about 10 Pb+Pb collisions while a J/ψ will be produced in one out of 500 Pb+Pb collisions. However, at the maximum RHIC energy of $\sqrt{S_{NN}} = 200$ GeV, the $c\bar{c}$ cross section is around a factor of 100 higher than at the SPS so that nearly 10 $c\bar{c}$ pairs will be produced in a typical Au+Au collision at RHIC. In this case, there are a significant number of c and $\bar{c}$ quarks in the medium which could come close enough in phase space to form a J/ψ through reactions like $c\bar{c} \to J/\psi$, providing a mechanism by which J/ψ suppression could become J/ψ enhancement. At the LHC,

$b\bar{b}$ production will be abundant enough for secondary $\Upsilon(1S)$ production to be possible.

These secondary charmonium states would have different kinematic distributions than those produced in the first nucleon-nucleon collisions. Since the center-of-mass energy of parton interactions within a quark-gluon plasma will be lower than in hadronic first collisions, their rapidity distributions will be narrower and the average p_T of the final-state charmonium would be smaller. Such kinematic changes could be observable, particularly if the initially-produced quarkonium has been broken up by color screening.

8.5 Quarkonium suppression by hadrons

The nuclear target dependence of hard processes is usually parameterized as a power law, based on empirical observations. In hA collisions,

$$\sigma_{hA} = \sigma_{hN} A^\alpha \; , \tag{8.48}$$

while in AB collisions,

$$\sigma_{AB} = \sigma_{NN} (AB)^\alpha \; , \tag{8.49}$$

where the exponent α characterizes all nuclear effects. The exponent α depends on x_F, p_T, the scale μ^2 and perhaps also $\sqrt{S}$.

Measurements of the J/ψ A dependence show that, in the integrated A dependence, α is typically less than unity. It decreases for $x_F > 0.25$ [124] but seems to be somewhat x_F independent for $x_F < 0$ [158]. The value of α increases with p_T. Indeed the $\langle p_T^2 \rangle$ of the J/ψ increases with A, an effect known as p_T-broadening. Similar trends have been observed for both Drell-Yan dileptons and Υ production [123]. The broadening effect is smaller for Drell-Yan than J/ψ while the Υ broadening is larger. See the review in Ref. [159] for more discussion and references.

The concept of p_T-broadening in hadron-nucleus collisions can be described rather simply. In an hA collision, there exists a fixed probability for the parton to scatter elastically in each nucleon-nucleon subcollision. The partons thus undergo a random walk in momentum space, increasing $\langle p_T^2 \rangle$ in proportion to the number of sub-collisions suffered by the parton [160]. A nucleon at impact parameter b in the target can undergo

$$n_A = \sigma_{NN} \int_{-\infty}^{z} dz' \; \rho_A(\vec{b}, z'), \tag{8.50}$$

sub-collisions prior to the hard interaction at longitudinal position z, where ρ_A is the nuclear density and σ_{NN} is the inelastic NN cross section. Thus in the hA collision, the $\langle p_T^2 \rangle$ is increased by

$$\Delta p_T^2 \equiv \langle p_T^2 \rangle - \langle p_T^2 \rangle_{NN} = \lambda^2(\langle n_A \rangle - 1) \; . \tag{8.51}$$

The factor λ^2, proportional to the elastic scattering probability, determines the incremental increase of the projectile parton p_T^2 due to each sub-collision. The difference, $\langle n_A \rangle - 1$, is taken to eliminate the final, hard sub-collision where

$$\langle n_A \rangle = [\int d^2b dz \rho_A]^{-1} \int d^2b dz \rho_A n_A \; . \tag{8.52}$$

The average p_T^2 in NN collisions is independent of A but increases with energy [161]. For a hard sphere nucleus, $\rho_A = \rho_0 \Theta(R_A - b)$ with radius $R_A \approx 1.2 \; A^{1/3}$ fm,

$$\langle n_A \rangle \approx 3\sigma_{NN}\rho_0 R_A/2 \approx 0.77 \; A^{1/3} \; . \tag{8.53}$$

If there are no nuclear effects, $\alpha = 1$, as determined in Chapter 3. However, it is not really true that, if $\alpha \equiv 1$, $\sigma_{hA} \equiv A\sigma_{hN}$ for every process. For example, the Drell-Yan process is an exception, as we already showed in Figs. 2.23 and 2.24 of Chapter 2. Thus the neutron and proton parton densities have to be weighted correctly by the neutron and proton numbers. In the case of quarkonium, where production is dominated by gg interactions out to very far forward x_F and p_T, the linear A dependence would hold rather precisely if counting neutrons and protons was all that was involved.

It was also shown in Chapter 2 that the parton densities in the nucleus do not behave like those in the free proton but are modified as a function of x and scale. While this effect is present, it is not strong enough alone to produce the observed x_F dependence of α for the J/ψ. In fact, at the fixed-target energies of the CERN SPS, the average x values are such that $x_F = 0$ is near the antishadowing region, leading to enhancement rather than suppression for nuclear parton density effects alone.

Another mechanism that can affect the A dependence is for the incoming parton to lose energy in the longitudinal direction by scattering in the target before the hard interaction that produces the quarkonium state. The projectile parton, assumed to have no initial transverse momentum, loses energy in the z direction (decreasing its x value relative to that before entering the target) but gains some p_T, leading to p_T-broadening, as described earlier. Thus energy loss and increasing $\langle p_T^2 \rangle$ can be thought of as arising from the same process. Note also that while shadowing can change the overall yield, energy loss shifts the final-state phase space without affecting the total yield.

All these effects can depend on the initial state, before the hard interaction occurs. However, the quarkonium states can also interact with hadrons in the final state, either the nucleons already in the system, 'nuclear absorption' or hadrons produced along with the quarkonium and comoving with it, 'comovers'. Here we will show one simplistic way of calculating a general absorption cross section in a pA interaction. For an in-depth discussion of the effects on J/ψ and ψ' production, see Ref. [162]. For details of the absorption models, see Ref. [163].

8.5.1 Nuclear Absorption

The effect of nuclear absorption alone on the quarkonium production cross section in hA collisions may be expressed as

$$\sigma_{hA \to C} = \sigma_{hN \to C} \int d^2b \, T_A^{\rm eff}(b) \; , \tag{8.54}$$

where C is the quarkonium state, b is the impact parameter, $T_A^{\rm eff}(b)$ is the effective nuclear profile function,

$$T_A^{\rm eff}(b) = \int_{-\infty}^{\infty} dz \, \rho_A(b,z) \exp\left\{ - \int_z^{\infty} dz' \rho_A(b,z') \sigma_{CN}(z'-z) \right\} \tag{8.55}$$

and σ_{CN} is the cross section for quarkonium absorption by nucleons. The nuclear density profile is $T_A(b) = \int_{-\infty}^{\infty} dz \rho_A(b,z)$. The exponential factor is the probability for the $Q\overline{Q}$ pair to avoid nuclear absorption and form quarkonium, called the nuclear absorption survival probability.

EXAMPLE: Calculate α and thus σ_{CN} from the integral of the survival probability over impact parameter, as in Eq. (8.55).

There is a simple way to relate σ_{CN} to a fitted value of α. For illustrative purposes, assume that the target is a hard sphere so that $\rho_A = \rho_0 = 3/(4\pi r_0^3)$ and neglect any spatial dependence of σ_{CN}. We expand the exponential factor in Eq. (8.55), keeping only terms linear in σ_{CN}, and integrate $T_A^{\rm eff}$ over impact parameter. Let

$$\int d^2b T_A^{\rm eff}(b) = A_{\rm eff} \tag{8.56}$$

so that

$$A_{\rm eff} = \int d^2b \int dz\, \rho_0 \left[1 - \int_z^{z_{\rm m}} dz' \rho_0 \sigma_{CN} + \cdots \right] \tag{8.57}$$

where $z_{\rm m} = \sqrt{R_A^2 - b^2}$. The integral over the first term is defined to give A. The second term is simply

$$\begin{aligned} -\sigma_{CN}\rho_0^2 \int d^2b \int_{-z_{\rm m}}^{z_{\rm m}} dz \int_z^{z_{\rm m}} dz' &= -\sigma_{CN}\rho_0^2 \int d^2b \int_{-z_{\rm m}}^{z_{\rm m}} dz (z_{\rm m} - z) \\ &= -\sigma_{CN}\rho_0^2 \int d^2b \left(z_{\rm m} z - \frac{1}{2} z^2\right)\Big|_{-z_{\rm m}}^{z_{\rm m}} \\ &= -2\pi\sigma_{CN}\rho_0^2 \int_0^{R_A} dbb \left(\sqrt{R_A^2 - b^2}\right)^2 \\ &= -\pi\sigma_{CN}\rho_0^2 R_A^4 \,. \end{aligned} \tag{8.58}$$

After substituting for ρ_0, the effective value of A is

$$A_{\rm eff} = A\left(1 - \frac{9\sigma_{CN}}{16\pi r_0^2} A^{1/3}\right) . \tag{8.59}$$

If the second term is small, we can re-exponentiate, giving [164]

$$A_{\rm eff} = A \exp\left[- \frac{9\sigma_{CN}}{16\pi r_0^2} A^{1/3} \right] . \tag{8.60}$$

For large targets with $A > 50$, $A^{1/3} \approx \ln A$. Then

$$\frac{\sigma_{hA\to C}}{\sigma_{hN\to C}} = A_{\rm eff} = A^{1-\eta} = A^{\alpha} \; . \tag{8.61}$$

The relationship between σ_{CN} and α is then

$$\sigma_{CN} = \frac{16\pi r_0^2}{9}(1-\alpha) \; . \tag{8.62}$$

Thus, if the measured α is 0.94, the quarkonium absorption cross section would be 4.8 mb.

There are a lot of assumptions that go into this association, many of them obvious. First, the absorption cross section, while small, is not necessarily small enough to justify keeping only the first term in the expansion of Eq. (8.55). Next, the approximate equality of $A^{1/3}$ and $\ln A$ holds only for sufficiently large A. However, this form is generally applied to all $A > 1$. The still approximate but less simplistic form,

$$\frac{\sigma_{hA\to C}}{\sigma_{hN\to C}} = \exp[-\sigma_{CN}\rho_0 L] \; , \tag{8.63}$$

has also been used where the A dependence enters through the nuclear path length traversed by the charmonium state, $L \equiv (3/4)R_A$, obtained when the integral over the nuclear volume is averaged over the impact parameter b. More advanced models use tabulated nuclear charge density distributions, tabulated in Table 3.1 of Chapter 3 directly in Eq. (8.55) and integrate it numerically.

The three ways of obtaining the nuclear absorption A dependence are compared in Fig. 8.9. The calculations use $\sigma_{CN} = 4.8$ mb, equivalent to $\alpha = 0.94$ in Eq. (8.62). When plotted on a logarithmic scale, the calculation with A^{α} is linear. The exponential form in Eq. (8.63) has concave curvature. The symbols show the results of the full numerical calculation using the measured nuclear charge density distributions, using Eqs. (8.54) and (8.55). While these last results agree fairly well with those of Eq. (8.63), there is some variation since the central density is not fixed to the same value for all nuclei.

All too often, a constant σ_{CN} is assumed, with no additional effects. In a hA interaction, there are enough effects possible that it is not likely

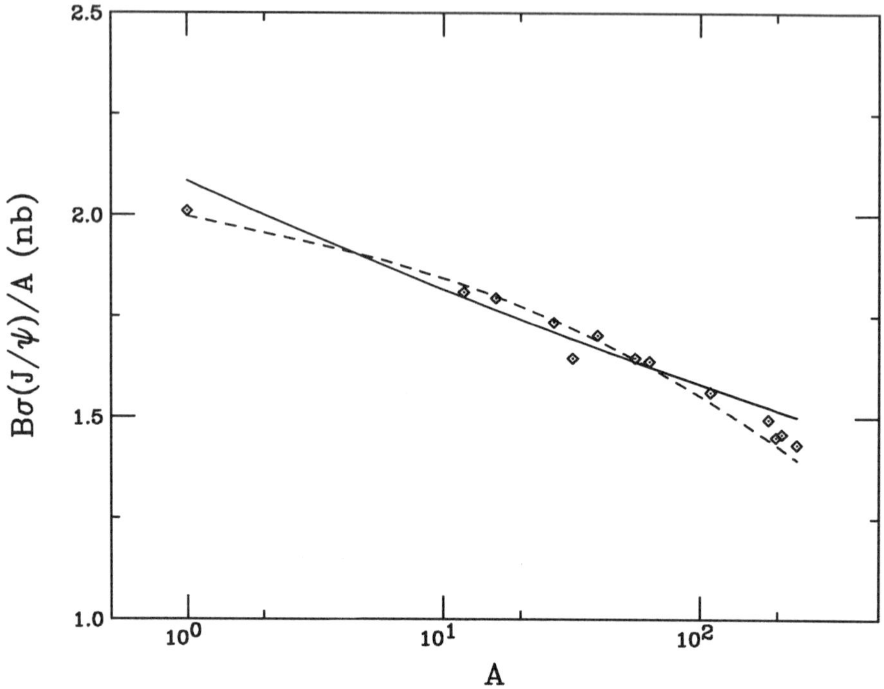

Figure 8.9: The A dependence of J/ψ production is compared for A^α (solid curve), the formula $\exp(-\sigma_{CN}\rho_0 L)$ (dashed curve) and the full numerical result (symbols). All results are for $\sigma_{CN} = 4.8$ mb. The per nucleon production cross section times branching ratio to lepton pairs is shown on the y axis.

that a universal value of a constant cross section can fit all the data as functions of A and $\sqrt{S}$. Instead, in such an analysis, all other effects are included in this 'effective' cross section.

There are also different models of quarkonium formation which depend on whether the $Q\overline{Q}$ is produced in a color octet or color singlet state or some admixture of the two, as in the NRQCD approach. The production mechanism affects not only the time of bound-state formation for each state, as already discussed, but also the strength of the final-state absorption. Most models of color octet production assume that the absorption cross section for all members of the quarkonium family, *e.g.* J/ψ, ψ' and χ_c, have the same asymptotic absorption cross section. However, if color singlet production is dominant, the asymptotic absorption cross sections are also assumed to be different. The difference in the production mechanism is not easy to sort out, particularly when the χ_c A dependence has not been measured with precision [163].

8.5.2 Hadronic comovers

Comoving secondaries, formed after $\tau_0 \sim 1-2$ fm, may also scatter with the $Q\overline{Q}$ pair or the quarkonium state. Because the typical quarkonium formation time is less than τ_0, the final-state meson is assumed to interact with the comovers. A spectator hadron moving with a velocity similar to the quarkonium state enhances the probability of heavy flavor hadron production at the expense of quarkonium production.

The A dependence due to comovers alone is determined from

$$\sigma_{hA\to C} = \sigma_{hN\to C} \int d^2b\, S(b) \ , \tag{8.64}$$

where the probability that the bound state survives its interactions with comovers is [164]

$$S(b) \approx \exp\left\{ - \int d\tau \, \langle \sigma_{C\mathrm{co}} v \rangle n(\tau, b) \right\} \ . \tag{8.65}$$

The parameters are $\sigma_{C\mathrm{co}}$, the quarkonium-comover absorption cross section, v is the velocity of the state relative to the comovers and $n(\tau, b)$ is the density of comovers at time τ and impact parameter b. Integrating over time τ and relating the initial density of the system to the final hadron rapidity density through the use of scaling dynamics with $n_0\tau_0 = (\pi R^2)^{-1}(dN/dy)$ [74], one finds [164]

$$\int d\tau\, n(\tau, b) \approx \frac{1}{\pi R^2} \ln\left(\frac{\tau_I}{\tau_0}\right) \frac{dN}{dy}|_{y=0} \sigma_{hN} T_A(b) \ , \tag{8.66}$$

where $dN/dy|_{y=0}$, the central rapidity density in a *pp* collision is scaled up to *pA* interactions by $\sigma_{hN}T_A(b)$, the number of participant nucleons in the collision. (In an *AB* collision, the multiplicity would be scaled up by the number of participants, $N_{\rm part}$. The effective proper lifetime τ_I over which the comovers, formed at time τ_0, interact with the quarkonium is $\tau_I \sim r_p/v$ and $r_p \sim 0.8$ fm. The shape of the rapidity density from inclusive quarkonium production in pA collisions is unknown. If one assumes a rapidity plateau, as in the longitudinal scaling expansion described in Chapter 5, in inclusive pA data [165], the central rapidity density can be assumed to be constant over the reduced rapidity range of quarkonium production. The rapidity density grows with center-of-mass energy [166].

Since the transverse area over which comovers are produced, πR^2, is approximately equal to σ_{hN}, the survival probability in hA collisions may be recast as

$$S(b) \approx \exp\left\{-\langle\sigma_{\psi {\rm co}} v\rangle \frac{dN}{dy}|_{y=0} \ln\left(\frac{\tau_I}{\tau_0}\right) T_A(b)\right\} . \tag{8.67}$$

Following the expansion and re-exponentiation needed to arrive at Eq. (8.60), the comover A dependence is

$$\begin{aligned} \frac{\sigma_{hA\to C}}{A\sigma_{hN\to C}} &\approx \exp\left\{-\frac{9}{8\pi r_0^2}\langle\sigma_{C{\rm co}} v\rangle \frac{dN}{dy}|_{y=0} \ln\left(\frac{\tau_I}{\tau_0}\right) A^{1/3}\right\} \\ &\equiv \exp\left\{-\beta A^{1/3}\right\} , \end{aligned} \tag{8.68}$$

suggesting that quarkonium-comover interactions have an A dependence very similar to nuclear absorption.

8.5.3 Absorption by Nucleons and Comovers

The total effect of nuclear absorption and comover scattering on the A dependence of quarkonium production may be obtained from the product of the two survival probabilities,

$$\sigma_{hA\to C} = \sigma_{hN\to C} \int d^2b\, T_A^{\rm eff}(b) S(b) . \tag{8.69}$$

Following the same procedure as for each effect alone,

$$\frac{\sigma_{hA\to C}}{A\sigma_{hN\to C}} \approx \exp\left\{-(\eta+\beta)A^{1/3}\right\} \approx A^\alpha \tag{8.70}$$

the true value of α, including both effects, is

$$\alpha = 1 - \eta - \beta \ . \tag{8.71}$$

An effective nucleon absorption cross section is commonly deduced from the A dependence, assuming that $\sigma_{C\text{co}}$ is negligible [159]. While this ultimately may be the case, it is difficult to determine with any certainty since the A dependencies are so similar. However, it seems that the CERN SPS data are difficult to describe by hadronic mechanisms alone and most models that do agree with the data include some contribution due to color screening [167].

8.6 Nucleus-nucleus collisions

Quarkonium suppression is expected in more central collisions where quark-gluon plasma production is more likely since the energy density is higher in these collisions. The centrality dependence may be quantified differently in different experiments.

At the CERN SPS, the J/ψ data were presented as a function of transverse energy, E_T. Rather than presenting the J/ψ E_T dependence alone, it was given relative to Drell-Yan production in the same mass region to eliminate systematic uncertainties. The Drell-Yan rate under the J/ψ peak in the dimuon invariant mass distributions was obtained by measuring the Drell-Yan yield above the J/ψ mass and extrapolating to the lower mass region by calculation, assuming a linear A dependence for Drell-Yan production: a rather good assumption [126].

The RHIC J/ψ data have so far been presented as a function of the number of participants. Instead of the J/ψ to Drell-Yan ratio, the ratio R_{AA} has been shown, the ratio of the J/ψ rate in AA relative to the rate in pp collisions at the same energy multiplied by the nuclear overlap function in the centrality bin [168]. RHIC is able to collide pp, d+Au and Au+Au at the same energy, not possible for the CERN SPS where the data at different energies were sometimes compared by scaling by a parameterization of the energy dependence [159].

8.6.1 The Drell-Yan and J/ψ E_T distributions

Recall that the Drell-Yan cross section as a function of E_T given in Chapter 3 is

$$\frac{d\sigma_{AB\to\mu^+\mu^-}}{dE_T} = \sigma_{NN\to\mu^+\mu^-}\int d^2b\int d^2s T_A(\vec{s})T_B(|\vec{b}-\vec{s}|)p(E_T;b) \quad (8.72)$$

where the cross section labeled $\sigma_{NN\to\mu^+\mu^-}$ is more correctly the AB cross section per nucleon to properly account for any initial-state effects on the parton distributions.

Since the J/ψ interacts in matter, the nuclear profile functions, T_A and T_B, should be replaced by $T_A^{\rm eff}$ and $T_B^{\rm eff}$ when nuclear absorption alone is included. If all the charmonium states share the same nuclear absorption cross section [169], the factor $T_A^{\rm eff}$ is the same for all of the states. If the asymptotic absorption cross sections are all different, the inclusive effective nuclear profile function for J/ψ production would be

$$T^{\rm eff}_{A_{J/\psi\,{\rm inc}}}(s) = 0.62\,T^{\rm eff}_{A_{J/\psi\,{\rm dir}}}(s) + 0.3\,T^{\rm eff}_{A_{\chi_c}}(s) + 0.08\,T^{\rm eff}_{A_{\psi'}}(s)\ . \quad (8.73)$$

In addition, the survival probabilities for interactions with hadronic secondaries and for color screening is included. Thus [164]

$$\begin{aligned}\frac{d\sigma_{AB\to J/\psi}}{dE_T} &= \sigma_{pp\to J/\psi}\int d^2b\int d^2s\,T_A^{\rm eff}(\vec{s})\,T_B^{\rm eff}(\vec{b}-\vec{s}) \\ &\quad\times S_{AB}(E_T;b,s)\,p(E_T;b)\ , \end{aligned} \quad (8.74)$$

The unnormalized E_T distributions calculated with Eq. (8.74) are shown in Fig. 8.10. Results are shown for no nuclear effects (solid curves) and nucleon absorption alone with $\sigma_{J/\psi N} = 4.8$ mb (dashed curves) for S+U and Pb+Pb collisions. Absorption tends to flatten the E_T distribution.

The ratios of the E_T distributions with and without absorption, comovers and color screening can be directly compared to data after multiplying by the correct normalization factor. The J/ψ E_T distribution is normalized by the cross section multiplied by the branching ratio to muon pairs. The Drell-Yan cross section in the interval $2.9 < M < 4.5$

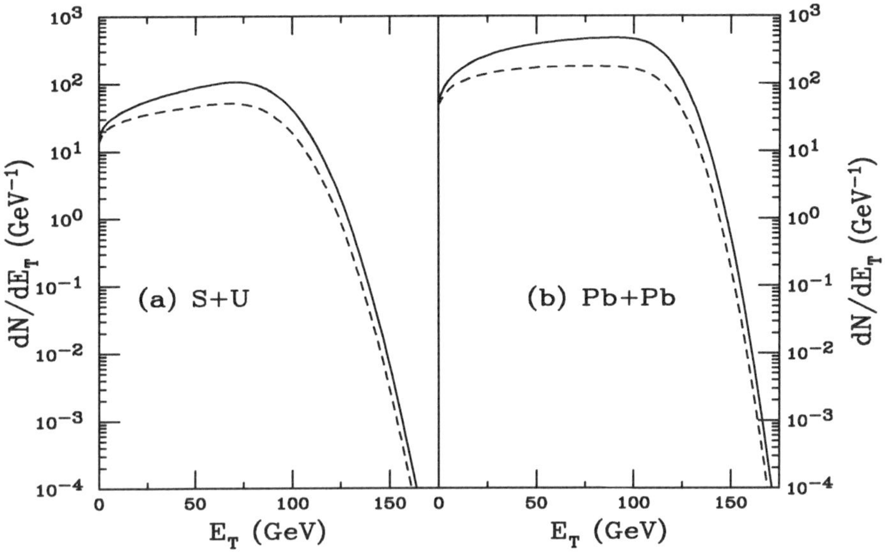

Figure 8.10: The E_T distributions in (a) S+U and (b) Pb+Pb interactions. Nuclear absorption alone is shown in the dashed curves for an absorption cross section of 4.8 mb.

GeV is calculated for the angular acceptance of the experiment using the same parton densities as in the experimental analysis for consistency. The left-hand side of Fig. 8.11 compares the S+U data to calculations with absorption cross sections of 4.8 mb (with a small contribution from hadronic comovers) and 7.3 mb (with no comover contribution). The data are not able to distinguish between the two results and are, in fact, able to accommodate other, smaller values of the absorption cross section obtained from fits to pA data [126]. However, the Pb+Pb data, shown on the right-hand side of Fig. 8.11 could not be fit without recourse to either larger comover densities (dashed curve) or inclusion of a quark-gluon plasma contribution (dot-dashed curve). The point at which the quark-gluon plasma contribution sets in, where the dot-dashed curve departs from the solid curve, is where the ψ' and χ_c contributions are suppressed. There is no sharp change at that E_T value because the range of impact parameters accessed within a given E_T bin washes out any threshold effect. The combined comover and color screening survival probability is

$$S_{AB}(\epsilon, b, s) = 0.58\, S_{\psi}^{\rm co}(b, s)\, S_{\psi}^{\rm QGP}(\epsilon) \\ +0.3\, S_{\chi_c}^{\rm co}(b, s)\, S_{\chi_c}^{\rm QGP}(\epsilon) + 0.12\, S_{\psi'}^{\rm co}(b, s)\, S_{\psi'}^{\rm QGP}(\epsilon) \; . \qquad (8.75)$$

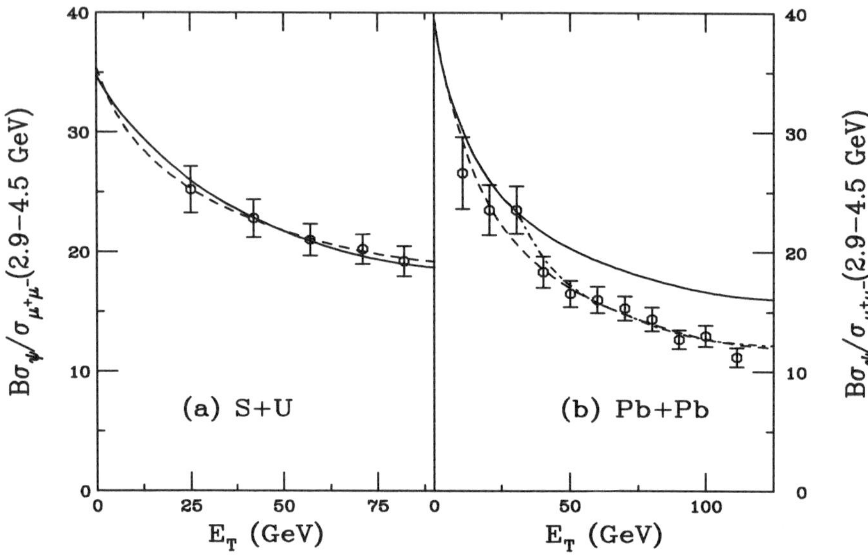

Figure 8.11: The calculated J/ψ to Drell-Yan ratios as a function of E_T are compared with data from (a) S+U [170] and (b) Pb+Pb [171] interactions. The solid curves show $\sigma_{J/\psi N} = 4.8$ mb and $\sigma^{\rm co}_{J/\psi,{\rm dir}} = 0.67$ mb. The dashed curve in (a) shows nuclear absorption alone with $\sigma_{J/\psi N} = 7.3$ mb. Two additional calculations are shown in (b): increasing the participant density by a factor of two (dashed) and a quark-gluon plasma with $n_f = 3$ (dot-dashed).

We note that already in comparison to these data some model calculations also included regeneration of the J/ψ through c and $\bar{c}$ coalescence, on the level of a few percent at these energies. However, at higher collision energies, these models suggest that most of the observed J/ψ's will be due to coalescence production [167]. At the LHC, secondary J/ψ's will also be produced through B meson decays. These states will have to be separated to obtain a truer picture of J/ψ suppression at the LHC. In addition, at the LHC, Υ measurements should provide an additional handle on the color screening phenomena [172].

We have not gone further in this chapter since the field is still new and yet evolving, particularly true of the RHIC data. Thus it is difficult to draw any lasting, generalized conclusions. However, this admission demonstrates the level of activity and interest in the problem.

Exercises

1. Calculate the breakup values of T_D for charmonium and bottomonium using Eq. (8.35) for $T_c = 170$ MeV with $n_f = 2$ and $n_f = 4$.

2. Calculate the values of t_D and $p_{T_{\rm m}}$ for the cases in Table 8.3 with $T(\tau_0) = 550$ MeV, $t_0 = 0.7$ fm and $T(\tau_0) = 820$ MeV, $t_0 = 0.5$ fm.

3. Use Eq. (8.52) to obtain Eq. (8.53) for a hard-sphere nucleus.

4. Show that $L = \langle b \rangle = (3/4)R_A$ by calculating $\langle b \rangle$ for a hard-sphere volume.

5. Using Eq. (8.62), calculate σ_{CN} for $\alpha = 0.97$ and 0.91. Use the results for σ_{CN} to calculate the A dependence in Eq. (8.63) and plot it with A^α as in Fig. 8.9.

Chapter 9

Hadronization

9.1 Introduction

In this final chapter, we discuss how final-state hadrons are formed from the partons produced in hard partonic scattering. Since no single partons are observed, only hadrons, the produced partons must hadronize before observation. In fact, because the partons are subject to confinement, they must form hadrons on the confinement length scale of about 1 fm. The process of hadron formation is often referred to as fragmentation and should be independent of the initial state, as suggested by the factorization theorem [19].

We will first discuss the typical fragmentation process in QCD, using the angular coverage of the BRAHMS experiment at RHIC to describe the changes in the final-state hadron distributions in *pp* collisions over the broad rapidity range measured by BRAHMS. We will then briefly discuss some possible changes in the hadronization mechanism due to the presence of a medium including recombination and energy loss by fast partons.

9.2 Fragmentation in *pp* collisions

The way hadrons are produced from partons can be empirically formulated by fragmentation functions. Fragmentation functions in the final state can be described analogously to the parton distributions in the initial state. It is simplest to start our discussion with e^+e^- collisions since electrons are point particles, providing a clean initial state. Let us say that in an e^+e^- collision, a quark-antiquark pair is produced, $e^+e^- \to q\overline{q}$, as shown in Fig. 9.1. After the quark is produced, it fragments into a hadron, h, which is ultimately observed in the detector. If we identify the center-of-mass energy of the e^+e^- collision as Q, the electron beam energy, $E_{\rm beam}$, is $Q/2$. The produced quark always has energy E_q equal to the beam energy of the electron or positron in the center-of-mass frame. Thus if the final-state hadron has energy E_h, the fraction of the quark energy it carries away is

$$z = \frac{E_h}{E_q} = \frac{2E_h}{Q} \, . \tag{9.1}$$

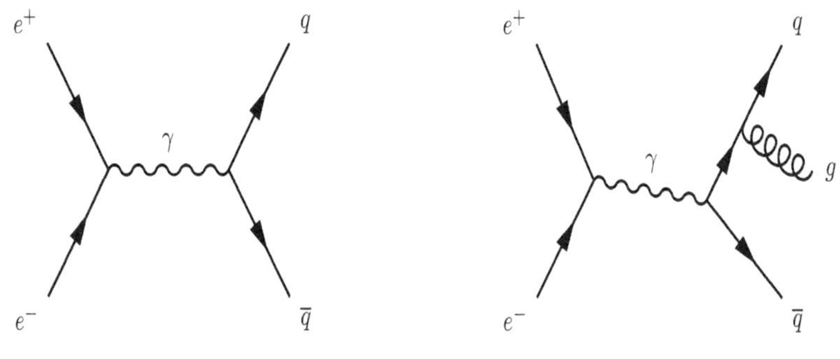

Figure 9.1: The diagrams for $e^+e^- \to q\overline{q}$ (left) and $e^+e^- \to q\overline{q}g$ (right).

The differential cross section for inclusive hadron production as a function of z is

$$\frac{d\sigma(e^+e^- \to hX)}{dz} = \sum_q \sigma(e^+e^- \to q\overline{q})[D_q^h(z) + D_{\overline{q}}^h(z)] \, . \tag{9.2}$$

Note that a quark or antiquark of any flavor can produce the hadron. The function $D_q^h(z)$ is the fragmentation function and represents the probability that final-state hadron h carries a fraction z of the initial quark momentum. The fragmentation functions describe the parton-to-hadron transition in the same way as the parton densities describe the parton structure of the hadron. Like the parton densities, the fragmentation functions are subject to momentum and probability constraints. The sum of the energies of all produced hadrons has to add up to the energy of the parent quark,

$$\sum_h \int_0^1 dz z D_q^h(z) = 1 \ . \tag{9.3}$$

The relation also holds if q is replaced by $\overline{q}$ in Eq. (9.3). In addition, the multiplicity of h is given by the sum of probabilities for producing h from all possible parent quarks and antiquarks,

$$\sum_q \int_{z_{\min}}^1 dz [D_q^h(z) + D_{\overline{q}}^h(z)] = n_h \ . \tag{9.4}$$

The lower limit on the integral over z is the threshold energy for producing a hadron of mass m_h, $z_{\min} = 2m_h/Q$, since the threshold energy is equal to the rest energy. The fragmentation functions are often parameterized as

$$D_q^h(z) = N\frac{(1-z)^n}{z} \tag{9.5}$$

where N and n are constants specific to a given h.

EXAMPLE: Show that the multiplicity grows as the log of the center-of-mass energy.
To determine the approximate energy dependence of the multiplicity, we can assume that $z \ll 1$ so that $(1-z)^n \approx 1$. Then $D_q^h(z) \approx N/z$. If we assume the same form for both the quark and antiquark fragmentation functions, Eq. (9.4) becomes

$$n_h \propto \int_{z_{\min}}^1 \frac{dz}{z} \propto \ln\left(\frac{1}{z_{\min}}\right) \approx \ln\left(\frac{Q}{2m_h}\right) \ . \tag{9.6}$$

Thus the dominant behavior of the multiplicity is its growth with $\ln Q$ or center-of-mass energy since $Q \equiv \sqrt{S}$.

Because the initial energy is known exactly, it is easy to determine the scale dependence of the fragmentation functions at leading order. The scale is simply the electron energy $Q/2$. At leading order, the $q\overline{q}$ pair is produced back-to-back and materializes into two jets of hadrons that are moving out toward the detectors with a momentum approximately collinear to that of the initial quarks. By measuring the momenta of the final-state hadrons in a jet, a range of z's are measured at a given interaction scale. In addition to inclusive hadron production, the fragmentation functions for individual hadrons can be measured by looking at different final states separately.

We can postulate some relationships between fragmentation functions based on charge and isospin. Take, for example, charged pion production with $\pi^+(u\overline{d})$ and $\pi^-(\overline{u}d)$. One would expect that the fragmentation functions of u, $\overline{u}$, d and $\overline{d}$ to charged pions would be

$$D_u^{\pi^+}(z) = D_{\overline{u}}^{\pi^-}(z) = D_d^{\pi^-}(z) = D_{\overline{d}}^{\pi^+}(z) \,, \tag{9.7}$$

$$D_u^{\pi^-}(z) = D_{\overline{u}}^{\pi^+}(z) = D_d^{\pi^+}(z) = D_{\overline{d}}^{\pi^-}(z) \,. \tag{9.8}$$

The first set of equalities are for charged partons that become valence quarks of the final-state pions while the second set are for those that do not. Thus, *e.g.* $D_u^{\pi^+}(z) \neq D_u^{\pi^-}(z)$. Finally, for strange quarks, we expect

$$D_s^{\pi^+}(z) = D_s^{\pi^-}(z) \tag{9.9}$$

and the same for the $\overline{s}$.

Note that the gluon has not yet been introduced. There is no gluon in e^+e^- annihilation at leading order. It must enter at the next order when either the produced quark or antiquark emits a gluon after production, $e^+e^- \to q\overline{q}g$, as shown on the right hand side of Fig. 9.1. The final-state momenta has to balance and the quark, antiquark and gluon all carry some fraction of the beam energy $Q/2$. We can write the fraction of the beam energy carried by each of the final-state partons as

$$x_q = \frac{2E_q}{Q} \,, \qquad x_{\overline{q}} = \frac{2E_{\overline{q}}}{Q} \,, \qquad x_g = \frac{2E_g}{Q} \,. \tag{9.10}$$

Typically, the direction of the most energetic jet is defined as the longitudinal axis, referred to as the 'thrust axis'. All three jets lie in the same plane, as shown in Fig. 9.2. The other two partons have combined

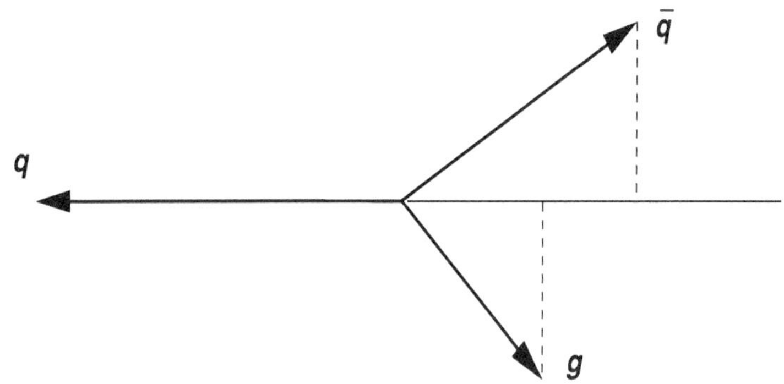

Figure 9.2: The final state of a three-jet event. The quark carries the largest momentum, making its direction the 'thrust' axis. The antiquark and gluon have equal and opposite transverse momentum while their longitudinal momentum balances that of the quark.

longitudinal momentum equal and opposite the longitudinal momentum along the thrust axis and momentum transverse to the thrust axis, p_T, so that

$$x_T = \frac{2p_T}{Q} \ . \tag{9.11}$$

Energy conservation tells us that the quark, antiquark and gluon energies have to add up to the total available energy, Q. Thus the sum of the momentum fractions gives

$$x_q + x_{\overline{q}} + x_g = 2 \ . \tag{9.12}$$

The fractional four-momenta of the final state partons is

$$q \longrightarrow (x_q, 0, -x_q) \ , \tag{9.13}$$
$$\overline{q} \longrightarrow (x_{\overline{q}}, x_T, x_L) \ , \tag{9.14}$$
$$g \longrightarrow (x_g, -x_T, x_q - x_L) \ . \tag{9.15}$$

The masses of the partons are constrained to be zero, giving us

$$0 = x_{\overline{q}}^2 - x_T^2 - x_L^2 \ , \tag{9.16}$$

$$0 = x_g^2 - x_T^2 - (x_L - x_q)^2 \ . \tag{9.17}$$

We can solve for x_T to find

$$x_T^2 = \frac{4}{x_q^2}(1 - x_q)(1 - x_{\overline{q}})(1 - x_g) \ . \tag{9.18}$$

EXAMPLE: Derive Eq. (9.18).
By subtracting Eq. (9.17) from Eq. (9.16), we solve for x_L alone,

$$x_L = \frac{1}{2x_q}(x_g^2 - x_q^2 - x_{\overline{q}}^2) \ . \tag{9.19}$$

Then, factoring Eq. (9.16) so that

$$x_T^2 = (x_{\overline{q}} - x_L)(x_{\overline{q}} + x_L) \ , \tag{9.20}$$

we have

$$\begin{aligned} x_T^2 &= \frac{1}{4x_q^2}(2x_{\overline{q}}x_q - x_g^2 + x_q^2 + x_{\overline{q}}^2)(2x_{\overline{q}}x_q + x_g^2 - x_q^2 - x_{\overline{q}}^2) \\ &= \frac{1}{4x_q^2}((x_q + x_{\overline{q}})^2 - x_g^2)(x_g^2 - (x_q - x_{\overline{q}})^2) \ . \end{aligned} \tag{9.21}$$

Using Eq. (9.12), the first term in Eq. (9.21) becomes

$$(x_q + x_{\overline{q}})^2 - x_g^2 = (2 - x_g)^2 - x_g^2 = 4(1 - x_g) \tag{9.22}$$

while the second becomes

$$\begin{aligned} x_g^2 - (x_q - x_{\overline{q}})^2 &= (2 - x_q - x_{\overline{q}})^2 - (x_q - x_{\overline{q}})^2 \\ &= 4 - 4(x_q + x_{\overline{q}}) + (x_q + x_{\overline{q}})^2 - (x_q - x_{\overline{q}})^2 \\ &= 4 - 4(x_q + x_{\overline{q}}) + 4x_q x_{\overline{q}} \\ &= 4(1 - x_q)(1 - x_{\overline{q}}) \ , \end{aligned} \tag{9.23}$$

leading to Eq. (9.18).

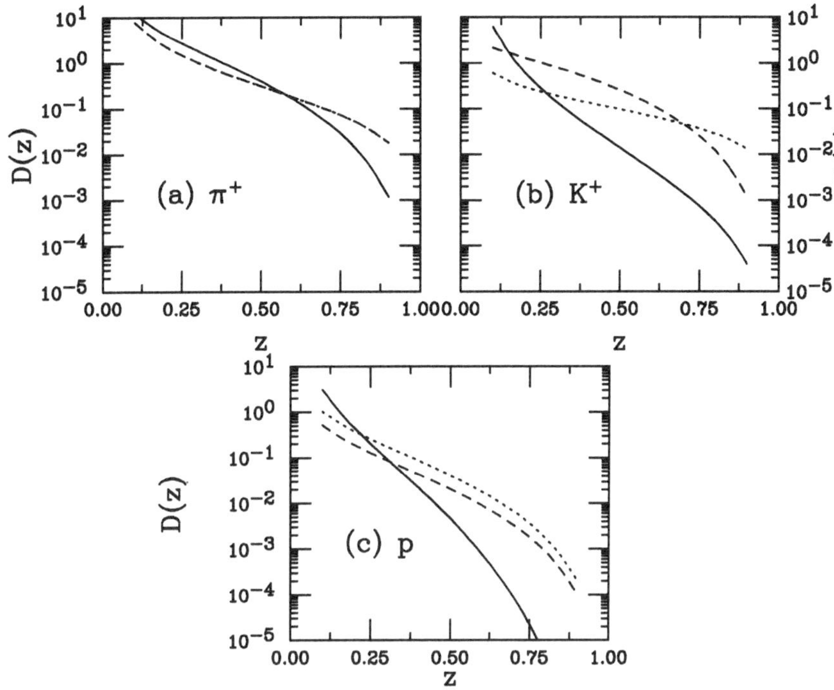

Figure 9.3: The KKP fragmentation functions for (a) charged pions, (b) charged kaons and (c) protons and antiprotons as a function of z, the fraction of the parton momentum carried by the final-state hadron. The curves show the contribution from u quarks (dotted), $\overline{u}$ quarks (dashed) and gluons (solid) with initial momentum of 3 GeV.

With measurements of hadron production in e^+e^- annihilation over a wide range of energies and calculations of the annihilation cross section for two and three jet production, the fragmentation functions can be parameterized from the data, similar to global fitting of the parton densities. Sets of fragmentation functions at leading and next-to-leading order have been prepared by several groups as the data improves. Most assume that the same fragmentation functions can describe the production of a particle and its antiparticle simultaneously such as π^+ and π^-, K^+ and K^-, p and $\overline{p}$, see Ref. [173] for an example. However, since the quarks and antiquarks that produce the jets fragment independently, there is no reason why *e.g.* the π^+ and π^- fragmentation functions should be identical as long as global charge and baryon number are conserved. The set by Kretzer [174] separates the hadron charge states for pions and kaons.

Unfortunately not all final-state hadron fragmentation functions have been extracted. The most commonly available are for pions and kaons as well as inclusive charged hadrons (mostly pions). The set by Kniehl, Kramer and Potter (KKP) [173] also includes protons and neutrons while the set by Kretzer [174] does not include any baryons. Fragmentation functions have also been extracted for the lightest heavy flavor mesons, D [175] and B [176].

Figure 9.3 shows the KKP fragmentation functions for charged pions, charged kaons and protons/antiprotons as functions of z for $Q = 3$ GeV, a typical value for RHIC. The hadron production by produced u (dotted), $\overline{u}$ (dashed) and g (solid) are shown separately. Note that the charged pions from u and $\overline{u}$ are identical, as anticipated in Eq. 9.7. While production by gluons is greatest at small z, at higher z production by quarks and antiquarks soon dominates, particularly for kaons and protons. Thus, while gluon production may dominate parton production in hadronic collisions, production of final states by gluons may be smaller, especially in certain regions of phase space where z is large. While some changes are expected as a function of scale, the general trends do not change substantially.

As we already mentioned, the fragmentation functions parameterize properties intrinsic to the partons and are thus universal, regardless of the parton production mechanism. Thus once the fragmentation functions are fixed in e^+e^- collisions, they can be applied to other collisions such as *ep*, *pp* and $p\overline{p}$. Thus, schematically, the production of hadron h in *ep* collisions as a function of z is

$$\frac{d\sigma(ep \to hX)}{dz} \propto \sum_q e_q^2 f_q^p(x) D_q^h(z) \ . \tag{9.24}$$

In *pp* collisions, since there are two partons in the initial state, many more processes open up in $2 \to 2$ scatterings. These are

$$qq' \rightarrow qq' , \tag{9.25}$$
$$qq \rightarrow qq , \tag{9.26}$$
$$q\overline{q} \rightarrow q\overline{q} , \tag{9.27}$$
$$q\overline{q} \rightarrow q'\overline{q}' , \tag{9.28}$$
$$q\overline{q} \rightarrow gg , \tag{9.29}$$
$$qg \rightarrow qg , \tag{9.30}$$
$$gg \rightarrow q\overline{q} , \tag{9.31}$$
$$gg \rightarrow gg . \tag{9.32}$$

The 'prime' in Eqs. (9.25) and (9.28) is to denote that q' is a different flavor than q. Thus *e.g.* Eq. (9.26) includes $uu \to uu$ while Eq. (9.25) does not. Note that processes like qq or qq' scattering, indicated with two quarks in the initial state, also include antiquarks: $\overline{qq}$ and $\overline{qq}'$. The replacement of q by $\overline{q}$ may also be made in Eq. (9.30).

In pp collisions, the inclusive p_T distribution of produced hadron h is [16]

$$\frac{d\sigma(pp \to hX}{dp_T} = 2p_T \int_{\theta_{\rm min}}^{\theta_{\rm max}} \frac{d\theta_{\rm cm}}{\sin\theta_{\rm cm}} \int dx_1 \int dx_2 \times f_i^p(x_1, Q^2) f_j^p(x_2, Q^2) \frac{D_q^h(z, Q^2)}{z} \frac{d\sigma_{ij}}{dt} . \tag{9.33}$$

The momentum of the other produced parton in the $2 \to 2$ scattering is integrated away. Thus only one fragmentation function appears in Eq. (9.33). The integral over center-of-mass scattering angle $\theta_{\rm min} \leq \theta_{\rm cm} \leq \theta_{\rm max}$ corresponds to an integral over the rapidity region covered by the detector. For an integral over all rapidity, $\theta_{\rm min} = 0$ and $\theta_{\rm max} = \pi$. The $2 \to 2$ hard scattering partonic cross sections in Eqs. (9.25)-(9.32) are given by $d\sigma_{ij}/dt$. We use the MRST leading order parton densities [177] and the leading order KKP parameterization of the fragmentation functions, both with $Q^2 = p_T^2$.

For a concrete example of parton production and fragmentation in pp collisions, we consider the BRAHMS experiment at RHIC which had wide pseudorapidity, η, coverage. BRAHMS made measurements in four η bins: $|\eta| \leq 0.2$; $0.8 \leq \eta \leq 1.2$ ($\eta = 1$); $1.9 \leq \eta \leq 2.35$ ($\eta = 2$) and $2.9 \leq \eta \leq 3.5$ ($\eta = 3.2$). The values of η in parentheses are the average values shown on the plots. These η ranges correspond to center-of-mass scattering angles of $101.4^\circ \leq \theta_{\rm cm} \leq 78.6^\circ$, $48.4^\circ \leq \theta_{\rm cm} \leq 33.5^\circ$, $17.01^\circ \leq \theta_{\rm cm} \leq 10.9^\circ$ and $6.3^\circ \leq \theta_{\rm cm} \leq 3.5^\circ$ respectively.

EXAMPLE: Calculate the correspondence between η and $\theta_{\rm cm}$.
In Chapter 1 we learned that

$$\eta = -\ln\tan(\theta_{\rm cm}/2) \ . \tag{9.34}$$

From this we can infer that

$$\theta_{\rm cm} = 2\tan^{-1}(\exp(-\eta)) \ . \tag{9.35}$$

The largest values of pseudorapidity are the most forward in phase space and correspond to the smallest angles in the center-of-mass frame. For example, in the pseudorapidity interval $0.8 \leq \eta \leq 1.2$, $\eta = 0.8$ gives $\theta_{\rm cm} = 48.4^\circ$ while $\eta = 1.2$ gives $\theta_{\rm cm} = 33.5^\circ$.

The η coverage is not continuous because the BRAHMS experiment consisted of two movable magnetic spectrometers, one that could be moved to cover the interval $-0.1 \leq \eta \leq 1.3$ and the other covering the forward interval, $1.3 \leq \eta \leq 4$. The spectrometers had to be mobile because the solid angles they subtended were too small to cover the full interval at once [178]. They reported results for $(h^+ + h^-)/2$ in the two most central bins while only h^- measurements were reported in the two forward bins. (Here h^+ and h^- refer to positive and negative charged hadrons.) The most important contributions to charged hadron production are pions, kaons and protons. The KKP fragmentation functions do not distinguish between h^+ and h^- but do predict more pions than protons in the final state.

Figure 9.4 shows the contributions to final-state pion production due to final-state quarks, antiquarks and gluons respectively. The results are shown only for the central η bin. The p_T distributions from the individual processes in Eqs. (9.25)-(9.32) are shown separately. In the central η bin, gluons are produced almost equally in the $gg \rightarrow gg$ and $qg \rightarrow qg$ channels. The qg channel is somewhat larger for $p_T > 5$ GeV.

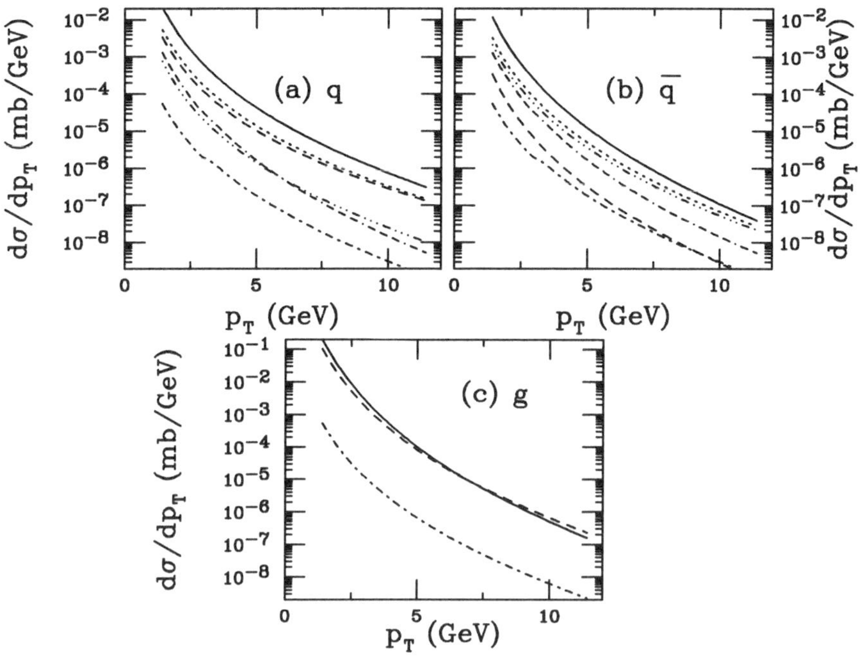

Figure 9.4: The p_T distribution of charged pions at $|\eta| < 0.2$ from (a) produced quarks, (b) produced antiquarks and (c) produced gluons. In (a) and (b) the curves correspond to $(q/\overline{q})g \to (q/\overline{q})g$ (solid), $qq \to qq$ (dashed), $q\overline{q} \to q'\overline{q}'$ (dot-dashed), $qq' \to qq'$ (dotted), $q\overline{q} \to q\overline{q}$ (dot-dot-dot-dashed), $gg \to q\overline{q}$ (dash-dash-dash-dotted). In (c) the curves are $gg \to gg$ (solid), $qg \to qg$ (dashed) and $q\overline{q} \to gg$ (dot-dashed).

There is a negligible contribution from $q\overline{q} \to gg$. Pion production by quarks and antiquarks proceeds mainly through the $qg \to qg$ channel for quarks and $\overline{q}g \to \overline{q}g$ for antiquarks. The next largest contribution to pion production by quarks are the $qq' \to qq'$ and $qq \to qq$ channels which are of very similar strength, followed by $q\overline{q} \to q\overline{q}$ and $gg \to q\overline{q}$ with a negligible contribution from $q\overline{q} \to q'\overline{q}'$. The contributions to antiquark production after $\overline{q}g \to \overline{q}g$ are $\overline{q}q' \to \overline{q}q'$, $\overline{q}q \to \overline{q}q$ and $gg \to q\overline{q}$, followed by smaller contributions from $\overline{qq} \to \overline{qq}$ and $q\overline{q} \to q'\overline{q}'$. Similar results are found for kaon and proton production. However, the proton distributions fall off more steeply with p_T.

The relative contributions from the production channels remains similar as rapidity increases. The most important change is in gluon production where the $qg \to qg$ channel grows more dominant, finally becoming larger than the $gg \to gg$ channel for all p_T for $\eta = 3.2$. Indeed, at the most forward rapidity, the $q\bar{q} \to gg$ channel becomes comparable to the qg channel at $p_T \sim 8$ GeV. This may seem counterintuitive since x_2 decreases as η grows, increasing the gluon density, see Chapter 2. However, x_1 increases more rapidly and, at large η, we are in a region where the gluon density is dropping steeply while the quark density, particularly that of the valence quarks, is still significant. Thus the qg channel is more important than the gg channel at large η, particularly when p_T and x_1 are large. Also, at high η and p_T antiquarks are predominantly produced by valence quark induced processes, since the valence densities are large at high x.

The p_T distributions of charged pions, kaons, protons and the sum of all charged hadrons are shown in the four η intervals in Fig. 9.5. At leading order and, for the given set of fragmentation functions, pion production dominates over all p_T, even in the most forward bin. The minimum p_T shown is the minimum scale of the fragmentation function, 1.4 GeV. The distributions steepen as η increases since we begin to approach the edge of phase space. The effect is most visible in the forward distributions where the edge of phase space is reached at lower p_T.

The edge of available phase space is roughly reached at $x_1 \to 1$. Since we have integrated away the other parton in the $2 \to 2$ scattering, we estimate the parton p_T at which $x_1 \to 1$ for a $2 \to 1$ process. In this case, since the partons are massless, $x_1 = (p_T/\sqrt{S}) \exp(\eta) \to 1$ is equivalent to $p_T = \sqrt{S} \exp(-\eta)$ for fixed η. At $\eta = 0$ then, we could take $p_T \to \sqrt{S}$. By $\eta = 2.2$, the maximum p_T is 22 GeV and by $\eta = 3.2$, the maximum p_T is reduced to 8.2 GeV. Of course, the hadron p_T is reduced relative to that of the parton by the fraction z so that unless the average value of z is close to unity, the maximum parton p_T will not be manifested by the final-state hadron.

We have calculated the projectile and target momentum fractions, $\langle x_1 \rangle$ and $\langle x_2 \rangle$ respectively. They are shown in Figs. 9.6 and 9.7. Note that these are averages over x_1 and x_2 in a specific pseudorapidity

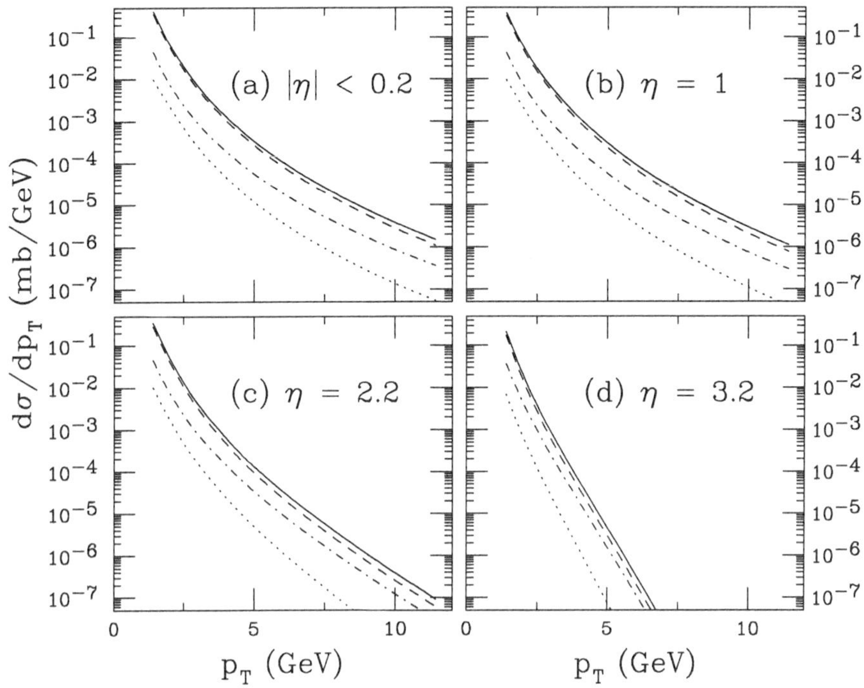

Figure 9.5: The p_T distributions of produced charged pions (dashed), charged kaons (dot-dashed), protons and antiprotons (dotted) and the total (solid) in (a) $|\eta| < 0.2$, (b) $\eta = 1$, (c) $\eta = 2.2$ and (d) $\eta = 3.2$.

window. Thus the actual x_1 and x_2 for a given interaction can be smaller or larger than the average. The averages shown in Figs. 9.6 and 9.7 are for combined quark, antiquark and gluon production of pions. While there is some variation with final-state hadron, it is rather small and the trends are the same. The average x values do not depend strongly on the identity of the produced hadron. The averages are also not very sensitive to changes in the parton densities or the choice of scale Q^2.

In the central η bin, $\langle x_1 \rangle = \langle x_2 \rangle$, as expected. At higher η, $\langle x_1 \rangle$ increases with η while $\langle x_2 \rangle$ decreases. The average values of x_1 increase rather rapidly with η while $\langle x_2 \rangle$ decreases more slowly. Both $\langle x_1 \rangle$ and $\langle x_2 \rangle$ increase with p_T. As $\langle x_1 \rangle$ increases, the valence quark distribu-

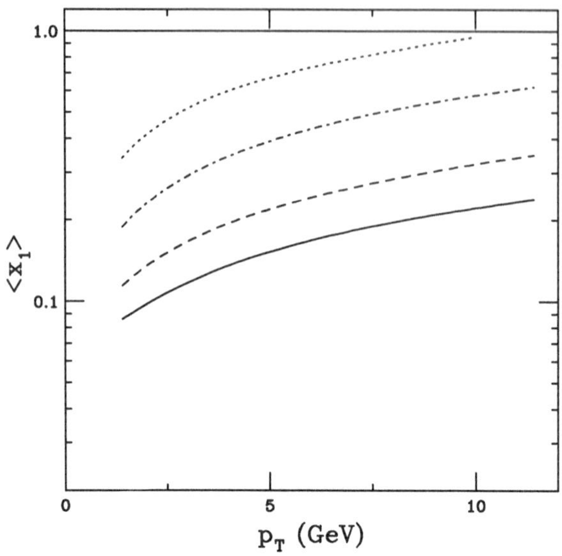

Figure 9.6: The average value of the 'projectile' parton, x_1, as a function of p_T for $|\eta| < 0.2$ (solid), $\eta = 1$ (dashed), $\eta = 2.2$ (dot-dashed) and $\eta = 3.2$ (dotted).

tions become more important. Above $x \sim 0.05$ the valence quarks are the largest contribution to the u and d quark densities. Thus the valence quarks will play a dominant role in quark-initiated scatterings, increasing the importance of quark production relative to antiquarks and gluons at high p_T, especially when combined with high η. In the bin with $\eta = 3.2$, $\langle x_1 \rangle \sim 1$ at $p_T \approx 10$ GeV.

The increase of $\langle x_1 \rangle$ with p_T is rather smooth and the results in the η bins are almost parallel to each other. This is not the case for $\langle x_2 \rangle$. The slope of $\langle x_2 \rangle$ as a function of p_T changes slightly. In the most forward η bin $\langle x_2 \rangle$ is smaller than in the more central bins at low p_T but changes slope and increases rather rapidly with p_T until the edge of phase space. This increase occurs because we run out of kinematic room for x_2 to remain small while x_1 is close to unity.

We have noted that these average x_1 and x_2 values are insensitive to most details of the calculation. However, one factor they are sensitive to is the proton number. In pp collisions we have more u quarks than d

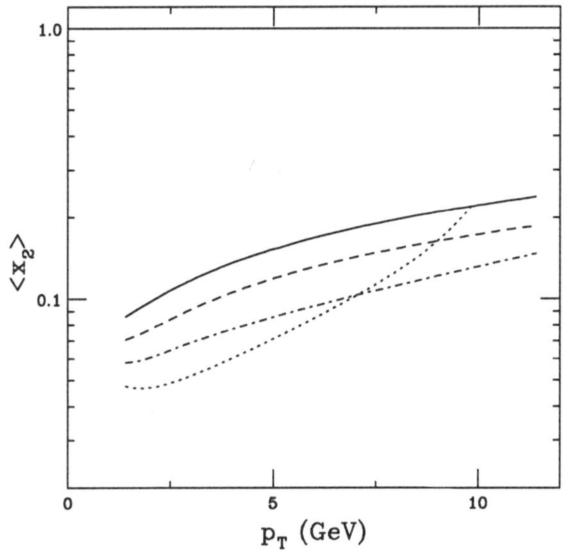

Figure 9.7: The average value of the 'target' parton, x_2, as a function of p_T for $|\eta| < 0.2$ (solid), $\eta = 1$ (dashed), $\eta = 2.2$ (dot-dashed) and $\eta = 3.2$ (dotted).

quarks (4:2). The valence parton densities are larger for u quarks than d quarks and the u quark densities peak at somewhat higher x values. Thus the average x values are biased more toward the behavior of the u densities. In nuclear collisions, the u/d ratio is altered. In a typical large nucleus, neutrons outnumber protons, making more valence d quarks available than u quarks. Since the average x of the d valence density is smaller than for u quarks, the average values of x_1 and x_2 are smaller in nucleus-nucleus collisions than in pp collisions.

The average z in the fragmentation functions is large, ≈ 0.5 at midrapidity, and increasing rather rapidly with η and p_T, as seen in Fig. 9.8. These large values mean that, on average, the parton typically loses less than half of its momentum to the produced hadron. Since the fragmentation functions are best determined for smaller z, the high z fragmentation functions have more uncertainties. Thus high p_T and high η hadron production contains important theoretical uncertainties due to the fragmentation functions. There is more variation in $\langle z \rangle$ due

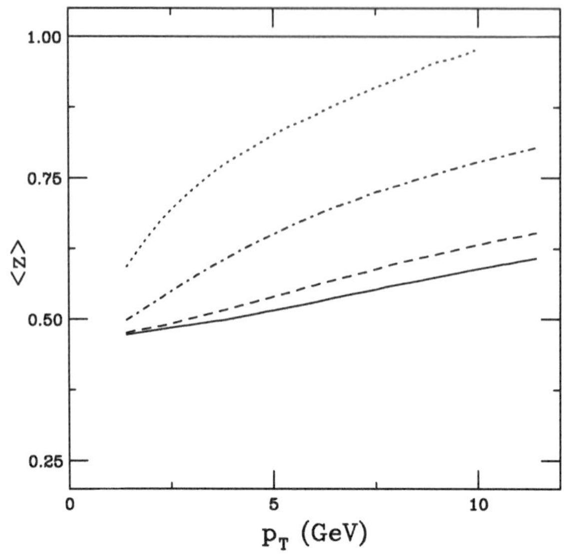

Figure 9.8: The average value of the fragmentation variable z, as a function of p_T for $|\eta| < 0.2$ (solid), $\eta = 1$ (dashed), $\eta = 2.2$ (dot-dashed) and $\eta = 3.2$ (dotted).

to parton type than in the average x values. The proton $\langle z \rangle$ tends to be somewhat smaller than for pions or kaons.

At such large z, kaon and proton production by gluons is suppressed. The gluon production cross sections need to be larger than the quark and antiquark production cross sections by several orders of magnitude to make up for the steep slope of the gluon fragmentation functions at $z > 0.25$, see Fig. 9.3. It is clear that kaon fragmentation is dominated by antiquarks over most of phase space while proton production will be dominated by quarks. The situation is less clear for pions since gluon fragmentation to pions is a much harder function of z and $D_g^\pi(z) > D_q^\pi(z)$ for $z < 0.6$.

These fragmentation effects are clearly reflected in the relative contributions to pion, kaon and proton production by quarks, antiquarks and gluons, shown in Figs. 9.9-9.11. We first discuss pion production as a function of η, shown in Fig. 9.9. At $|\eta| \leq 0.2$, pion production is dominated by produced gluons up to $p_T > 9$ GeV where pion pro-

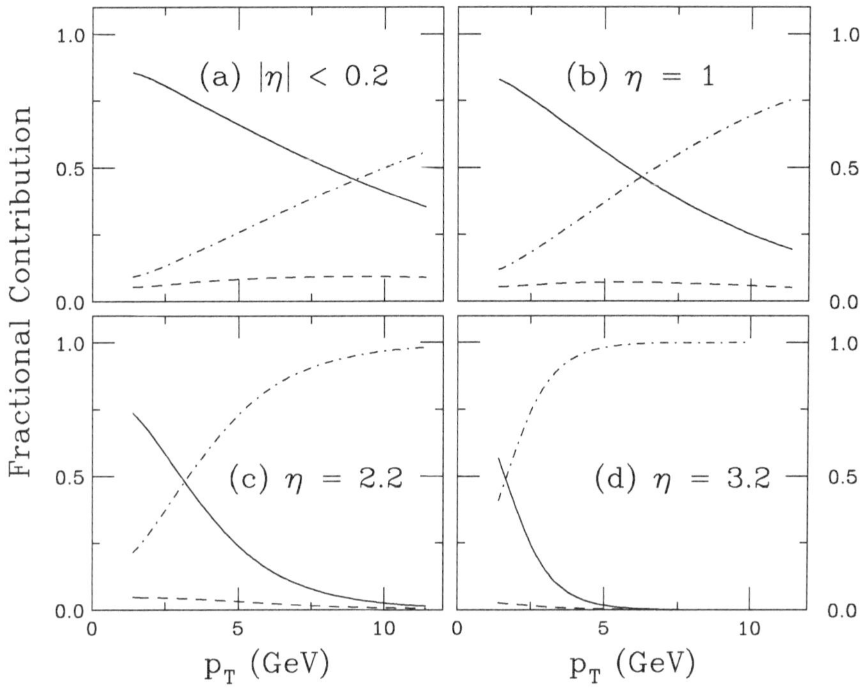

Figure 9.9: The contributions to charged pion production from quarks (dot-dashed), antiquarks (dashed) and gluons (solid) in (a) $|\eta| < 0.2$, (b) $\eta = 1$, (c) $\eta = 2.2$ and (d) $\eta = 3.2$.

duction by quarks becomes larger. From Figs. 9.3 and 9.8, we see that this occurs at the point where the quark fragmentation function to pions becomes harder than the gluon fragmentation function. Since $\langle z \rangle \sim 0.6$ is reached at lower values of p_T, the crossover between gluon and quark production of pions occurs at lower values of p_T with increasing η. Thus quarks dominate pion production for $p_T > 6.5$, 3 and 1.5 GeV with $\eta = 1$, 2.2 and 3.2 respectively. Note that although the quark and antiquark fragmentation functions are the same for pions, quark production is always larger than antiquark production for pions so that the fractional contribution to pion production by antiquarks is small over all phase space.

Figures 9.10 and 9.11 show the fractional contributions of quarks,

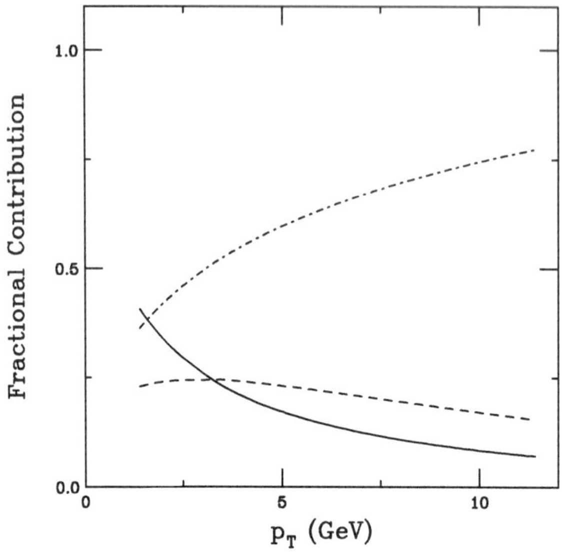

Figure 9.10: The contributions to charged kaon production from quarks (dot-dashed), antiquarks (dashed) and gluons (solid) in $|\eta| < 0.2$.

antiquarks and gluons to kaon and proton production in the central η bin only. It is clear from Fig. 9.10 that the gluon contribution to kaon and proton production is already small at rather low p_T. Gluon production of kaons is rather small, similar to the quark contribution at the minimum p_T of the fragmentation functions and dropping below the antiquark contribution at $p_T \sim 3.5$ GeV. Although Fig. 9.3 shows that $D_{\overline{u}}^K(z) > D_u^K(z)$ for $z < 0.65$, the antiquark contribution to kaon production remains smaller than the quark contribution since the quark production cross section is larger than the antiquark production cross section. The same effects are seen for proton production in Fig. 9.11 although here quark production is most important for protons up to $p_T \sim 3.5$ GeV.

As η increases, the quark contribution increasingly dominates kaon and proton production. Already at $\eta = 1$, more than half of kaons and protons are produced by quarks for $p_T > 2.5$ GeV. At higher η, antiquarks and gluons make negligible contributions to low p_T kaon and proton production.

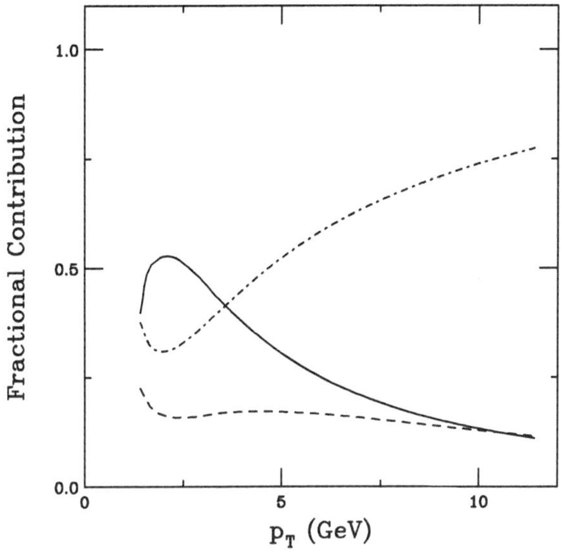

Figure 9.11: The contributions to proton/antiproton production from quarks (dot-dashed), antiquarks (dashed) and gluons (solid) in $|\eta| < 0.2$.

9.3 Nuclear effects

We have spent considerable time discussing the pp results for the specific circumstances of the BRAHMS detector because the results are not necessarily intuitive. In addition, the data show that nuclear effects are important. Already in d+Au collisions at $\sqrt{S_{NN}} = 200$ GeV, important differences are seen between the low p_T d+Au and pp results as a function of η [179]. BRAHMS observed a depletion of the d+Au to pp ratio per nucleon, $R_{\rm dAu}$, similar to the R_{AA} defined for nucleus-nucleus collisions in Chapter 5. The depletion increases at more forward rapidities where x_2, the momentum fraction in the gold ion, is smallest. In addition, calculations of particle production differ in their assessment of the x regime reached by BRAHMS. Calculations such as the one described in the previous section suggest that the average x_2 values in the gold nucleus are larger than calculations including low x saturation effects as in the color-glass condensate, briefly mentioned in

Chapter 2. The difference comes about because the saturation calculations are based on $2 \to 1$ interactions which reach lower values of x than the $2 \to 2$ kinematics assumed in the previous section [180]. Other models, involving recombination of quarks rather than fragmentation, also claim to describe the same d+Au data without invoking small x effects, see *e.g.* Ref. [181]. No conclusions have yet been reached.

In addition, other effects are seen in nucleus-nucleus collisions. An even stronger depletion than BRAHMS observed at low p_T in d+Au collisions has been observed in AA collisions by the other RHIC experiments. As stated at the end of Chapter 5, measurements of R_{AA} show a strong suppression of high p_T particle production. Instead of increasing toward $R_{AA} = 1$ at high p_T, the suppression is still strong at the highest p_T values measured at RHIC. None of the RHIC experiments measure the full jet. Instead they measure 'leading' hadrons with relatively high p_T compared to the rest of the event [83].

Finally, the proton/pion ratio in low impact parameter Au+Au collisions is nearly unity for $p_T \sim 2-4$ GeV [182], considerably larger than either the fragmentation functions or the pp measurements can account for. A similar effect is seen for $\overline{p}/\pi$ albeit smaller in magnitude. These ratios decrease to the pp level as the impact parameter increases. This cannot be understood in terms of the e^+e^- fragmentation functions.

The rest of this section will briefly describe some of the possible nuclear effects that may be present in heavy-ion collisions.

9.3.1 Recombination

Recombination models were first used to describe fast hadron production in pp collisions. The idea is rather simple: quarks and antiquarks can combine with comoving partons to form mesons and baryons. If some of the valence quarks of the final-state hadron are shared with those of the projectile, this hadron typically carries more longitudinal momentum than other produced hadrons. This fast hadron, called the leading hadron, can be produced entirely from the projectile wavefunction and come on shell by soft interactions with the target. One such model for leading charm production, the intrinsic charm model [183], provides a simple example of recombination. We use it here to explain the basic concepts of recombination.

In the proton wavefunction, short-lived $c\overline{c}$ fluctuations may exist. The lowest possible such state is $|uudc\overline{c}\rangle$. Charm hadrons that can be produced from such a state are the $\Lambda_c^+(udc)$, $D^-(\overline{c}d)$ and $\overline{D}^0(\overline{c}u)$. The dominant configurations are not far off mass shell and thus have minimal invariant mass, $M^2 = \sum_i^n \widehat{m}_i^2/x_i$ where $\widehat{m}_i^2 = m_i^2 + \langle \vec{k}_{T,i}^2 \rangle$ is the square of the average transverse mass of parton i. The general, frame-invariant form of the wavefunction is

$$\propto \frac{1}{m_h^2 - M^2} \tag{9.36}$$

where a slowly-varying vertex function has been ignored. The particle distributions are then controlled by the energy denominator and phase space. States with minimum invariant mass correspond to configurations with equal rapidity constituents. Thus charm quarks from such a state carry a larger fraction of the projectile momentum than the light quarks [183].

Momentum conservation demands $\sum_{i=1}^n x_i = 1$ and $\sum_{i=1}^n \vec{k}_{T,i} = 0$ where n is the number of partons in the state. Assuming it is sufficient to use $\langle k_T^2 \rangle$ for the transverse momentum, the probability distribution as a function of x in a general n–particle intrinsic $c\overline{c}$ state is

$$\frac{dP_{\rm IC}}{dx_i \cdots dx_n} = N_n \, \frac{\delta(1 - \sum_{i=1}^n x_i)}{(m_h^2 - \sum_{i=1}^n (\widehat{m}_i^2/x_i))^2} \,, \tag{9.37}$$

where N_n normalizes the n-particle Fock state probability. When $n = 5$ and $P_{\rm IC} \equiv 1\%$, $N_5 = 36$. The delta function in Eq. (9.37) takes care of longitudinal momentum conservation. The charm quark distribution is when

$$\begin{aligned} c(x_c) &= \int dx_1 \cdots dx_{\overline{c}} \frac{dP_{\rm IC}}{dx_i \cdots dx_{\overline{c}} dx_c} \\ &= \frac{N_5}{2} x_c^2 [\frac{1}{3}(1 - x_c)(1 + 10x_c + x_c^2) \\ &\quad + 2x_c(1 + x_c)\ln x_c] \,. \end{aligned} \tag{9.38}$$

The second equality holds if $n = 5$ and $\widehat{m}_c$, $\widehat{m}_{\overline{c}} \gg m_h$, $\widehat{m}_q$.

The model has been most widely applied to the longitudinal, x_F, distributions. To calculate the x_F distribution of a final-state charm hadron, we need a recombination function. In practice, an additional delta function is used to sum up the momentum fractions of the charm hadron valence quarks. Thus for the Λ_c^+, we include $\delta(x_{\Lambda_c^+} - x_u - x_d - x_c)$ in Eq. (9.37). The effect of the recombination function is to move the charm hadron forward to higher x_F relative to standard perturbative QCD calculations of charm production. The model predictions for the charm quark, D^- and Λ_c^+ distributions from the $|uudc\bar{c}\rangle$ state are shown in Fig. 9.12. All the distributions are normalized to unit probability. The charm quark distribution alone has an average x_F of 0.2 while the D^- and Λ_c^+ distributions peak at $x_F \sim 0.4$ and 0.6 respectively. Thus the intrinsic charm contribution to the total cross section naturally produces leading charm hadrons with valence quarks in common with the projectile at forward x_F, see Ref. [21] for more details and references.

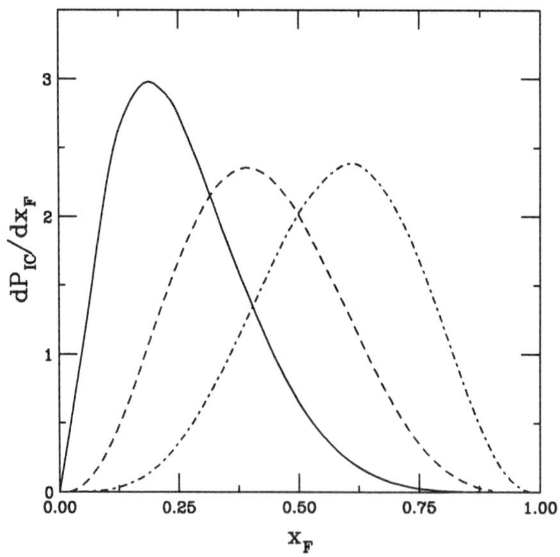

Figure 9.12: The intrinsic charm probability distributions for the charm quark (solid), D meson (dashed) and Λ_c^+ baryon (dot-dashed) in the $|uudc\bar{c}\rangle$ proton state.

In heavy-ion collisions, things are more complicated and it is not possible to write down a simple wavefunction for the quark-gluon plasma that includes all the quarks and antiquarks present in the medium. However, the idea is the same as in the intrinsic charm model. Two or three comoving partons in the quark-gluon plasma combine their transverse momentum to produce a final-state meson or baryon with higher p_T than the partons themselves. In addition to the hard partons produced by initial interactions governed by perturbative QCD, the other partons in the medium have 'soft' thermal distributions. The partons that combine can all be hard, all soft, or a mixture of the two [181]. There is thus a competition between the effects of fragmentation and recombination. At central impact parameters, recombination effects should be most important while, in more peripheral collisions, fragmentation should be a better description of hadronization. A schematic picture of the two effects, leading to the same p_T of the final-state hadron is shown in Fig. 9.13. While fragmentation functions demand $z < 1$, reducing the parton momentum, the recombination function, $\delta(p_{T_h} - p_{T_1} - p_{T_2})$ sums the momenta of two partons to obtain the final-state hadron momentum. For a review of some early successes of this approach in heavy-ion collisions, see Ref. [184].

9.3.2 Energy loss by fast partons

In this section, we briefly discuss a mechanism that could cause the depletion of particle production in nucleus-nucleus collisions to rather high p_T. The mechanism is often referred to as 'energy loss' or 'jet quenching'. Note that 'energy loss' is a bit of a misnomer: total energy must still be conserved. If the produced parton has 'lost' some energy, the medium it travels through gains that energy.

The basic idea can be linked to the same mechanisms that produce 'stopping', discussed in Chapter 3, and broaden the J/ψ p_T distribution, discussed in Chapter 8: multiple scattering. In the quarkonium picture, the broadening occurs in the initial state because the interacting parton, with no initial transverse momentum, suffers elastic collisions that transfers some of its longitudinal momentum into transverse motion, leading to a larger J/ψ p_T in the final state. While the effect

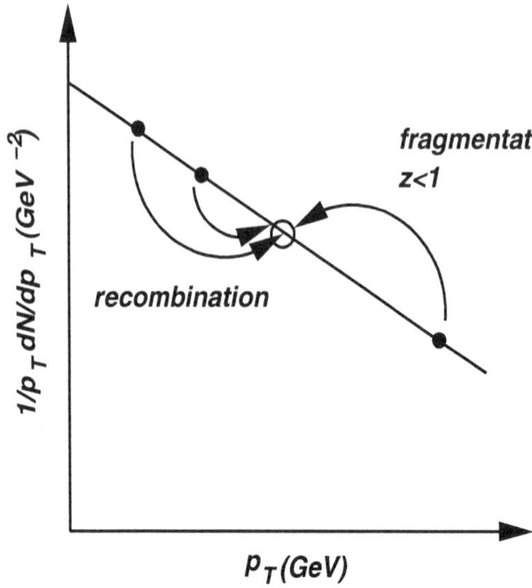

Figure 9.13: The competing effects of recombination and fragmentation on a final-state p_T distribution are shown, after [184].

on the J/ψ happens in the initial state, before the hard scattering, jet quenching occurs in the final state, after the hard scattering.

In the $2 \rightarrow 2$ scattering kinematics, two partons are produced back to back. In a pp collision, as in e^+e^- collisions, they hadronize into two back-to-back jets. However, after the partons are produced in a heavy-ion collision, they move apart through the medium on the way to the surface. Even though the partons are produced with high momentum, they may still either radiate gluons within the medium or scatter elastically with other partons in the medium, losing some of their initial momentum to the medium they are traveling through. The final-state jets may then not emerge with equal and opposite p_T but, instead, the jets may be observed with unequal momentum and may no longer be oppositely directed. In fact, some partons may lose so much energy that they will no longer appear as jets and, instead of two jets, only a single jet, called a 'monojet', will be detected [185].

Whether one, two or no jets emerge from a hard scattering depends on the path length of the parton in the medium. If the partons are produced at the edge of the medium, they may travel tangentially to it and emerge relatively unscathed. However, if the two partons are produced near the surface but one is directed outwards, away from the medium, while the other is directed into it, only one jet may be observed. If the partons are produced at the center of the medium, both may lose enough energy such that neither are observed as jets. These scenarios are depicted in Fig. 9.14.

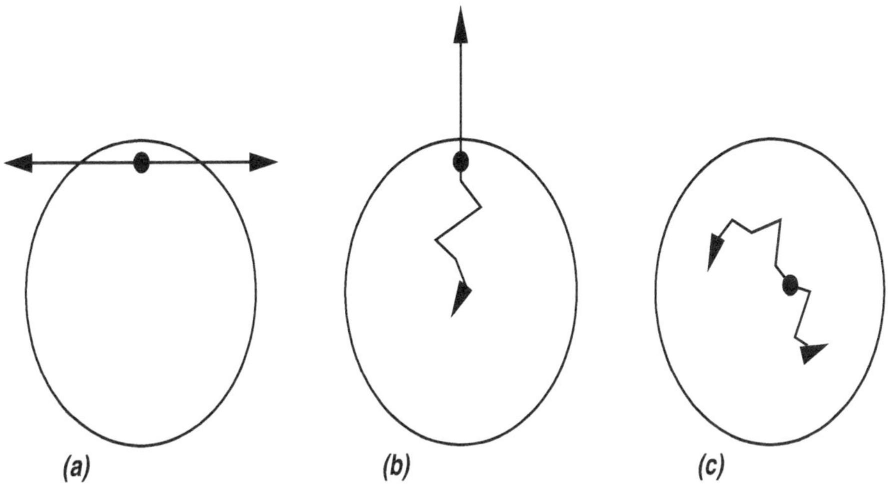

Figure 9.14: Energy loss scenarios. In (a) and (b), the partons are produced at the surface of the medium. In (a), both partons move tangential to the medium and hadronize as two back-to-back jets. In (b), one parton moves out of the medium and is unaffected by it while the other is directed into it and loses its energy. Only one jet is observed. In (c), both partons are produced in the center of the medium and do not escape it. The point of origin of all three $2 \to 2$ scatterings is indicated by the dot.

This energy loss mechanism is likely the source of the high p_T hadron suppression in R_{AA}. Evidence that jet quenching is at work at RHIC is found in other measurements besides ratios of single inclusive particle spectra. Although no jets are observed, the effects of energy loss can be quantified through leading-hadron correlations. In peripheral heavy-ion collisions, two high p_T hadrons are detected back to back, similar to the measurements of pp interactions at the same energy. However, in central collisions, the opposite side hadron has disappeared and the azimuthal correlation is similar to that of the underlying soft hadrons in the event [186]. This jet suppression is absent in d+Au collisions at RHIC, strongly suggesting that the suppression is a final-state effect arising from the dense medium. Both measurements and analysis are proceeding apace, including also three particle correlations. For the most recent results, consult the proceedings of the Quark Matter conferences.

It is difficult to use dijets as a probe because the original jet energy is unknown – it is impossible to tell a priori how much matter the probe has traveled through before reaching the detector. Thus it is nontrivial to determine how much of the initial parton energy was imparted to the medium. For a better calibrated probe, it would be preferable to use a known energy 'tag' for the jet [187]. Quark jets of known energy can be produced in reactions such as $qg \to q\gamma$ or $qg \to qZ^0$. In these cases, the energy is known since, at leading order, the quark jet is produced with transverse momentum equal and opposite to that of a gauge boson (the photon or the Z^0) unaffected by the presence of the medium. While the reaction $qg \to qZ^0$ is a small contribution to the total Z^0 yield [188], it is a more distinctive signature since the Z^0 is free from the high background of hadronic decays contributing to the direct photon spectrum [189]. At the LHC, the Z^0+jet yield is large enough to extract physics signals. Any energy loss suffered by the jet can thus be more cleanly identified than in the dijet channel.

In addition, since the jet 'tagged' by a gauge boson is a quark jet, it may be possible to determine any difference between quark and gluon energy loss through comparison to dijet studies. Since the two largest contributions to the dijet cross sections are $gg \to gg$ and $qg \to qg$, at least one of the final-state dijets is a gluon.

Energy loss calculations suggest that the nucleus-nucleus results, whether quantified by R_{AA} or by azimuthal correlations, or, at higher energies, by observed jets, provide information about the density of the medium [190, 191]. Other models assume that the fragmentation functions extracted in e^+e^- annihilation no longer apply and focus on modifications of the fragmentation functions in the medium [192]. There are interesting times ahead in this field, both for theorists and experimentalists.

Exercises

1. Write down the equivalent relations of Eqs. (9.7)-(9.9) for charged kaons.

2. If we let $N_p^\pi(z) = d\sigma(ep \to \pi X)$, use Eq. (9.24) together with Eqs. (9.7) and (9.8) to calculate

$$\frac{\int dz[N_n^{\pi^+}(z) - N_n^{\pi^-}(z)]}{\int dz[N_p^{\pi^+}(z) - N_p^{\pi^-}(z)]} = \frac{2}{7} \tag{9.39}$$

   if the proton and neutron consist only of valence quarks.

3. The CMS experiment covers the largest part of full phase space of all the LHC detectors. It has coverage for muons in the region $|\eta| \leq 2.4$, hadronic calorimeter coverage for $|\eta| < 5$ and specialized calorimeters at $5 \leq \eta \leq 7$ and $|\eta| > 8$. Calculate the corresponding $\theta_{\rm cm}$ covered by the various detector components.

4. Calculate the charm quark distribution from the intrinsic charm model to derive Eq. (9.38).

Bibliography

[1] http://www.bnl.gov/RHIC

[2] http://lhc-machine-outreach.web.cern.ch/lhc-machine-outreach/

[3] S. Eidelman *et al.*, Phys. Lett. B **592** (2004) 1.

[4] F. Halzen and A. D. Martin, *Quarks and Leptons: An Introductory Course in Modern Particle Physics*, 1984, Wiley, New York.

[5] The data may be found on the web, courtesy of the COMPAS group, IHEP, Protvino, Russia at http://wwwppds.ihsp.su:8001/ppds.html or at http://pdg.lbl.gov/xsect/contents.html. The data is also plotted in the Particle Data Book, S. Eidelman *et al.*, Phys. Lett. B **592** (2004) 1.

[6] C. Avila *et al.* (E811 Collaboration), Phys. Lett. B **445** (1999) 419; F. Abe *et al.* (CDF Collaboration), Phys. Rev. D **50** (1994) 5550; N. Amos *et al.* (E710 Collaboration), Phys. Rev. Lett. **68** (1992) 2433.

[7] V. Berardi *et al.* (TOTEM Collaboration), CERN-LHCC-2004-002; M. Deile (TOTEM Collaboration), arXiv:hep-ex/0410084.

[8] C. Adler *et al.* (STAR Collaboration), Phys. Rev. Lett. **86** (2001) 4778 [Erratum-ibid. **90** (2003) 119903] [arXiv:nucl-ex/0104022]; J. J. Gaardhøje *et al.* (BRAHMS Collaboration), Nucl. Phys. A **734** (2004) 13 [arXiv:nucl-ex/0401026]; J. Adams, *et al.* (STAR

Collaboration), Phys. Rev. Lett. **92** (2004) 112301 [arXiv:nucl-ex/0310004].

[9] F. Abe *et al.* (CDF Collaboration), Phys. Rev. Lett. **73** (1994) 225 [arXiv:hep-ex/9405005]; Phys. Rev. D **50** (1994) 2966; Phys. Rev. Lett. **74** (1995) 2626 [arXiv:hep-ex/9503002]; S. Abachi *et al.* (D0 Collaboration), Phys. Rev. Lett. **72** (1994) 2138; Phys. Rev. Lett. **74** (1995) 2422 [arXiv:hep-ex/9411001]; Phys. Rev. Lett. **74** (1995) 2632 [arXiv:hep-ex/9503003].

[10] See the review in Ref. [3] on "The Electroweak model and constraints on new physics.

[11] http://www.lbl.gov/Science-Articles/Archive/sabl/2005/October/01-antiproton.html

[12] Yu. Dokshitzer, Sov. Phys. JETP **46** (1977) 1649; V. N. Gribov and L. N. Lipatov, Sov. Nucl. Phys. **15** (1972) 438, V. N. Gribov and L. N. Lipatov, Sov. Nucl. Phys. **15** (1972) 675; G. Altarelli and G. Parisi, Nucl. Phys. B **126** (1977) 298.

[13] F. Abe *et al.* (CDF Collaboration), Phys. Rev. Lett. **67** (1991) 2418.

[14] A. Bodek (CDF Collaboration), *Presented at 28th International Conference on High-energy Physics (ICHEP 96), Warsaw, Poland, 25-31 Jul 1996*, edited by Z. Ajduk and A. K. Wroblewski, World Scientific (1997) p. 1438.

[15] F. Abe *et al.* (CDF Collaboration), Phys. Rev. Lett. **74** (1995) 850 [arXiv:hep-ex/9501008].

[16] R. D. Field, *Applications of Perturbative QCD* Addison-Wesley, 1989.

[17] G. Sterman *et al.* (CTEQ Collaboration), Rev. Mod. Phys. **67** (1995) 157.

[18] S. D. Drell and T. M. Yan, Ann. Phys. **66** (1971) 578.

[19] J. C. Collins, D. E. Soper, and G. Sterman, Nucl. Phys. B **261** (1985) 104; Nucl. Phys. B **308** (1988) 833; *Perturbative QCD*, edited by A.H. Mueller, World Scientific, (1989) 1; G. T. Bodwin, Phys. Rev. D **31** (1985) 2616; erratum **D34** (1986) 3932; J. Qiu and G. Sterman, Nucl. Phys. B **353** (1991) 105; *ibid.*, Nucl. Phys. B **353** (1991) 137.

[20] E. A. Hawker *et al.* (E866 Collaboration), Phys. Rev. Lett. **80** (1998) 3715 [arXiv:hep-ex/9803011].

[21] R. Vogt, Prog. Part. Nucl. Phys. **45** (2000) S105 [arXiv:hep-ph/0011298].

[22] J. Pumplin, D. R. Stump, J. Huston, H. L. Lai, P. Nadolsky and W. K. Tung, JHEP **0207** (2002) 012 [arXiv:hep-ph/0201195].

[23] A. D. Martin, R. G. Roberts, W. J. Stirling and R. S. Thorne, Phys. Lett. B **604** (2004) 61 [arXiv:hep-ph/0410230].

[24] M. Glück, E. Reya, and A. Vogt, Eur. Phys. J. C **5** (1998) 461 [arXiv:hep-ph/9806404].

[25] J. F. Gunion, Phys. Lett. B **88** (1979) 150.

[26] J. Breitweg *et al.* (ZEUS Collaboration), Eur. Phys. J. C **7** (1999) 609 [arXiv:hep-ex/9809005]; A. Ahmed *et al.* (H1 Collaboration), Nucl. Phys. B **439** (1995) 471 [arXiv:hep-ex/9503001].

[27] M. Glück, E. Reya and I. Schienbein, Eur. Phys. J. C **10** (1999) 313 [arXiv:hep-ph/9903288].

[28] L. V. Gribov, E. M. Levin and M. G. Ryskin, Phys. Rep. **100** (1982) 1; A. H. Mueller and J. Qiu, Nucl. Phys. B **268** (1986) 427.

[29] K. J. Eskola, H. Honkanen, V. J. Kolhinen, J. w. Qiu and C. A. Salgado, Nucl. Phys. B **660** (2003) 211 [arXiv:hep-ph/0211239].

[30] E. Iancu, A. Leonidov and L. McLerran, Lectures given at Cargese Summer School on QCD Perspectives on Hot and Dense Matter, Cargese, France, 6-18 Aug 2001, arXiv:hep-ph/0202270.

[31] M. Arneodo, Phys. Rept. **240** (1994) 301.

[32] K. J. Eskola, V. J. Kolhinen and P. V. Ruuskanen, Nucl. Phys. B **535** (1998) 351 [arXiv:hep-ph/9802350]; K. J. Eskola, V. J. Kolhinen and C. A. Salgado, Eur. Phys. J. C **9** (1999) 61 [arXiv:hep-ph/9807297].

[33] G. Baur, K. Hencken, D. Trautmann, S. Sadovsky and Yu. Kharlov, Phys. Rept. **364** (2002) 359 [arXiv:hep-ph/0112211].

[34] C. A. Bertulani, S. R. Klein and J. Nystrand, Ann. Rev. Nucl. Part. Sci. **55** (2005) 271 [arXiv:nucl-ex/0502005].

[35] E. Witten, Nucl. Phys. B **120** (1977) 189.

[36] D. Brandt, LHC Project Report 450, December, 2000.

[37] K. Hencken and G. Baur, private communication.

[38] S. R. Klein, J. Nystrand and R. Vogt, Phys. Rev. C **66** (2002) 044906 [arXiv:hep-ph/0206220].

[39] M. Strikman, R. Vogt and S. White, Phys. Rev. Lett. **96**, 082001 (2006) [arXiv:hep-ph/0508296].

[40] D. Griffiths, *Introduction to Elementary Particles*, 1987, Wiley, New York.

[41] K. S. Krane, *Introductory Nuclear Physics*, 1987, Wiley, New York.

[42] C. W. deJager, H. deVries, and C. deVries, Atomic Data and Nuclear Data Tables **14** (1974) 485.

[43] D. Kharzeev, E. M. Levin and M. Nardi, hep-ph/0212316; L. Hulthén and M. Sagawara, in *Handbüch der Physik*, **39** (1957).

[44] R. J. Glauber and G. Matthiae, Nucl. Phys. B **21** (1970) 135; R. J. Glauber, arXiv:nucl-th/0604021.

[45] W. Busza and A. Goldhaber, Phys. Lett. B **139** (1984) 235.

[46] K. Kinoshita, A. Minaka and H. Sumiyoshi, Prog. Theo. Phys. **63** (1980) 1268; S. Daté, M. Gyulassy and H. Sumiyoshi, Phys. Rev. D **32** (1985) 619.

[47] C. Y. Wong, Phys. Rev. Lett. **52** (1985) 1393; C. Y. Wong, Phys. Rev. D **32** (1985) 94; C. Y. Wong, Phys. Rev. C **33** (1986) 1340; C. Y. Wong and Z. D. Lu, Phys. Rev. D **39** (1989) 2606.

[48] H. Appelshauser *et al.* (NA49 Collaboration), Phys. Rev. Lett. **82** (1999) 2471 [arXiv:nucl-ex/9810014].

[49] M. C. Abreu *et al.* (NA50 Collaboration), Phys. Lett. B **410** (1997) 327.

[50] J. Ftáčnik, K. Kajantie, N. Pišútová and J. Pišút, Phys. Lett. B **188** (1987) 279.

[51] M. Plümer and R. M. Weiner, Phys. Rev. D **37** (1988) 3136.

[52] G. Baym, in proceedings of the IX Autumn School, "The Physics of the Quark-Gluon Plasma", edited by J. Dias de Deus and S. Costa Ramos (World Scientific, Singapore, 1988), p. 209.

[53] G. Baym, G. Friedman and I. Sarcevic, Phys. Lett. B **219** (1989) 205.

[54] S. Sorensen *et al.* (WA80 Collaboration) Z. Phys. C **38** (1988) 3.

[55] D. Kharzeev and M. Nardi, Phys. Lett. B **507** (2001) 121 [arXiv:nucl-th/0012025].

[56] R. Vogt, Heavy Ion Phys. **9** (1999) 339.

[57] P. L. McGaughey *et. al.*, Int. J. Mod. Phys. A **10** (1995) 2999 [arXiv:hep-ph/9411438].

[58] http://wine1.sb.fsu.edu/chm1045/notes/Forces/Phase/Forces06.htm.

[59] K. Rajagopal and F. Wilczek, arXiv:hep-ph/0011333.

[60] M. Alford, P. Jotwani, C. Kouvaris, J. Kundu and K. Rajagopal, Phys. Rev. D **71** (2005) 114011 [arXiv:astro-ph/0411560].

[61] F. Weber, *Pulsars as Astrophysical Laboratories for Nuclear and Particle Physics*, 1999, Institute of Physics Pub., Bristol.

[62] http://www-new.gsi.de/zukunftsprojekt/index_e.html.

[63] C. Kittel and H. Kroemer, *Thermal Physics*, 2 Edition, 1980, W. H. Freeman and Company, New York.

[64] See *e.g.* E. Shuryak, Prog. Part. Nucl. Phys. **53** (2004) 273 [arXiv:hep-ph/0312227]; U. W. Heinz, arXiv:nucl-th/0512051.

[65] J. Cleymans, R. V. Gavai and E. Suhonen, Phys. Rept. **130** (1986) 217.

[66] A. Chodos, R. L. Jaffe, K. Johnson, C. B. Thorn and V. F. Weisskopf, Phys. Rev. D **9** 3471 (1974); T. DeGrand, R. L. Jaffe, K. Johnson and J. Kiskis, Phys. Rev. D **12** 2060 (1975).

[67] R. L. Jaffe, MIT-CTP-814 *Lectures presented at Erice School of Subnuclear Physics, Erice, Sicily, Jul 31 - Aug 10, 1979*

[68] J. Cleymans, H. Oeschler, K. Redlich and S. Wheaton, arXiv:hep-ph/0511094.

[69] L. P. Csernai, *Introduction to Relativistic Heavy Ion Collisions*, 1994, Wiley, New York.

[70] See *e.g.*, D. Teaney, Phys. Rev. C **68** (2003) 034913 [arXiv:nucl-th/0301099].

[71] L. D. Landau, Izv. Akad. Naul. SSSR, Ser. Fiz. **17** (1953) 51.

[72] E. Fermi, Prog. Theor. Phys. **5** (1950) 570; Phys. Rev. **81** (1951) 683.

[73] P. Carruthers and M.-D. Van, Phys. Rev. D **8** (1973) 859; F. Cooper, G. Frye and E. Schonberg, Phys. Rev. D **11** (1974) 192.

[74] J. D. Bjorken, Phys. Rev. D **27** (1983) 140.

[75] K. Alpøard *et al.* (UA5 Collaboration), Phys. Lett. B **107** (1981) 310; 315.

[76] A. Bujak *et al.*, Phys. Rev. D **23** (1981) 1895; J. Burg *et al.*, Nucl. Phys. B **187** (1981) 205.

[77] K. Kajantie, P. V. Landshoff and J. Lindfors, Phys. Rev. Lett. **59** (1987) 2527; K. J. Eskola, K. Kajantie and J. Lindfors, Nucl. Phys. B **323** (1989) 37.

[78] K. J. Eskola and K. Kajantie, Z. Phys. C **75** (1997) 515 [arXiv:nucl-th/9610015].

[79] K. J. Eskola, Nucl. Phys. A **698** (2002) 78 [arXiv:hep-ph/0104058].

[80] V. Emel'yanov, A. Khodinov, S. R. Klein and R. Vogt, Phys. Rev. C **61** (2000) 044904 [arXiv:hep-ph/9909427].

[81] K. J. Eskola, K. Kajantie, P. V. Ruuskanen, K. Tuominen, Nucl. Phys. B **570** (2000) 379 [arXiv:hep-ph/9909456].

[82] G. Baym, B. L. Friman, J.-P. Blaizot, M. Soyeur and W. Czyz, Nucl. Phys. A **407** (1983) 541.

[83] I. Arsene *et al.* (BRAHMS Collaboration), Nucl. Phys. A **757** (2005) 1 [arXiv:nucl-ex/0410020]; B. B. Back *et al.* (PHOBOS Collaboration), Nucl. Phys. A **757** (2005) 28 [arXiv:nucl-ex/0410022]; J. Adams *et al.* (STAR Collaboration) Nucl. Phys. A **757** (2005) 102 [arXiv:nucl-ex/0501009]; K. Adcox *et al.* (PHENIX Collaboration), Nucl. Phys. A **757** (2005) 184.

[84] W. Reisdorf and H. G. Ritter, Ann. Rev. Nucl. Part. Sci. **47** (1997) 663.

[85] C. Pinkenburg *et al.* (E895 Collaboration), Phys. Rev. Lett. **83** (1999) 1295.

[86] C. Adler *et al.* (STAR Collaboration), Phys. Rev. Lett. **87** (2001) 182301 [arXiv:nucl-ex/0107003].

[87] B. B. Back *et al.* (PHOBOS Collaboration), Phys. Rev. Lett. **94** (2005) 122303 [arXiv:nucl-ex/0406021].

[88] A. C. Mignerey (PHOBOS Collaboration), arXiv:nucl-ex/0602008.

[89] M. Creutz, *Quarks, Gluons and Lattices*, 1983, Cambridge University Press, Cambridge.

[90] J. B. Kogut, Rev. Mod. Phys. **51** (1979) 659; Rev. Mod. Phys. **55** (1983) 775.

[91] K. G. Wilson, Phys. Rev. D **14** (1974) 2455.

[92] T.-P. Cheng and L.-F. Li, *Gauge Theory of Elementary Particle Physics*, 1984, Clarendon Press, Oxford.

[93] R. P. Feynman and A. R. Hibbs, *Quantum Mechanics and Path Integrals*, 1965, McGraw-Hill Book Company, New York.

[94] http://en.wikipedia.org/wiki/Path_integral_formulation

[95] C. Y. Wong, *Introduction to High-Energy Heavy-Ion Collisions*, 1994, World Scientific, Singapore.

[96] F. Karsch, Lect. Notes Phys. **583** (2002) 209 [arXiv:hep-lat/0106019].

[97] P. Ramond, *Field Theory: A Modern Primer*, 1981, Benjamin Cummings, Reading, Massachusetts.

[98] R. Gupta, arXiv:hep-lat/9807028.

[99] R. D. Pisarski and F. Wilczek, Phys. Rev. D **29** (1984) 338.

[100] F. Karsch, private communication.

[101] F. Karsch, A. Peikert and E. Laermann, Phys. Lett. B **478** (2000) 447 [arXiv:hep-lat/0002003].

[102] F. Karsch, Nucl. Phys. A **698** (2002) 199 [arXiv:hep-ph/0103314].

[103] Z. Fodor and S. D. Katz, JHEP **0203** (2002) 014 [arXiv:hep-lat/0106002]; Phys. Lett. B **534** (2002) 87 [arXiv:hep-lat/0104001]; Z. Fodor, Nucl. Phys. A **715** (2003) 319 [arXiv:hep-lat/xxxx].

[104] C. R. Allton, S. Ejiri. S. J. Hands, O. Kaczmarek, F. Karsch, E. Laermann, C. Schmidt and L. Scorzato, Phys. Rev. D **66** (2002) 074507 [arXiv:hep-lat/0204010].

[105] P. deForcrand and O. Philipsen, Nucl. Phys. B **642** (2002) 290 [arXiv:hep-lat/0205016].

[106] Z. Fodor, S. D. Katz, K. K. Szabo, Phys. Lett. B **568** (2003) 73 [arXiv:hep-lat/0208078].

[107] L. D. McLerran and T. Toimela, Phys. Rev. D **31** (1985) 545.

[108] M. I. Gorenstein and O. P. Pavlenko, Z. Phys. C **37** (1988) 611.

[109] K. Kajantie and P. V. Ruuskanen, Z. Phys. C **44** (1989) 167.

[110] J. I. Kapusta, L. McLerran, and D. K. Srivastava, Phys. Lett. B **283** (1992) 145.

[111] R. Vogt, B. V. Jacak, P. L. McGaughey and P. V. Ruuskanen, Phys. Rev. D **49** (1994) 3345 [arXiv:hep-ph/9309213].

[112] A. Shor, Phys. Lett. B **215** (1988) 375; B **233** (1989) 231.

[113] S. Gavin, P. L. McGaughey, P. V. Ruuskanen and R. Vogt, Phys. Rev. C **54** (1996) 2606 [arXiv:hep-ph/9604369].

[114] M. Cacciari, P. Nason and R. Vogt, Phys. Rev. Lett. **95** (2005) 122001 [arXiv:hep-ph/0502203].

[115] J. Adams *et al.* (STAR Collaboration), Phys. Rev. Lett. **94** (2005) 062301 [arXiv:nucl-ex/0407006]; A. Tai (STAR Collaboration), J. Phys. G **30** (2004) S809 [arXiv:nucl-ex/0404029].

[116] J. J. Aubert *et al.*, Phys. Rev. Lett. **33** (1974) 1404.

[117] J. E. Augustin *et al.*, Phys. Rev. Lett. **33** (1974) 1406.

[118] S. L. Glashow, J. Iliopoulos and L. Maiani, Phys. Rev. D **2** (1970) 1285.

[119] S. W. Herb *et al.*, Phys. Rev. Lett. **39** (1977) 252.

[120] T. Matsui and H. Satz, Phys. Lett. B **178** (1986) 416.

[121] R. L. Anderson *et al.*, Phys. Rev. Lett. **38** (1977) 263.

[122] J. Badier *et al.* (NA3 Collab.), Z. Phys. **C20** (1983) 101.

[123] D. M. Alde *et al.* (E772 Collab.), Phys. Rev. Lett. **66** (1991) 133.

[124] M. Leitch *et al.* (E866 Collaboration), Phys. Rev. Lett. **85** (2000) 4012 [arXiv:nucl-ex/9909007].

[125] S. S. Adler *et al.* (PHENIX Collaboration), arXiv:nucl-ex/0507032.

[126] L. Kluberg, Eur. Phys. J C **43** (2005) 145.

[127] M. Bedjidian *et al.*, arXiv:hep-ph/0311048.

[128] F. Karsch, M. T. Mehr, and H. Satz, Z. Phys. C **37** (1988) 617.

[129] D. H. Perkins, *Introduction to High Energy Physics*, Addison-Wesley, 1987.

[130] F. Abe *et al.* (CDF Collaboration), Phys. Rev. Lett. **79** (1997) 572; 578.

[131] T. Affolder *et al.* (CDF Collaboration), Phys. Rev. Lett. **84** (2000) 2094.

[132] S. Okubo, Phys. Rev. Lett. **5** (1963) 165; G. Zweig, CERN-TH-401 (1964); J. Iizuka, Suppl. Prog. Theor. Phys. **37** (1966) 21.

[133] R. Baier and R. Rückl, Z. Phys. **C19** (1983) 251.

[134] G. T. Bodwin, E. Braaten, and G. P. Lepage, Phys. Rev. D **51** (1995) 1125; **55** (1997) 5853(E) [arXiv:hep-ph/9407339].

[135] M. B. Einhorn and S. D. Ellis, Phys. Rev. D **12** (1975) 2007; H. Fritzsch, Phys. Lett. B **67** (1977); M. Glück, J. F. Owens and E. Reya, Phys. Rev. D **17** (1978) 2324; J. Babcock, D. Sivers and S. Wolfram, Phys. Rev. D **18** (1978) 162.

[136] P. Hoyer and S. Peigné, Phys. Rev. D **59** (1999) 034011 [arXiv:hep-ph/9806424].

[137] N. Brambilla *et al.* (Quarkonium Working Group), arXiv:hep-ph/0412158.

[138] R. V. Gavai *et al.* (Hard Probes Collaboration), Int. J. Mod. Phys. A **10** (1995) 3043.

[139] V. Barger, W. Y. Keung, and R. N. Phillips, Phys. Lett. B **91** (1980) 253; Z. Phys. C **6** (1980) 169.

[140] G. A. Schuler and R. Vogt, Phys. Lett. B **387** (1996) 181.

[141] M. L. Mangano, P. Nason and G. Ridolfi, Nucl. Phys. **B405** (1993) 507.

[142] S. Digal, P. Petreczky and H. Satz, Phys. Rev. D **64** (2001) 094015 [arXiv:hep-ph/0106017].

[143] T. Affolder *et al.* [CDF Collaboration], Phys. Rev. Lett. **84** (2000) 2094 [arXiv:hep-ex/9910025].

[144] W. E. Caswell and G. P. Lepage, Phys. Lett. B **167** (1986) 437.

[145] B. A. Thacker and G. P. Lepage, Phys. Rev. D **43** (1991) 196.

[146] See G. A. Schuler, arXiv:hep-ph/9403387, and references therein.

[147] M. Beneke and I. Z. Rothstein, Phys. Rev. D **54** (1996) 2005 [Erratum-ibid. D **54** (1996) 7082] [arXiv:hep-ph/9603400].

[148] S. Digal, O. Kaczmarek, F. Karsch and H. Satz, Eur. Phys. J C **43** (2005) 71 [arXiv:hep-ph/0505193].

[149] H. Satz, J. Phys. G **32** (2006) R25 [arXiv:hep-ph/0512217].

[150] E. Shuryak, Phys. Rep. **61** (1980) 71; D. J. Gross, R. D. Pisarski, and L. G. Yaffe, Rev. Mod. Phys. **53** (1981) 43.

[151] P. Levai and U. Heinz, Phys. Rev. C **57** (1998) 1879.

[152] G. Boyd *et al.*, Nucl. Phys. B **469** (1996) 419.

[153] J. Fingberg, U. M. Heller and F. Karsch, Nucl. Phys. B **392** (1993) 493.

[154] B. Petersson, Nucl. Phys. A **525** (1991) 237c.

[155] M.-C. Chu and T. Matsui, Phys. Rev. D **37** (1988) 1851.

[156] F. Karsch and H. Satz, Z. Phys. C **51** (1991) 209.

[157] J.-P. Blaizot and J.-Y. Ollitrault, Phys. Lett. B **199** (1987) 499; F. Karsch and R. Petronizio, Phys. Lett. B **212** (1988) 255.

[158] U. Husemann, DESY-THESIS-2005-005.

[159] R. Vogt, Phys. Rep. **310** (1999) 197.

[160] S. Gavin and M. Gyulassy, Phys. Lett. B **214** (1988) 241.

[161] N. S. Craigie, Phys. Rep. **47** (1978) 1.

[162] R. Vogt, Phys. Rev. C **61** (2000) 035203 [arXiv:hep-ph/9907317].

[163] R. Vogt, Nucl. Phys. A **700** (2002) 539 [arXiv:hep-ph/0107045].

[164] S. Gavin and R. Vogt, Nucl. Phys. B **345** (1990) 104.

[165] R. J. Ledoux, Nucl. Phys. A **498** (1989) 205c.

[166] H. Satz, Nucl. Phys. A **544** (1992) 371c.

[167] L. Grandchamp and R. Rapp, Nucl. Phys. A **715** (2003) 545 [arXiv:hep-ph/0209141]; L. Grandchamp, R. Rapp and G. E. Brown, Phys. Rev. Lett. **92** (2004) 212301 [arXiv:hep-ph/0306077].

[168] H. Periera (PHENIX Collaboration), arXiv:nucl-ex/0510051.

[169] D. Kharzeev, C. Lourenço, M. Nardi and H. Satz, Z. Phys. C **74** (1997) 307.

[170] M. C. Abreu *et al.* (NA38 Collaboration), Phys. Lett. B **449** (1999) 128.

[171] B. Alessandro *et al.* (NA50 Collaboration), Eur. Phys. J. C **39** (2005) 335.

[172] J. F. Gunion and R. Vogt, Nucl. Phys. B **492** (1997) 301.

[173] B. A. Kniehl, G. Kramer and B. Pötter, Nucl. Phys. B **582** (2000) 514 [arXiv:hep-ph/0010289].

[174] S. Kretzer, Phys. Rev. D **62** (2000) 054001 [arXiv:hep-ph/0003177].

[175] R. Seuster *et al.* [Belle Collaboration], Phys. Rev. D **73** (2006) 032002 [arXiv:hep-ex/0506068].

[176] G. Abbiendi *et al.* [OPAL Collaboration], Eur. Phys. J. C **29** (2003) 463 [arXiv:hep-ex/0210031].

[177] A. D. Martin, R. G. Roberts, W. J. Stirling and R. S. Thorne, Phys. Lett. B **443** (1998) 301 [arXiv:hep-ph/9808371].

[178] I. G. Bearden *et al.* [BRAHMS Collaboration], Nucl. Phys. A **698** (2002) 29c.

[179] I. Arsene *et al.* [BRAHMS Collaboration], Phys. Rev. Lett. **93** (2004) 242303 [arXiv:nucl-ex/0403005].

[180] A. Accardi, Acta Phys. Hung. A **22** (2005) 289 [arXiv:nucl-th/0405046].

[181] R. C. Hwa, C. B. Yang and R. J. Fries, Phys. Rev. C **71** (2005) 024902 [arXiv:nucl-th/0410111].

[182] K. Adcox *et al.* [PHENIX Collaboration], Phys. Rev. Lett. **88** (2002) 242301 [arXiv:nucl-ex/0112006]; S. S. Adler *et al.* [PHENIX Collaboration], Phys. Rev. Lett. **91** (2003) 172301 [arXiv:nucl-ex/0305036].

[183] S. J. Brodsky, P. Hoyer, C. Peterson and N. Sakai, Phys. Lett. B **93** (1980) 451; S. J. Brodsky, C. Peterson and N. Sakai, Phys. Rev. D **23** (1981) 2745.

[184] R. J. Fries, J. Phys. G **30** (2004) S853 [arXiv:nucl-th/0403036].

[185] M. Gyulassy and M. Plumer, Phys. Lett. B **243** (1990) 432.

[186] C. Adler *et al.* [STAR Collaboration], Phys. Rev. Lett. **90** (2003) 082302.

[187] X. N. Wang, Z. Huang, and I. Sarcevic, Phys. Rev. Lett. **77** (1996) 231 [arXiv:hep-ph/9605213].

[188] R. Vogt, Phys. Rev. C **64** (2001) 044901 [arXiv:hep-ph/0011242].

[189] G. Baur *et al.*, CMS/2000-60, 2000.

[190] M. Gyulassy, P. Levai and I. Vitev, Phys. Rev. Lett. **85** (2000) 5535 [arXiv:nucl-th/0005032].

[191] U. A. Wiedemann, Nucl. Phys. B **588** (2000) 303 [arXiv:hep-ph/0005129].

[192] A. Accardi *et al.*, arXiv:hep-ph/0310274.

Index

Kolibri 360 GmbH
Ferdinand-Jühlke-Straße 7
99095 Erfurt, Deutschland
produktsicherheit@kolibri360.de